Lecture Notes in Electrical Engineering

Volume 79

Marian Adamski, Alexander Barkalov,
and Marek Węgrzyn (Eds.)

Design of Digital Systems and Devices

Springer

Prof. Marian Adamski
University of Zielona Góra
Institute of Computer Eng. &
Electronics
ul. Podgórna 50
65-246 Zielona Góra, Poland
E-mail: M.Adamski@iie.uz.zgora.pl

Dr. Marek Węgrzyn
University of Zielona Góra
Institute of Computer Eng. &
Electronics
ul. Podgórna 50
65-246 Zielona Góra, Poland
E-mail: M.Wegrzyn@iie.uz.zgora.pl

Prof. Alexander Barkalov
University of Zielona Góra
Institute of Computer Eng. &
Electronics
ul. Podgórna 50
65-246 Zielona Góra, Poland
E-mail: A.Barkalov@iie.uz.zgora.pl

ISBN 978-3-642-17544-2

e-ISBN 978-3-642-17545-9

DOI 10.1007/978-3-642-17545-9

Typeset & Coverdesign: Scientific Publishing Services Pvt. Ltd., Chennai, India.

Printed on acid-free paper

9 8 7 6 5 4 3 2 1

springer.com

Contents

Part III: Testing, Modeling and Signal Processing

About the Editors

Marian Andrzej Adamski received the M.Sc. degree in Electrical Engineering (specialty of Control Engineering) from Poznan Technical University, Poland, in 1970, the Ph.D. degree in Control and Computer Engineering from Silesian Technical University, Gliwice, Poland, in 1976, and Habilitated Doctor (D.Sc.) degree in Computer Engineering from Warsaw University of Technology, Poland, in 1991.

After M.Sc. study (1970) he joined Research Laboratory in Nuclear Electronics Company in Poznan. In 1973 he became a Senior Lecturer at Technical University of Zielona Gora, Poland. From 1976 until 1991 he was employed as an Assistant Professor, and later from 1991 to 1992 as an Associate Professor. From 1993 until 1996 he was a Visiting Professor at University of Minho, in Braga and Guimaraes, Portugal. Currently he is Full, Tenured Professor of Computer Engineering at University of Zielona Gora, Poland. He is a Chairman (Head) of Computer Engineering and Electronics Institute at University of Zielona Góra.

Prof. Adamski's research includes mathematical logic and Petri nets in digital systems design, formal development of Logic Controller programs, VHDL, FPLD and FPGA in industrial applications.

Prof. M. Adamski is an author of more than 180 publications, including 6 books, and he holds 5 patents. He is a member of several international and national societies, including Committees of Polish Academy of Sciences, Polish Computer Science Society, ACM and IEEE. He has supervised more than 100 M.Sc. theses and several Ph.D.dissertations. He has been a principal investigator for government-sponsored research projects and a consultant to industry. He is a member of editorial board of *International Journal* of *Applied Mathematics and Computer Science*, and a referee of international conferences and journals. He has been involved as a program and organizing committee member of several international workshops and conferences. He obtained the Scientific Award from Ministry of Higher Education and won several times the University Distinguished Teaching and Research Awards.

Alexander Barkalov received the M.Sc. degree in Computer Engineering from the Donetsk Politechnical Institute (now Donetsk National Technical University), Ukraine, in 1976, Ph.D. degree in Computer Science from the Leningrad Institute of Fine Mechanics and Optics, Russia, in 1983. In 1995 he received Doctor of Technical Sciences degree in Computer Science from Institute of Cybernetics named after V.M.Glushkov (Kiev, Ukraine). He has been an Assistant Professor (since 1976), an Associate Professor (since 1984) and a Professor (since 1996) at the Institute of Computers, Donetsk National Technical University. From 2003 he is a Professor of Computer Engineering at the Institute of Informatics and Electronics, University of Zielona Góra, Poland, and he still is a Professor at the Institute of Computers, Donetsk National Technical University. His current research interests include theory of digital automata, especially the methods of synthesis and optimization of control units implemented with field-programmable logic devices. He has taken part in the realization of a number of research projects sponsored by different institutions of former USSR. These researches were connected with development of computer-aided design tools for implementation of control units.

Alexander Barkalov has published more than 400 papers in international journals and conference proceedings. He is an author two and co-author of 8 monographs and one book chapter.

Marek Węgrzyn received the M.Sc. of Electrical Engineering degree (*summa cum laude*) from the Technical University of Zielona Gora, Poland, in 1991. Since 1991 he has been a lecturer of digital systems in Computer Engineering and Electronics Department, Faculty of Electrical Engineering at the University. He spent one academic year (1992/93) at UMIST, Manchester, UK, working on VLSI design and HDLs (Verilog and VHDL). He has been a visiting Research Fellow in the Department of Industrial Electronics of University of Minho, Braga and Guimaraes, Portugal (in 1996). He received his Ph.D. in Computer Engineering from the Faculty of Electronics and Information Techniques at Warsaw University of Technology, Poland, in 1999. Currently, Dr. Marek Węgrzyn is an Assistant Professor, and a Head of Computer Engineering Division at University of Zielona Gora, Poland. He is a chair of IFAC Technical Committee 3.1. Computer for Control. He is a member of IEEE, Polish Information Processing Society (PTI) and Polish Society for Measurement, Automatic Control and Robotics (POLSPAR).

His research interests focus on Hardware Description Languages, Petri Nets, concurrent controller designs, and information technology. Recent work includes design of dedicated FPGA-based digital systems and tools for the automatic synthesis of programmable logic. He is a referee of international conferences and journals.

Dr. Marek Węgrzyn was the 1991 Recipient of the Best Young Electrical Engineer award from District Branch of Electrical Engineering Society. As the best student he obtained in 1989 the Golden Medal *Maxima cum Laude* from the Rector-Head of the Technical University of Zielona Góra, *Primus Inter Pares* diploma and *Nicolaus Copernicus Award* from the National Student Association. He won the National Price from Ministry of Education for the distinguished Ph.D. Dissertation (2000) and as book co-author (2006). He obtained several awards from the Rector-Head of the University of Zielona Góra. He published more than 100 papers on conferences and journals. He was a co-editor several books and post-conference proceedings.

Foreword

Logic design of digital devices is a very important part of the Computer Science. It deals with design and testing of logic circuits for both data-path and control unit of a digital system. Design methods depend strongly on logic elements using for implementation of logic circuits. Different programmable logic devices such as CPLD and FPGA are wide used for implementation of logic circuits. Nowadays, we can see the rapid growth of new and new chips, but there is a strong lack of new design methods. To diminish this gap, we created a group of experts in Computer Science and Electronics, and they present here their new design and test approaches.

This book includes a variety of design and test methods targeted on different digital devices, as well as different logic elements. The authors of the book represent such countries as Israel, Poland, Russia, and Ukraine. The book is divided by three main parts, including thirteen different Chapters. The following problems are discussed in these Chapters.

The Chapter 1 is written by Professor Samary Baranov and is devoted to original methods of digital system design. It represents some approach for design a central processor unit of a computer. It presents an example of a simple processor design. The proposed design procedure included such steps as combination of separate algorithmic state machines (ASM); synthesis of data-path; control unit design, and composition of data-path and control unit into whole processor. The main advantage of proposed procedure is formalization of the design process, where all steps are formalized and automated in EDA tool "Abelite" designed by the author of this chapter.

The Chapter 2 is prepared by Professor Edward Hrynkiewicz and is devoted to application of the rectangular function $\Pi(x)$ for description of the operation for some logical devices. This function describes the carrying out of the logic operations on pulses, as well as pulse series. Such functions as logic sum, logic product, logic negation and Ex-OR are investigated in the Chapter. Using these functions, it is possible to describe the frequency multiplying. Moreover, this Chapter deals with the problem of rectangular function $\Pi(x)$ utilization for description of operation of such logical devices as digital sine wave generators and for nonlinear distortions analyzing in such generators.

The Chapter 3 is written by Alexander Palagin and Vladimir Opanasenko. It gives the theoretical base for construction and designing of the PLD–based reconfigurable devices, including the new formalized design techniques of construction and dynamic reconfiguration of architecture and structure of digital devices with a high degree of reconfiguration, corresponding with properties of performing algorithms, constructive and technological features PLD, and also tool means of their designing, are presented. Bases of the theory of adaptive logic networks, intended for the solution of a wide class of tasks by direct structural realization of algorithms of processing and direct representation of input data to output data by functional and structural customization for universal components of a network, are developed by the authors. Synthesis algorithms of adaptive logic networks on the classes of tasks set are developed by them, too.

The Chapter 4 is devoted to application of multilevel design on the base of UML for digital system developing. It is written by Professor Raisa Malcheva. In this chapter the features of image generation and performing systems' design are analyzed. Estimation of complexity of the standard rendering pipeline is done. The architectural decisions and algorithm approaches for the real-time rendering systems' creation are discussed. Adaptation of a multilevel designing method of the built-in systems with realization of separate modules on reconfigurable devices, based on application of architecture operated by models and the unified modeling language, is offered. The end of the Chapter shows the graphical application of the proposed by author modified method.

The Chapter 5 is written by Mariusz Rawski, Paweł Tomaszewicz, Grzegorz Borowik, and Tadeusz Łuba. It includes the new method of logic synthesis for digital circuits with FPGAs. The main feature of the method is a wide use of specialized embedded memory blocks (EMBs). Existing methods do not ensure effective utilization of the possibilities provided by such modules. The problem of efficient mapping of combinational and sequential parts of design can be solved using decomposition algorithms. The main issue of the Chapter is the application of decomposition based methods for efficient utilization of modern FPGAs. It shows that functional decomposition method allows for very flexible synthesis of the designed system onto heterogeneous structures of modern FPGAs composed of logic cells and EMBs. Finally there are some results of the experiments, which evidently show, that the application of functional decomposition algorithms in the implementation of typical signal and information processing systems greatly influences the performance of resulting digital circuits.

The Chapter 6 is prepared by Professor Dariusz Kania. It includes the original technology mapping method for PAL-based devices based on the analysis of graph of outputs. The method is oriented on CPLD chips having a PAL-based logic blocks which consist of a programmable AND matrix and a fixed OR matrix. The presented approach uses original method for illustrating a minimized form of a multi-output Boolean function. Graph node represents groups of multiple-output implicants with common output part. The essence of the method is the process of searching for appropriate multi-output implicants that can be shared by several functions. A new method for the description of cascaded feedback connections is presented in the Chapter. The experimental results show that the proposed algorithm leads to significant reduction of chip area used by resulting circuits.

The Chapter 7 is written by Nataliya Yakymets and Vyacheslav Kharchenko and is devoted to use of genetic algorithms (GAs) for the reliable design of finite state machines (FSMs). This chapter introduces the principles of multi-version digital system design and describes the concept of developing a reliable and robust system out of unreliable parts. The Authors started with the state of the art in the area of multi-version design and explore the motivations for using different approaches to development of digital projects. A few techniques to manage design diversity for FPGA-based systems are proposed. These techniques are based on the use of genetic algorithms, and partially correct and partially definite automata obtained with GAs. Finally, they suggested GA-based method of multi-version fault-tolerant systems synthesis and discuss case-study for on-board device implementation.

The Chapter 8 is written by Alexander Barkalov, Larysa Titarenko, Jacek Bieganowski, and Alexander Miroshkin, and proposed a new method for logic synthesis of a linear control algorithm. In this case, the control unit can be represented by the model of compositional microprogram control unit (CMCU) with dedicated area of inputs. The chapter is devoted to CMCU optimization, based on the modification of the microinstruction format. Proposed modifications are intended to eliminate code transformers from the CMCU and reduce the hardware amount of circuits used in the FSM for the microinstruction addressing, as compared with the CMCU basic structure. The reduction of the hardware amount is achieved at the cost of increasing the number of cycles needed for the execution of the control algorithms, and in some cases also at the cost of increasing control memory size.

The Chapter 9 is prepared by Marek Węgrzyn, and Agnieszka Węgrzyn. They present the CAD system dedicated for modeling, verification, and synthesis of concurrent logic controllers. The core of the proposed PeNLogic system is Petri net models. The Petri net can be prepared as graph or as textual form. Controllers specified by Petri nets can be analyzed and implemented using method suitable for such models. Results of verification are applied also for decomposition of net into several communicating state machines (as finite state machines, FSMs). After verification it is possible to transform Petri net model into HDLs model (VHDL and Verilog) and alternatively into EDIF or XNF netlist format. Such prepared models are also simulated and synthesized using other academic or commercial CAD systems.

The Chapter 10 is written by Professor Vladimir Popovskiy and includes some new methods of signals processing in radio access networks. The Chapter considers optimum stochastic methods of radio signal processing, including parameter assessment tasks and management of parameters of receiving and transmitting devices. Such tasks are formed by state variable methods using Kalman-Bucy optimum recursive procedures. It's recommended to solve the management problems basing on the division theorem. The Chapter includes analysis of steadiness and efficiency of state and management assessment procedures in steady-state and unsteady-state conditions. It gives recommendations regarding the choice of parameters and efficiency of processing devices taking into account statistics of signals and constraints attributable to certain telecommunication technologies. It analyzes a proposal, within which recursive procedures are used efficiently. The main tasks are united on Multiple-input/Multiple-output (MIMO) principle and are aimed at solving the access problems in mobile communication networks, Wi-Fi and Wi-Max systems, etc. Such tasks include: space-time encoding, multipath effect reduction, radio link power improvement, interference effect reduction, adaptation to channel parameter changes and current signal interference situation, possible repeated use of frequencies.

The Chapter 11 is prepared by Professor Alexander Ojiganov and discusses the recursive code scales for moving converters. The original methods of construction of recursive code scales (RCS), as well as the algorithms of placing on a scale of reading out elements (RE) are considered in the Chapter. It includes also some results of research of correcting possibilities of such scales. The recursive code scales for synthesis of drawing of an information path of a scale of sequence have received the name pseudo-random (PRCS) and composite code scales (CCS). Offered scales can be applied as the coded element in moving converters. Recursive scales at the expense of use in them of only one information code path more technologically than the known traditional scales. In traditional scales, code paths (CP) which are carried out, as a rule, in an ordinary binary code or in the Gray code. The Chapter shows that RCS allow only at the expense of redundancy introduction on number of reading out elements without use of additional control paths to form the codes which are correcting and (or) finding out errors of reading.

The Chapter 12 is written by professor Vladimir Hahanov and deals with the class of infrastructure Intellectual Property for System on Chip simulation and diagnosis service The models and methods for creating Infrastructure Intellectual Property (I-IP) service for the functionalities System on Chip (SoC), which has a minimum set of the real time Built-In Self Test (BIST) tools, are proposed in this Chapter. The means I-IP provide an opportunity to services: fault modeling and simulation for the functional primitives to evaluate the test quality and to build Fault Detection Table (FDT); diagnosis of a given defects search depth in the SoC; repairing embedded memory functionality, by using spare row and column components. High performance deductive-parallel fault analysis method for building FDT and tests quality assessment is offered. Algebra logical methods of fault diagnosis and embedded memory repair by synthesis Disjunctive Normal Form (DNF) completing all decisions for diagnosis SoC functionalities in the real time are represented in the Chapter.

The Chapter 13 is written by Yuriy A. Skobtsov and Vadim Y. Skobtsov. It discusses some new evolutionary test generation methods for digital devices. It is shown how evolutionary methods can be used for test generation of digital circuits. In present time it is strongly investigated the new direction in theory and practice of artificial intelligence and information systems, named as evolutionary computations. This term is used to generic description of the search, optimizing or learning algorithms, based on some formal principles of natural evolutional selection, which are sufficiently applied in solving various problems of machine learning, data mining, databases etc. Among these approaches following main paradigms can be picked out: genetic algorithms, evolutionary strategy (ES), evolutional programming (EP), and genetic programming (GP). The differences of these approaches mainly consist in the way of target solution representation and in different set of evolutional operators used in evolutional simulation. Classical GA uses the binary encoding of problem solution and basic genetic operators are crossover and mutation. In ES solution is represented by real numbers vector and basic operator is mutation. EP uses FSM as solution representation and mutation operator. In GP problem solution is represented by program, crossover and mutation operators are applied. The Chapter includes some new test methods for digital devices based on GA.

The editors of the book hope that it will be interesting and useful for experts in Computer Science and Electronics, as well as for students and postgraduates, who will be designers of future digital devices and systems.

M. Adamski
A. Barkalov
M. Węgrzyn

Part I
System Design

1 Digital System Design

Samary Baranov

Department of Computer Science, Holon Institute of Technology, 52 Golomb St., P.O.B. 305, Holon, 58102, Israel

Abstract. The most complicated stage of each design, namely the system design, is discussed. An example of the design for a rather simple processor is shown. A design procedure is proposed, which included such steps as combination of separate algorithmic state machines (ASM); synthesis of data-path; control unit design, and composition of data-path and control unit into whole processor. The main advantage of proposed procedure is formalization of the design process, where all steps are formalized and automated in EDA tool "Abelite" designed by the author of this chapter.

1.1 Main Processor Units and Instruction Sets

1.1.1 Main Units

Our processor has 16-bit word and contains main components common for such a device – memory, arithmetic logic unit, program counter, instruction register etc.

Memory (Fig. 1.1,a) has 16-bit address bus (*adr*) and two 16-bit data buses – input (*din*) and output (*dout*). Input *rdwr* (read-write) is the instruction input of the memory. If *rdwr = 0,* instruction *read* is implemented: *A := M[Adr]*. Here *A* is the word at the output bus *dout* and *M[Adr]* is the word of the memory with address *Adr*. This address should be set at the address bus *adr*. If *rdwr = 1,* instruction *write* is implemented: *M[Adr] := B*. Here *B* is the word at the input bus *din* and *M[Adr]* is the word of the memory with address *Adr*. This address should be set at the address bus *adr*. Our processor has two memories – instruction memory *M0* and operand memory *M1*.

Arithmetic logic unit (ALU) is used for implementation of arithmetic and logic operations (Fig. 1.1,b). Two 16-bit inputs *in1* and *in2* serve for operands of ALU, 5-bit input *ctr* serves for instruction codes of ALU operations. One bit input *rg_c* will be connected with the output of carry flag *cf* to implement circular shifts *cil* and *cir* (see below). The result of the operation appears at the output *dout* of ALU. Three additional one-bit outputs *c*, *z* and *v* are connected with the inputs of three flags cf (*carry*), zf (*zero*) and vf (*overflow*). Our ALU is a combinational circuit (circuit without memory) so Processor has *RALU* (the register of ALU) to save the results of arithmetic and logic operations.

M. Adamski et al. (Eds.): Design of Digital Systems and Devices, LNEE 79, pp. 3–41.
springerlink.com

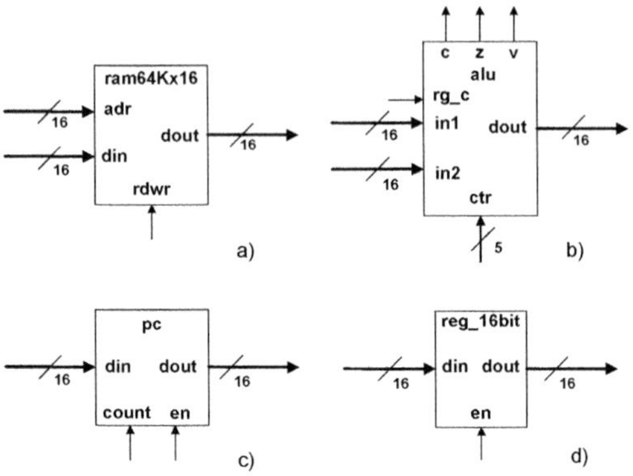

Fig. 1.1 Processor units

Program counter (*PC*) points to the next instruction which should be implemented (Fig. 1.1,c). It saves the address of such instruction. Input *en* is used to load 16-bit information from *din* to the counter when *en* = *1*. When *count* = *1* the content of the counter increases by one.

Instruction register contains the instruction which is being currently implemented. As most of processor registers, it has the structure shown in Fig. 1.1,d. Such a register has 16 clocked *D* flip-flops with input *rst* to reset it and input *en* (enable signal), this signal should be equal to one to write information from input *din* into the register. We skipped signals *clk* (clock) and *rst* (reset) in our pictures.

One more small memory – Block of Register *(BoR)* contains sixteen 16-bit registers (Fig. 1.2). For operation *write*, there must be some data at the bus *din* and an address at the bus adrW. When enable signal *en* = *1*, data is loaded through the input *din* at the rising edge of the clock signal. For example, if *adrW* = *0011* and *en* = *1*, then *R[3]* (register number 3) will get information from input *din*.

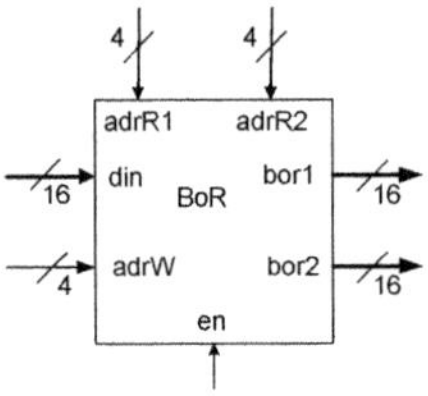

Fig. 1.2 Block of registers

In operation *read*, two registers can be read at the same time. Their addresses are specified with the *adrR1* and *adrR2* signals. To avoid unintentional storage of information during the read operation, the signal *en* should be set to '*0*'.

1.1.2 Instruction Set and Instruction Formats

The instruction set of Processor is presented in Table 1.1. It contains four subsets, or subgroups of operations: *aosh, load, branch* and *inout.* Assembly names of instructions are in the column *Name*, their machine codes are in the column *Code*. Column *Description* contains short description of each operation. You can use Table 1 as a reference table. We will discuss execution of some instructions in the following sections.

Table 1.1 Instruction set

Type	Name	Description	Code b0-b4	Format and addressing mode: Short b5 = 0 Dir	Long b5 = 1 Dir b7=0	Long b5 = 1 Imm b7=1
aosh	add	Add Op1 and Op2; store result in Op1	00 001	•		
	and	Bitwise logical AND between Op1 and Op2 store result in Op1	00 010	•		
	sub	Subtract Op2 from Op1; store result in Op1	00 011	•		
	shl	Shift Op2 one bit to the left; store result in Op1	00 100	•		
	shr	Shift Op2 one bit to the right; store in Op1	00 101	•		
	cil	Rotate Op2 to the left; store result in Op1	00 110	•		
	cir	Rotate Op2 to the right; store result in Op1	00 111	•		
load	lod	Copy Op2 into Op1	10 000	•	•	•
	str	Copy Op1 into Op2	10 001	•	•	
	inc	Increment Op1; store result in Op1	10 010	•		
	dec	Decrement Op1; store result in Op1	10 011	•		
	com	Complement Op1; store result in Op1	10 100	•		
branch	bcz	If z = 1, load new address into PC	01 000	•		•
	bcf	If v = 1, load new address into PC	01 001	•		•
	bcc	If c = 1, load new address into PC	01 010	•		•
	bun	Branch unconditionally to a new address	01 011	•		•
inout	ski	Skip next instruction (if flag of input fgi = 1, increment PC twice)	11 000			
	sko	Skip next instruction (if flag of output fgo = 1, increment PC twice)	11 001			
	inp	Copy input register into one of the registers of BOR and reset fgi	11 010	•		
	out	Copy one of the registers from BOR into output register and reset fgo	11 011	•		
	ion	Set flag interrupt enable (ien) to 1	11 100			
	iof	Set flag interrupt enable (ien) to 0	11 101			

Processor has two instruction formats (Fig. 1.3) – *short* (one word, 16 bits) and *long* (two words, 32 bits). The first twelve bits have the same mission in both formats. Bit 5 points to the *length* of instructions – *short* (bit 5 equal to zero) or *long* (bit 5 equal to one). The next two bits define the *addressing mode*. We will talk about it later.

Short instruction (needs IR1 only)

0	1	2	3	4	5	6	7	8	9	10	11	12	13	14	15
Instruction code					0	Addr mode		First operand address				Second operand address			

Long instruction (needs IR1 and IR2)

0	1	2	3	4	5	6	7	8	9	10	11	12	13	14	15
Instruction code					1	Addr mode		First operand address							

16	17	18	19	20	21	22	23	24	25	26	27	28	29	30	31
Second operand (data or address)															

Fig. 1.3 Instruction formats

The short instruction cannot use operands from operand memory M1. To access this memory, the 16 bit address should be in the instruction. So, the short instruction uses only operands from BoR – bits 8-11 define the address of the first operand in BoR and the address of the result, whereas bits 12-15 define the address of the second operand in BoR. In the long instruction, we will not use the field from bit 12 to bit 15, the second operand is in the memory M1 and its 16 bit address is in the second instruction word. Sometimes, we have operand itself in the second word – see the immediate addressing mode below.

Our Processor uses 16-bit words, so 16-bit *Instruction Register* cannot save the long instruction. Therefore, Processor has two instruction registers – *IR1* and *IR2*, sixteen bits each. In the program memory *M0*, the short instruction takes one word, the long one – two sequential words. Processor fetches the short instruction into *IR1* and the long instruction – into *IR1* and *IR2*. The first half of the long instruction, containing its code, length, addressing mode and two short addresses is fetched into *IR1*, the second half – into *IR2*.

1.1.3 Addressing Modes

Addressing mode defines how Processor finds operands using their addresses in the instruction. There are many different addressing modes, we will discuss most of them but will use only some of them in our Processor. In both formats – short and long, the address of the first operand and the address of the result are in bits 8-11 of the first instruction word. So, the different addressing modes will be essential only for the second operand. Now we shortly discuss several addressing modes:

- *Direct addressing* mode is the simplest one. In this mode, addresses of operands are written within the instruction. In the short format (Fig. 1.4,a), addresses of the first and the second operands *A1* and *A2* are in bits 8-11 and 12-15, operands themselves are in *BoR*. In the long format (Fig. 1.4,b), address *A1* is in the bits 8-11 and the long address *A3* of the second operand is in the second word of such instruction. The second operand is in the operand memory *M1* in the cell with address *A3*.

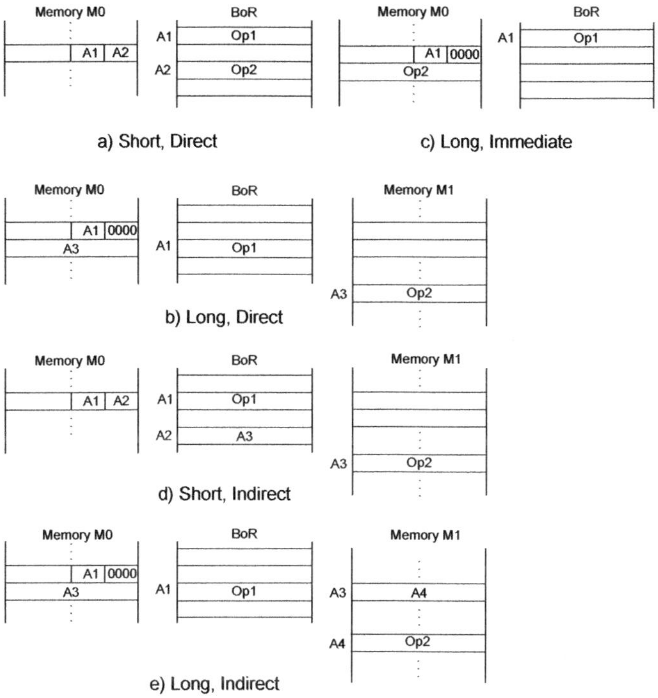

Fig. 1.4 Addressing modes

- In *immediate mode*, not the address of the operand, but the second operand itself is written in the second word of the instruction (Fig. 1.4,c). Thus, the immediate mode can be used only in the long instructions.
- In *indirect mode*, not the address of the operand, but the address of the address of the operand is written in the instruction. For the short format (Fig. 1.4,d), *A2* in the field 12-15 points to the register in *BoR*, where the memory address *A3* of the second operand is saved. For the long format (Fig. 1.4,c), address *A3* in the second word of the instruction points to the cell of the memory *M1* which contains address *A4* of the second operand. Thus, to

reach the second operand in the long instruction, Processor should access the memory twice: first time – for its address, second time – for the oper and itself.

- In the *indexed addressing mode*, Processor uses an index register to find the address of the operand. As a rule, any *BoR* register except BoR[0], can be used as an index register. For the long direct indexed mode the content *A2* of the second address field (bits 12-15) in the first instruction word points to the index register in *BoR*. Processor should add the content of this index register (*offset*) to the long address *A3* (base address) from the second instruction word to find the address of the second operand (*A3* + *offset*). Exactly in the similar way, we can define indirect indexed addressing mode.

We will use only two addressing modes in our Processor – direct and indirect. But I left here two bits (*6* and *7* in Fig. 1.3; *00* – direct mode, *01* – immediate mode) because my students use various addressing modes in their projects.

1.2 ASMs for Processor Instructions

How to describe instructions by ASMs?

Before we present ASMs for processor instructions let us discuss how to describe the instruction execution with ASM [7, 8]. As an example we will use ASM for instruction *lod* in Fig. 1.5. The goal of this instruction is to send the second operand to the place of the first one. For the long instruction, the second operand is in memory *M1*, for the short instruction it is in *BoR* with the second short address. The first operand in both cases is in *BoR* with the first address.

Look at Fig. 1.5. A waiting conditional vertex with the logical condition *S* is placed immediately after vertex *Begin*. You can look at *S* as a signal from a *Start-Stop* button, so the instruction will be executed only when *S* = *1*. Such conditional vertex should be at the beginning of ASM for each instruction. The next condition *DMA* (Direct Memory Access) corresponds to the special mode that doesn't execute an operation but connects its memory with the outside storage device to read or write information. *Ext_RdWr* is the signal from outside. When *Ext_RdWr* = *1*, external system writes information to memory *M0* (external signal *M* should be equal to zero) or to memory *M1* (*M* = *1*) from its output *Ext_out*. When *Ext_RdWr* = *0*, external system reads information from memory *M0* or *M1* to its input *Ext_in*.

Signal *R* in the next conditional vertex tells us about an interrupt. If *R* = *1*, the special mode of Processor – *Interrupt* takes place. When *R* = *0*, there is no interrupt and the instruction is executed in the regular mode. At the end of each instruction Processor checks the conditions for interrupt; we will discuss it later as well. So, now we will consider the implementation of instruction *lod* (long, direct) from *R* = *0* (from the beginning of block *Fetch1*) until the first condition *IEN* of block *CheckInt*.

The only information important prior to instruction implementation is the content of *PC*, which contains an address of the instruction in memory *M0*. Therefore, in the first step (the first operator vertex in Fetch1 in Fig. 1.5), the content of *PC* should be sent to the address bus of the memory *(Adr0 := PC)* and the content of the memory cell with this address will be sent to the first instruction register IR1 *(IR1 := M0[Adr0])*. Only when an instruction, or the first word of the long instruction is in *IR1*, Processor can analyze the instruction code *(IR1(0-4))*, length *(IR1(5))*, addressing mode *(IR1(6-7))* etc. In our notation, we use square brackets for addresses *(M0[Adr0])* and parenthesis for the fields of the register. For example, *IR1(0-4)* means bits *0-4* of *IR1*, *IR1(5)* – the fifth bit of *IR1* etc.

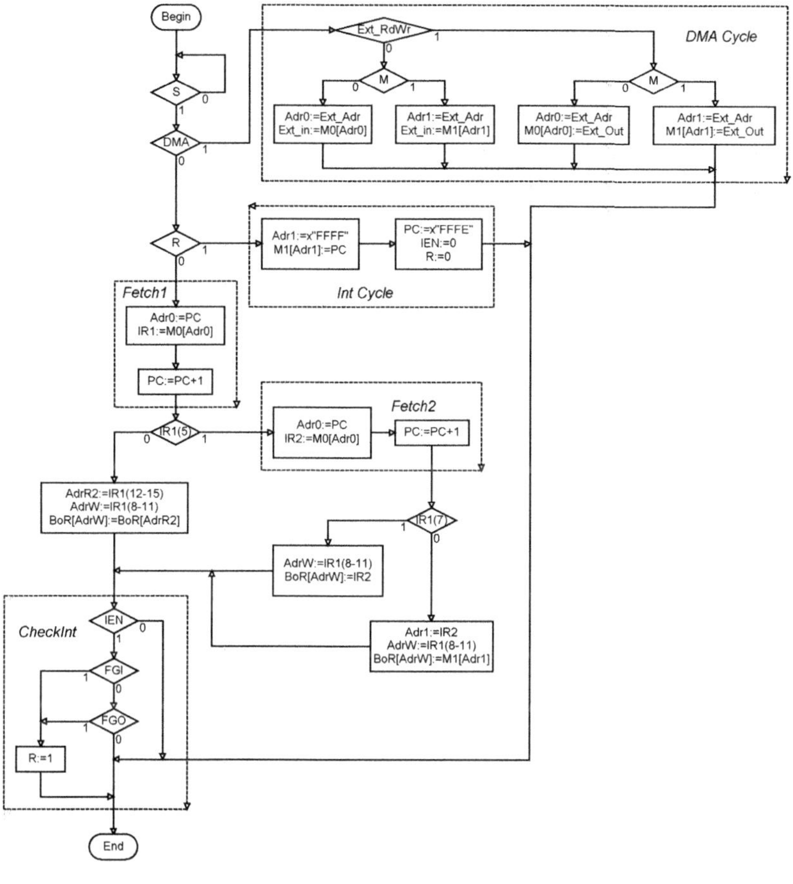

Fig. 1.5 ASM of instruction *lod*

In the second step (the second operator vertex in Fetch1 in Fig. 1.5), Processor increments *PC (PC := PC + 1)*. It is necessary to get an address of the next instruction in *PC* if the current instruction is short, or to give *PC* the address of the second word of the long instruction. So, if the current instruction is short, the fetch is finished and the instruction is in *IR1*. If the instruction is long, the fetch should continue to access memory *M0* again for the second part of this instruction – see the check of *IR1(5)* and the stage Fetch2 with two steps in Fig. 1.5.

After Fetch2, the long instruction is in two instruction registers *IR1* and *IR2*. In the case of the direct addressing mode *(IR(7) = 0)*, the second operand is in memory *M1* and this operand should be loaded into the place of the first operand in *BoR*:

$$Adr1 := IR2;\ AdrW := IR1(8\text{-}11);\ BoR[AdrW] := M1[Adr1].$$

If *IR1(5)* = *0*, we have the short direct mode (see the check of the conditional vertex containing *IR1(5)* immediately after Fetch1 in Fig. 1.5):

$$AdrR2 := IR1(12\text{-}15);\ AdrW := IR1(8\text{-}11);\ BoR[AdrW] := BoR[AdrR2].$$

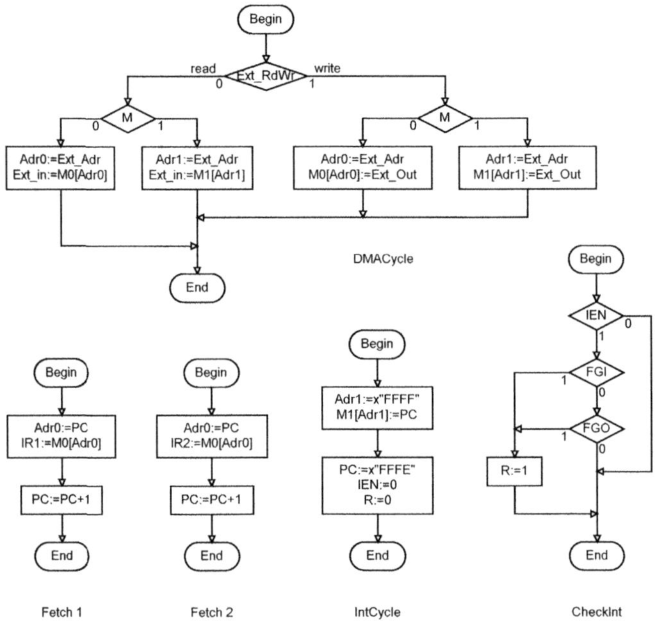

Fig. 1.6 Generalized Operators

For the long immediate mode (see *IR1(7)* = *1* after Fetch2 in Fig. 1.5) we have:

$$AdrW := IR1(8\text{-}11);\ BoR[AdrW] := IR2,$$

because the second operand itself, not its address, is in *IR2*. Note, that it is sufficient to check only bit 7 in *IR1* for the addressing mode since we use only two addressing modes – direct *(00)* and immediate *(01)*.

Blocks Fetch1, DMACycle, IntCycle and CheckInt should be in ASMs for all instructions Block Fetch2 will be only in ASMs for instructions with the long format. These blocks are drawn as separate ASMs in Fig. 1.6. If we replace dotted blocks in Fig. 1.5 by generalized operators we will get ASM in Fig. 1.7, which is simpler and clearer than ASM in Fig. 1.5.

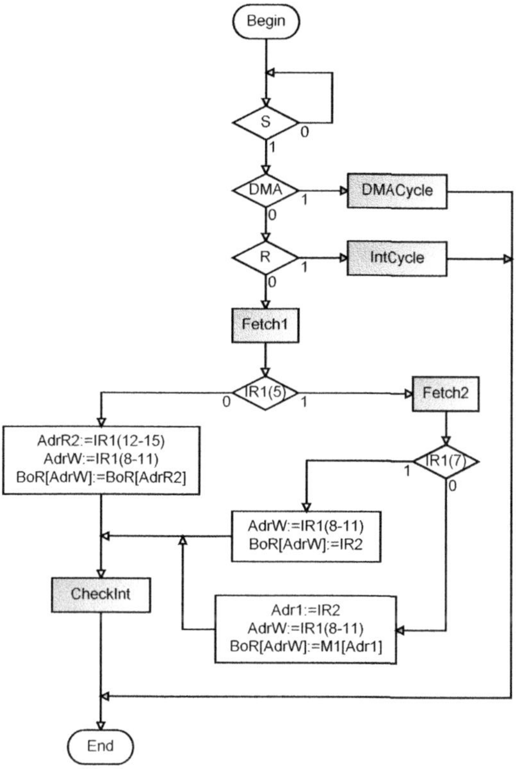

Fig. 1.7 ASM of instruction *lod* with generalized operators

In the next several figures, we will present at least one ASM from each group of instructions with generalized operators.

All instructions from group Aosh (see Table 1.1) have a short format and are implemented in *ALU*. Processor must send operands and instruction code to *ALU* and return the result from *RALU* to the place of the first operand with an address in *IR1(8-11)*. We can divide these instructions into two subsets

- *ao*, containing instructions *add, and* and *sub*. They use two operands from BoR with addresses in *IR1(8-11)* and *IR1(12-15)* – see Fig. 1.8;
- *sh*, containing other instructions. They use one operand from BoR with an address in *IR1(12-15)*.

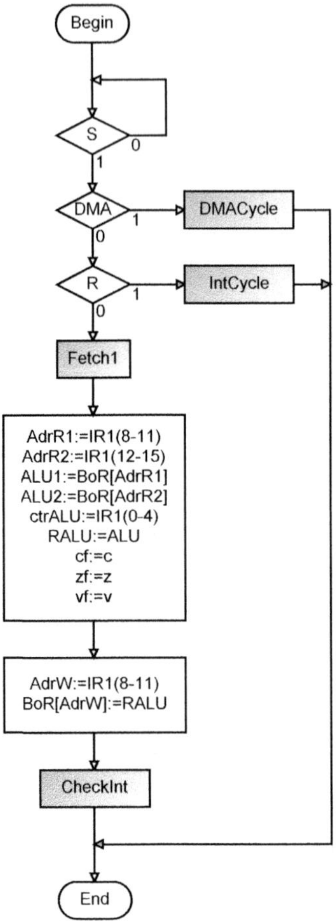

Fig. 1.8 ASM *ao*

ASM for three conditional branch instructions are presented in Fig. 1.9. In each of these instructions, the address of the next instruction depends on the values of flags – *zero* (flag *zf*), *carry* (flag *cf*) and *overflow* (flag *vf*), which are results of the previous operations in ALU. After fetch, short or long, which depends on *IR1(5)*, Processor checks the value of flag *zf* (see the shadowed conditional vertices in Fig. 1.9,a). If *zf* = *1*, a jump to the instruction with the new address takes place. If *zf* = *0*, Processor executes the instruction which is immediately after *bcz* in memory *M0*. To construct two other ASMs we should replace the shadowed vertices by a vertex with *cf* for instruction *bcc* (Fig. 1.9,b) or by a vertex with *vf* for *bcf* (Fig. 1.9,c).

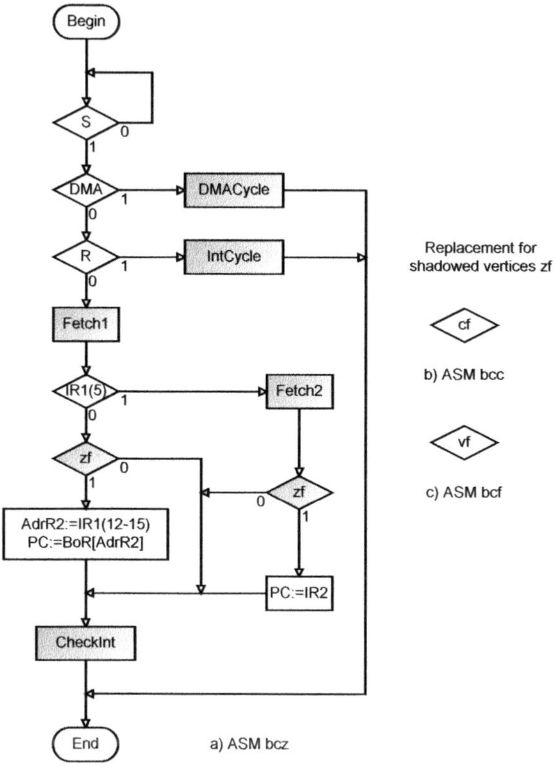

Fig. 1.9 ASMs *bcz, bcc, bcf*

All instructions from group *Inout* are short and only two of them – *inp* and *out* use an address of the operand written in *IR1(8-11)*. Other instructions are zero-address instructions. To explain ASMs of *inp* and *out* let us appeal to Fig. 1.10. An input device can send new information into input register *InpR* at any time. At the same time, it sets a special flag of input *FGI (FlaG of Input)* into *1*. The goal of instruction *inp* is to take the content of *InpR* into *BoR* with an address in *IR1 (8-11)* and to reset *FGI*. Thus, the implementation of this instruction is reduced to one shadowed operator vertex in Fig. 1.11,a:

AdrW := IR1(8-11); BoR[AdrW] := InpR; FGI := 0.

To construct an ASM for instruction *out* we should replace the shadowed vertex in Fig. 1.11,a by the vertex in Fig. 1.11,b. In this instruction Processor sends the word from *BoR* with an address in *IR1(8-11)* into output register *OutR* and resets *FGO (FlaG of Ouput)*:

AdrR1 := IR1(8-11); OutR := BoR[AdrR1]; FGO := 0.

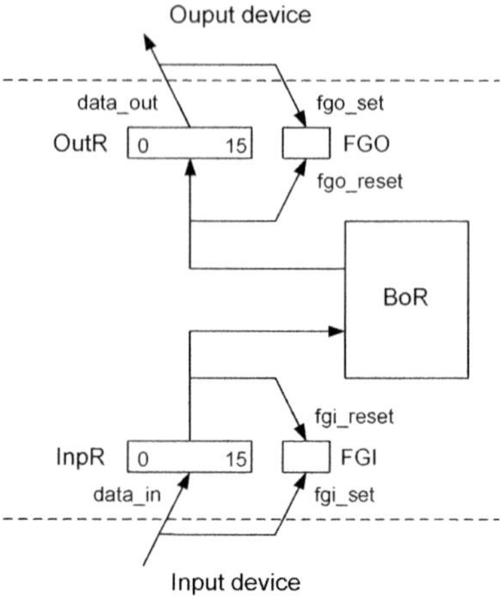

Fig. 1.10 Interface between Input-Output system and Processor

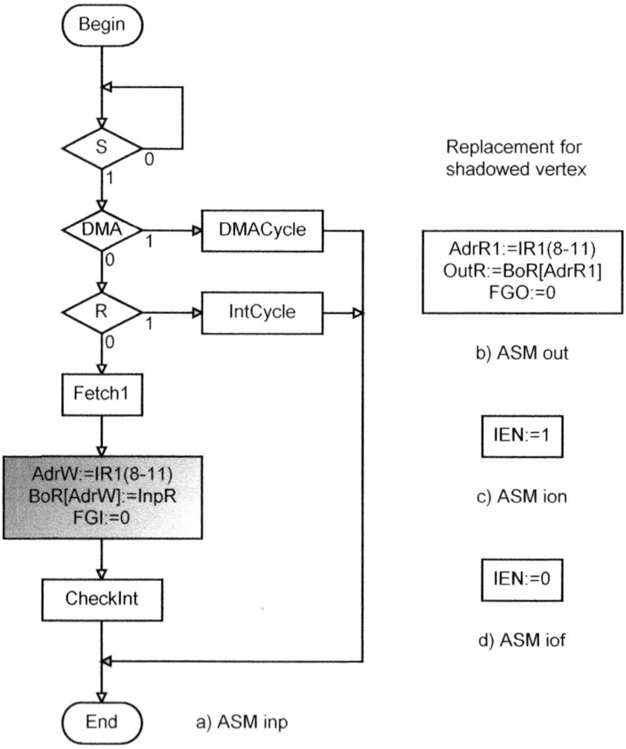

Fig. 1.11 ASMs inp, out, ion, iof

1.3 Data Path Design

1.3.1 Combined Functional ASM

After constructing functional ASMs for each instruction, our next step is to combine them into one ASM [8]. The combined and minimized functional ASM for our Processor is presented in Fig. 1.12. Combining and minimization was made by EDA tool *Abelite*.

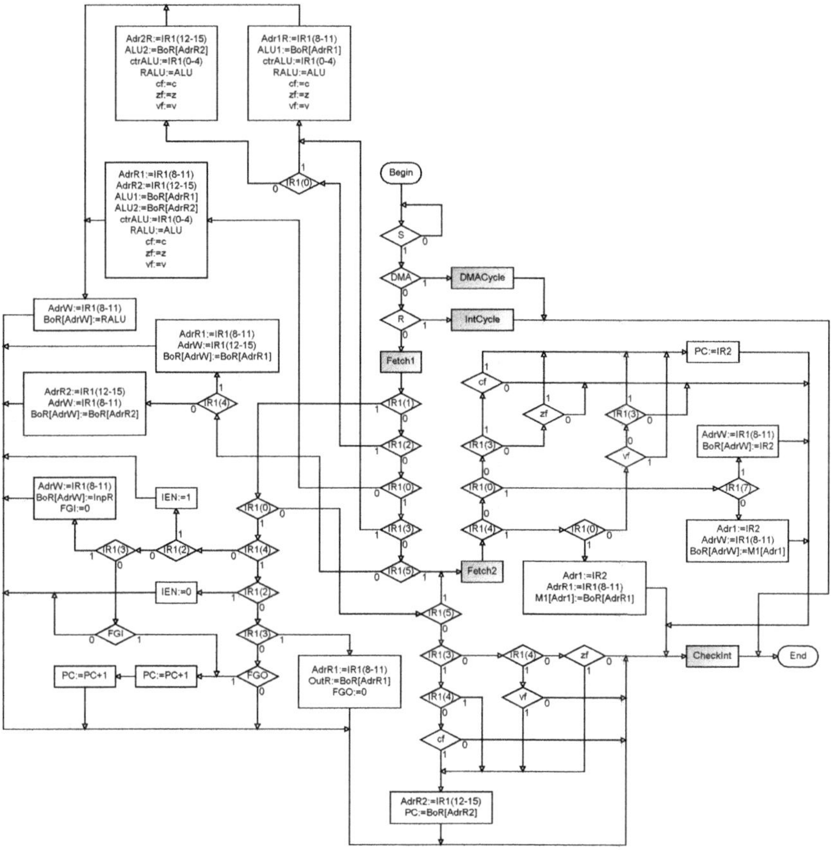

Fig. 1.12 Combined ASM with generalized operators

We use term *functional ASM* because this ASM represents the functional behavior of Processor. In this ASM, processor units in each assignment are used as variables. We do not have a real structure of Processor yet. For example, in assignments

$$Adr1 := IR2;\ AdrW := IR1(8\text{-}11);\ BoR[AdrW] := M1[Adr1],$$

we know nothing about how these units are connected – whether direct connections or buses (multiplexers) are used to provide information from the output of one unit to the input of another one. But for behavior representation and for understanding of this behavior, functional ASMs are very comprehensible and very compact, especially with the use of generalized operators, such as Fetch1, Fetch2, DMACycle, IntCycle and CheckInt. In Fig. 1.13 we replaced these blocks by subgraphs from Fig.1.6 and got total combined Functional ASM G_{Fn} with included generalized operators.

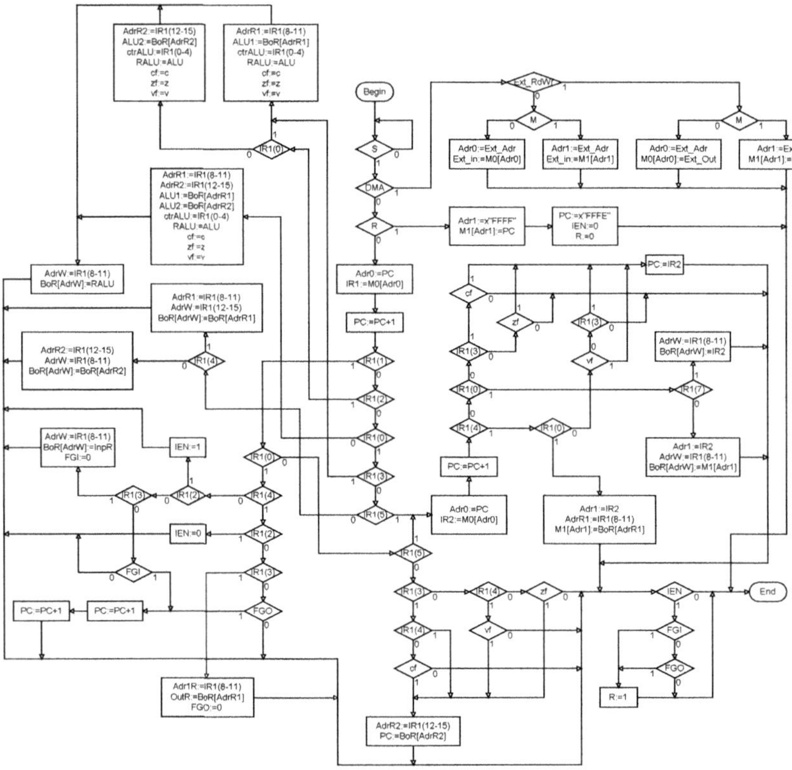

Fig. 1.13 Combined Functional ASM G_{Fn} with included replacements

To explain the difference between functional representation and structural representation (between the function and the structure in our context) let us use a very simple structure in Fig. 1.14. In this figure, we have sources *s0, … , s3* and targets (receivers) *t0, …, t4* (disregard target *t5* for now, we will use it later). For simplicity, you can think about these sources and targets as registers.

Let us suppose that we should send information from each source to each target except *t5*. Of course, we can connect each source with each target directly but the cost of such connection is very high. So, we use here a multiplexer (*MUX*) with four inputs. To send information from any input to the output of *MUX* we should supply 2-bit vector (control signal), corresponding to the input, to the *ctr_mux*. Suppose that the control signal is equal to the number of the input, so for *in0*, the signal at *ctr_mux* is equal to *00*, for *in1* – *01* etc.

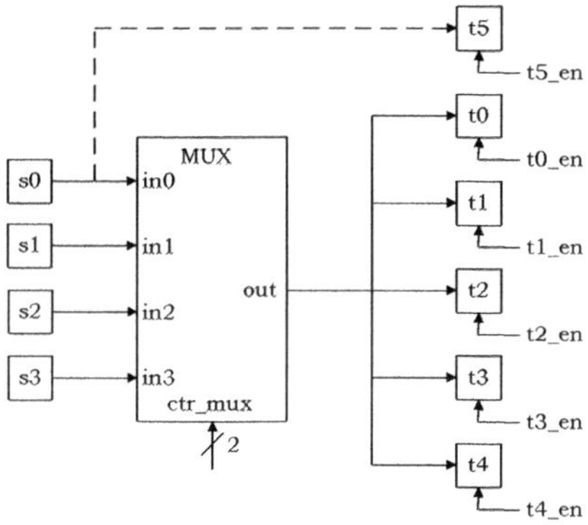

Fig. 1.14 Our first structure

We will discuss several examples in Fig. 1.15. In the first example, source *s2* transfers information to target *t3*. At the functional level, the description is very simple and clear: *t3 := s2*. At the structural level, to pass information from input *in2* to the output of *MUX*, *ctr_mux* should get control signal *10*. To write information from the output of *MUX* to the target *t3* the signal *enable* of *t3* should be equal to *1*: *t3_en := '1'*.

Functional	Structure
1. *t3 := s2*	*ctr_mux := '10'* *t3_en := '1'*
2. *t1 := s3* *t2 := s3*	*ctr_mux := '11'* *t1_en := '1'* *t2_en := '1'*
3. *t5 := s0*	*t5_en := '1'*
4. *t5 := s0* *t4 := s1*	*t5_en := '1'* *ctr_mux := '01'* *t4_en := '1'*

Fig. 1.15 Descriptions at the functional and structural levels

In the second example we send the same information from one source *s3* to two targets *t1* and *t2*. The control signal of *MUX* must be equal to *11* and two *enable* signals, equal to *1*, will write information into *t1* and *t2*. However, if we should like to pass information from several sources to several outputs at the same time, we need several MUXes because one *MUX* transfers only one input to its output. To put it more precisely, we need as many MUXes as the number of sources we wish to pass simultaneously.

One of the targets *t5* in our structure is connected only with one source, i.e. it can get information only from *s0*. We call such a connection a *direct connection*. In example 3 in Fig. 1.15, to send information from *s0* to *t5* it is sufficient to supply enable signal to this target: *t5_en := '1'*.

In the fourth example we have two simultaneous transfers – the first through the direct connection: *t5 := s0*, and the second – through the indirect connection: *t4 := s1*. Since only one indirect transfer takes place in this example, one MUX is sufficient in such a case.

Let us return to our Processor. Any digital system is usually regarded as a composition of *Control unit* and *Operational unit* (*Data Path*) – see Fig. 1.16. Data path contains such regular blocks as memory, registers, ALU, counters, coders, encoders, multiplexers, demultiplexers etc. A control unit produces a sequence of control signals that force implementation of microoperations in data path. Sometimes designers include cloud (non-regular) circuits in data path as well. In Fig. 1.17 we have a fragment of data path with two registers *R1*, *R2* and a cloud circuit.

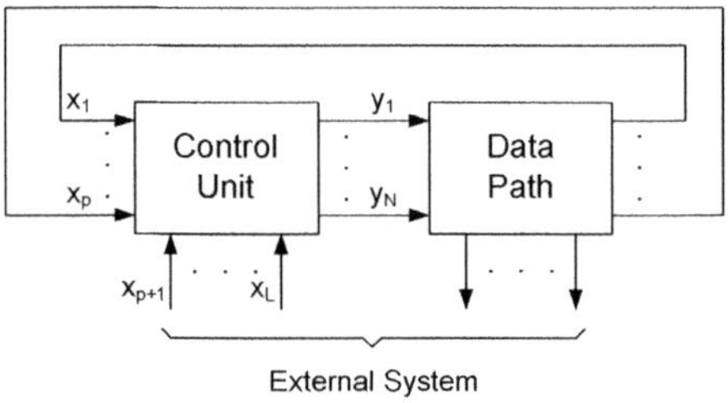

Fig. 1.16 Digital system as a composition of Control unit and Data path

Suppose that in the digital system there are transfers from *R1* to *R2* at different times with different conditions. Designers often use a cloud circuit to realize some Boolean function, and the output of this circuit is the signal for the transfer. So, this circuit defines when and under which logic conditions the transfer information from *R1* to *R2* takes place.

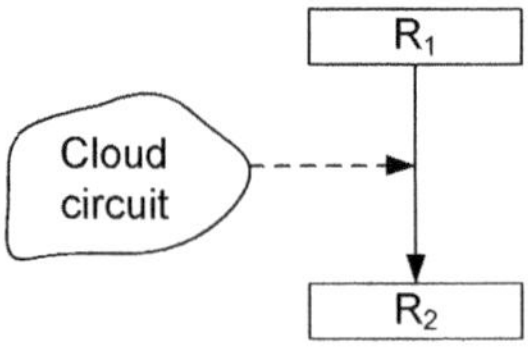

Fig. 1.17 Element of Data path with a cloud circuit

One of the main concepts in our design methodology is the construction of "*naked data path*". Naked data path doesn't contain any cloud circuits, only standard regular units with their inputs and outputs. Such units can be predesigned or even taken from libraries. We leave all check-ups of conditions to control unit. We can afford this because we know how to design very complicated FSM with hardly any constraints on their size, that is, the number of inputs, outputs and states. We will try to show that such design and its verification are very simple. Moreover, we will formalize a design of the digital system with naked data path.

1.3.2 Process Table and Connection Graph

To design the data path of our processor we will use the combined functional ASM G_{Fn} in Fig. 1.13. When we construct the data path we will transform functional ASM into structural ASM with microinstructions and microoperations corresponding to the data path. The last step of our design will consist in constructing a control unit (Finite state machine) and combining FSM with the data path.

We will begin with filling the left, functional part of the *Process table* (Table 1.2). To do this, we copy each microinstruction from combined ASM G_{Fn} (Fig. 1.13) into this table. If some microinstruction appears several times in combined ASM G_{Fn} we write it several times in Table 1.2. For example, microinstruction $PC := PC + 1$ is written in four vertices of ASM G_{Fn}, so it is written four times in Table 1.2. *The order of writing microinstructions in this table is absolutely unimportant.*

There are two classes of microoperations in this table. The first one presents the transfer information from the output of *one* unit to the input of *another* unit, for example, $Adr1 := IR2$, $M1[Adr1] := BoR[AdrW]$, $RALU := ALU$ etc. These microoperations are marked in the last column of the left part of Table 1.2 by the number equal to the number of bits in the source and in the target (1, 4, 5 or 16 in our example). The microoperations from the second class, such as $PC := PC + 1$, $R := 1$, $IEN := 0$, $R := 0$, $FGI := 0$ etc. are executed in one operational unit. They are marked by *0* in the same column. For some time we will continue to work only with microoperations from the first class marked by nonzero numbers in the process table. The microoperations of the second class will not be "our clients" for a while.

In the connection graph (Fig. 1.18a for the sixteen bit transfers) source A, located on the left, is connected by an arc with target B on the right, if there is microoperation $B := A$ in the set of microoperations marked by 16. In this graph, we use a dotted arc if a target has only one source (similarly, target $t5$ has only one source $s0$ in Fig. 1.14). Otherwise, we use a solid arc. Each solid arc has a weight, written over the arc. This weight is equal to the number of appearances of the corresponding microoperation in Table 1.2. For example, the arc with source *PC* and target *Adr0* has the weight equal 2, because $Adr0 := PC$ appears twice in Table 1.2.

We assume that each dotted arc has a weight equal to *0* and we do not write zero weights in our connection graph. Each source of the connection graph has a weight written on the left of this source. This weight is equal to the sum of weights of arcs outgoing from the source. We will use these weights later in the optimal encoding of the MUX inputs.

Table 1.2 Process table

	Functional ASM			Structural ASM		
Micro-instruction		Functional microoperations		Structural microoperations	Minimized structural microoperations	
Y1	y1	AdrW:=IR1(8-11)	4	ctr_mux3 := 0	ctr_mux2[1] := 1	y1
	y2	BoR[AdrW]:=RALU	16	ctr_mux2 := 0110	ctr_mux2[2] := 1	y2
				bor_en := 1	bor_en := 1	y3
Y2	y3	AdrR1:=IR1(8-11)	4	ralu_en := 1	ralu_en := 1	y4
	y4	AdrR2:=IR1(12-15)	4	cf_en := 1	cf_en := 1	y5
	y5	ALU1:=BoR[AdrR1]	16	zf_en := 1	zf_en := 1	y6
	y6	ALU2:=BoR[AdrR2]	16	vf_en := 1	vf_en := 1	y7
	y7	ctrALU:=IR1(0-4)	5			
	y8	RALU:=ALU	16			
	y9	cf:=c	1			
	y10	zf:=z	1			
	y11	vf:=v	1			
Y3	y4	AdrR2:=IR1(12-15)	4	ralu_en := 1	ralu_en := 1	y4
	y6	ALU2:=BoR[AdrR2]	16	cf_en := 1	cf_en := 1	y5
	y7	ctrALU:=IR1(0-4)	5	zf_en := 1	zf_en := 1	y6
	y8	RALU:=ALU	16	vf_en := 1	vf_en := 1	y7
	y9	cf:=c	1			
	y10	zf:=z	1			
	y11	vf:=v	1			
Y4	y12	Adr1:=IR2	16	ctr_mux1 := 001	ctr_mux1[2] := 1	y8
	y1	AdrW:=IR1(8-11)	4	ctr_mux3 := 0	ctr_mux2[3] := 1	y9
	y13	BoR[AdrW]:=M1[Adr1]	16	ctr_mux2 := 0001	bor_en := 1	y3
				rdwrM1 := 0		
				bor_en := 1		
Y5	y1	AdrW:=IR1(8-11)	4	ctr_mux3 := 0	ctr_mux2[0] := 1	y10
	y14	BoR[AdrW]:=IR2	16	ctr_mux2 := 1000	bor_en := 1	y3
				bor_en := 1		
Y6	y4	AdrR2:=IR1(12-15)	4	ctr_mux3 := 0	ctr_mux2[1] := 1	y1
	y1	AdrW:=IR1(8-11)	4	ctr_mux2 := 0101	ctr_mux2[3] := 1	y9
	y15	BoR[AdrW]:=BoR[AdR2]	16	bor_en := 1	bor_en := 1	y3
Y7	y12	Adr1:=IR2	16	ctr_mux1 := 001	ctr_mux1[2] := 1	y8
	y3	AdrR1:=IR1(8-11)	4	ctr_mux2 := 0000	rdwrm1 := 1	y11
	y16	M1[Adr1]:=BoR[AdrR1]	16	rdwrM1 := 1		
Y8	y3	AdrR1:=IR1(8-11)	4	ctr_mux3 := 1	ctr_mux3 := 1	y12
	y17	AdrW:=IR1(12-15)	4	ctr_mux2 := 0000	bor_en := 1	y3
	y18	BoR[AdrW]:=BoR[AdrR1]	16	bor_en := 1		
Y9	y3	AdrR1:=IR1(8-11)	4	ralu_en := 1	ralu_en := 1	y4
	y5	ALU1:=BoR[AdrR1]	16	cf_en := 1	cf_en := 1	y5
	y7	ctrALU:=IR1(0-4)	5	zf_en := 1	zf_en := 1	y6
	y8	RALU:=ALU	16	vf_en := 1	vf_en := 1	y7
	y9	cf:=c	1			
	y10	zf:=z	1			
	y11	vf:=v	1			
Y10	y4	AdrR2:=IR1(12-15)	4	ctr_mux1 := 100	ctr_mux1[0] := 1	y13
	y19	PC:=BoR[AdrR2]	16	pc_en := 1	pc_en := 1	y14
Y11	y20	PC:=IR2	16	ctr_mux1 := 001	ctr_mux1[2] := 1	y8
				pc_en := 1	pc_en := 1	y14

Table 1.2 (*continued*)

Y12	y21	PC:=PC+1	0	pc_count := 1	pc_count := 1	y15
Y12	y21	PC:=PC+1	0	pc_count := 1	pc_count := 1	y15
Y12	y21	PC:=PC+1	0	pc_count := 1	pc_count := 1	y15
Y12	y21	PC:=PC+1	0	pc_count := 1	pc_count := 1	y15
Y13	y1	AdrW:=IR1(8-11)	4	ctr_mux3 := 0	ctr_mux2[1] := 1	y1
	y22	BoR[AdrW]:=InpR	16	ctr_mux2 := 0100	bor_en := 1	y3
	y23	FGI:=0	0	bor_en := 1	fgi_reset := 1	y16
				fgi_reset := 1		
Y14	y3	AdrR1:=IR1(8-11)	4	outr_en := 1	outr_en := 1	y17
	y24	OutR:=BoR[AdrR1]	16	fgo_reset := 1	fgo_reset := 1	y18
	y25	FGO:=0	0			
Y15	y26	IEN:=1	0	ien_set := 1	ien_set := 1	y19
Y16	y27	IEN:=0	0	ien_reset := 1	ien_reset := 1	y20
Y17	y28	PC:=x"FFFE"	16	ctr_mux1 := 011	ctr_mux1[1] := 1	y21
	y27	IEN:=0	0	pc_en := 1	ctr_mux1[2] := 1	y8
	y29	R:=0	0	ien_reset := 1	pc_en := 1	y14
				r_reset := 1	ien_reset := 1	y20
					r_reset := 1	y22
Y18	y30	Adr1:=x"FFFF"	16	ctr_mux1 := 101	ctr_mux1[0] := 1	y13
	y31	M1[Adr1]:=PC	16	ctr_mux2 := 1001	ctr_mux1[2] := 1	y8
				rdwrM1 := 1	ctr_mux2[0] := 1	y10
					ctr_mux2[3] := 1	y9
					rdwrm1 := 1	y11
Y19	y32	Adr0:=Ext_Adr	16	ctr_mux1 := 000	rdwrm0 := 1	y23
	y33	M0[Adr0]:=Ext_Out	16	rdwrM0 := 1		
Y20	y34	Adr1:=Ext_Adr	16	ctr_mux1 := 000	ctr_mux2[2] := 1	y2
	y35	M1[Adr1]:=Ext_Out	16	ctr_mux2 := 0010	rdwrm1 := 1	y11
				rdwrM1 := 1		
Y21	y34	Adr1:=Ext_Adr	16	ctr_mux1 := 000	ctr_mux2[3] := 1	y9
	y36	Ext_in:=M1[Adr1]	16	ctr_mux2 := 0001		
				rdwrM1 := 0		
Y22	y32	Adr0:=Ext_Adr	16	ctr_mux1 := 000	ctr_mux2[2] := 1	y2
	y37	Ext_in:=M0[Adr0]	16	ctr_mux2 := 0011	ctr_mux2[3] := 1	y9
				rdwrM0 := 0		
Y23	y38	R:=1	0	r_set := 1	r_set := 1	y24
Y24	y39	Adr0:=PC	16	ctr_mux1 := 010	ctr_mux1[1] := 1	y21
	y40	IR1:=M0[Adr0]	16	rdwrM0 := 0	ir1_en := 1	y25
				ir1_en := 1		
Y25	y39	Adr0:=PC	16	ctr_mux1 := 010	ctr_mux1[1] := 1	y21
	y41	IR2:=M0[Adr0]	16	rdwrM0 := 0	ir2_en := 1	y26
				ir2_en := 1		

Sometimes, when a graphic representation of a connection graph is too complicated, we can present it as a list (Fig. 1.18,b). In this list, weights on the left are the weights of sources and weights on the right are the weights of arcs. Thus, the weight on the right which is equal to zero corresponds to the direct connection.

Next important information that we use in design of data path is the *list of parallel (concurrent) microoperations* (Fig. 1.19 for the sixteen-bit transfers). In this list, we include microinstructions, containing two or more microoperations, marked by 16 in the fourth column of Table 1.2. I remind you that if several microoperations are in one microinstruction, they are implemented concurrently (at the same clock).

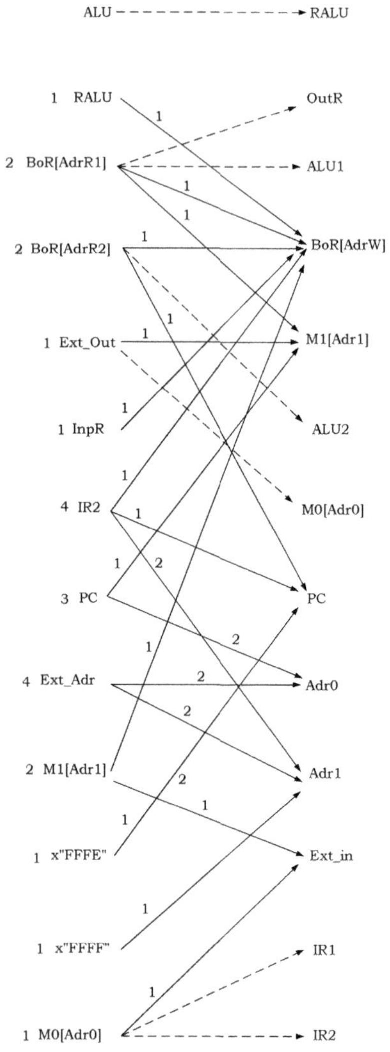

a)

weight	:	sources	targets	:	weight
0	:	ALU	RALU	:	0
2	:	BoR[AdrR1]	ALU1	:	0
			BoR[AdrW]	:	1
			M1[Adr1]	:	1
			OutR	:	0
2	:	BoR[AdrR2]	ALU2	:	0
			BoR[AdrW]	:	1
			PC	:	1
4	:	Ext_Adr	Adr0	:	2
			Adr1	:	2
1	:	Ext_Out	M0[Adr0]	:	0
			M1[Adr1]	:	1
4	:	IR2	Adr1	:	2
			BoR[AdrW]	:	1
			PC	:	1
1	:	InpR	BoR[AdrW]	:	1
1	:	M0[Adr0]	Ext_in	:	1
			IR1	:	0
			IR2	:	0
2	:	M1[Adr1]	BoR[AdrW]	:	1
			Ext_in	:	1
3	:	PC	Adr0	:	2
			M1[Adr1]	:	1
1	:	RALU	BoR[AdrW]	:	1
1	:	x"FFFE"	PC	:	1
1	:	x"FFFF"	Adr1	:	1

b)

Fig. 1.18 Connection Graph

Let us discuss construction of this list from Table 1.2 We didn't insert microinstruction Y_1 in this list because it contains only one 16 bit microoperation. Microinstruction Y_2 contains nine microoperations, but only three of them are marked by 16 in the forth column of Table 1.2.

We can compress the list of parallel microoperations (Fig. 1.19), if we remove microoperations corresponding to direct connections in the connection graph (dotted arcs in Fig. 1.18,a). We have six such microoperations; they are marked by *d* (*direct*) in Fig. 1.19 in the right column. There are 9 symbols *d* in this column because some microoperations occur several times there. After such reduction, three microinstructions – Y_{19}, Y_{24} and Y_{25} contain only one microoperation, so we can remove them as well (it is similar to example 4 in Fig. 1.15). The final list of parallel microoperations is shown in Fig. 1.20.

```
Y2    : y5    ALU1:=BoR[AdrR1]          d
        y6    ALU2:=BoR[AdrR2]          d
        y8    RALU:=ALU                 d

Y3    : y6    ALU2:=BoR[AdrR2]          d
        y8    RALU:=ALU                 d

Y4    : y12   Adr1:=IR2
        y13   BoR[AdrW]:=M1[Adr1]

Y7    : y12   Adr1:=IR2
        y16   M1[Adr1]:=BoR[AdrR1]

Y9    : y5    ALU1:=BoR[AdrR1]          d
        y8    RALU:=ALU                 d

Y18   : y30   Adr1:=x"FFFF"
        y31   M1[Adr1]:=PC

Y19   : y32   Adr0:=Ext_Adr
        y33   M0[Adr0]:=Ext_Out         d

Y20   : y34   Adr1:=Ext_Adr
        y35   M1[Adr1]:=Ext_Out

Y21   : y34   Adr1:=Ext_Adr
        y36   Ext_in:=M1[Adr1]

Y22   : y32   Adr0:=Ext_Adr
        y37   Ext_in:=M0[Adr0]

Y24   : y39   Adr0:=PC
        y40   IR1:=M0[Adr0]             d

Y25   : y39   Adr0:=PC
        y41   IR2:=M0[Adr0]             d
```

Fig. 1.19 Parallel microoperations before considering direct connections

```
Y4   : y12   Adr1:=IR2
       y13   BoR[AdrW]:=M1[Adr1]

Y7   : y12   Adr1:=IR2
       y16   M1[Adr1]:=BoR[AdrR1]

Y18  : y30   Adr1:=x"FFFF"
       y31   M1[Adr1]:=PC

Y20  : y34   Adr1:=Ext_Adr
       y35   M1[Adr1]:=Ext_Out

Y21  : y34   Adr1:=Ext_Adr
       y36   Ext_in:=M1[Adr1]

Y22  : y32   Adr0:=Ext_Adr
       y37   Ext_in:=M0[Adr0]
```

Fig. 1.20 Parallel microoperations after considering direct connections

1.3.3 Graph of Incompatibility. Main MUXes and Direct Connections

If we use MUXes for indirect connections between Processor units, we can connect the output of only one MUX with the input of the target. So, if we have two targets *A* and *C*, one MUX is sufficient to transfer information to this targets if we do not have parallel transfers to *A* and *C* from the different sources.

Let us now suppose that we have two transfers $A := B$ and $C := D$ that must be implemented concurrently. Then we must use two MUXes, the output of one of them will be connected to the input of target *A* and source *B* should be among the inputs of this MUX. The output of the second MUX will be connected to the input of target *C* and source *D* should be among the inputs of this MUX.

After this, our next steps are almost evident. First, we find all targets with non-zero weights of arcs from the connection graph in Fig. 1.18b. These weights are written in the last column of this figure. We go along this column from the top to the bottom and write targets without repetition. Here they are:

$$BoR[AdrW], M1[Adr1], PC, Adr0, Adr1, Ext\_in. \tag{1.1}$$

Next, we construct the *graph of incompatibility* (Fig. 1.21:

1. Each vertex of this graph is a target from (1).
2. We connect two vertices (targets) by edge (line) if these two targets are together in the same microinstruction in the set of parallel micro-operations (as *Adr1* and *BoR[AdrW]* in *Y4*, *Adr1* and *M1[Adr1]* in *Y7*, *Y18* and *Y20*, *Adr1* and *Ext_in* in *Y21* and *Adr0* and *Ext_in* in *Y22* in Fig 1.20).

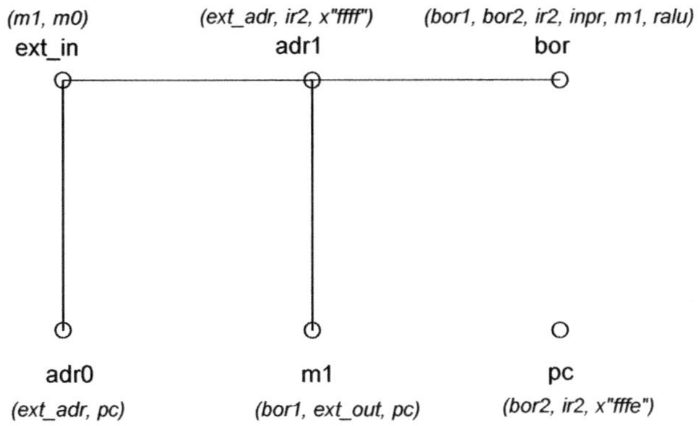

Fig. 1.21 Graph of incompatibility

From the connection graph (Fig. 1.18,b) write sources for each target (vertex). For example, we write *(bor1, bor2, ir2, inpr, m1, ralu)* next to vertex (target) *bor* because *bor* is written 6 times with the sources *bor1, bor2, ir2, inpr, m1, ralu* in the connection graph in Fig. 1.18,b. Here we use the abbreviations bo*r1, bor2* for BoR[AdrR1], BoR[AdrR2] and *bor* for BoR[AdrW].

If two vertices (targets) are connected by edge in this graph we cannot pass information to these targets through the same MUX because these targets are written together in some set of concurrent microoperations with *different* sources. For example, target *adr1* cannot be acquired from the same MUX with *ext_in, bor* and *m1* since *adr1* is connected with these vertices by arcs. However, *adr0* can be acquired from the same MUX with *adr1, m1, pc* or *bor* – *adr0* is not connected with them in the graph of incompatibility.

To find the minimal number of MUXes in our design we must color this graph with a minimal number of colors in such a way that each two connected vertices are colored by different colors. The targets (vertices) colored by the same colors will be received from the same MUXes and the number of MUXes will be equal to the number of colors. And we will use *mux1, mux2, …* as colors for such a coloring.

Table 1.3 Coloring process for our Processor

Vertices	Forbidden vertices	Colors
adr1	ext_in, m1, bor	mux1
ext_in	adr1, adr0	mux2
adr0	ext_in	mux1
bor	adr1	mux2
m1	adr1	mux2
pc	-	mux1

The coloring process is presented in Table 1.3. It is reasonable to order vertices in such table according to their ranks – to the decreasing number of edges connected with each vertex (three such edges for *adr1*, two edges for *ext_in*, one edge for *adr0, bor* and *ml* and zero edges for *pc).* We place these vertices in the column *Forbidden vertices.*

We color the first vertex *adr1* with color *mux1*. Since the second vertex *ext_in* is connected to *adr1 (ext_in* has *adr1* in the column *Forbidden vertices),* we cannot color *ext_in* with the same color *mux1*. We color *adr0* with the same color *mux1* because *adr0* does not contain *adr1* in the second column of Table 1.3. We cannot color *bor,* and *ml* with *mux1* because these vertices are connected with vertex *adr1*. Continue until the end of the list with color *mux1* we use this color for *pc*.

In the next step, taking color *mux2* for *ext_in,* we go down the list and color *bor* and *ml* with *mux2*. Now all vertices are colored. The total number of MUXes (colors) is equal to two.

Thus, we got the outputs of MUXes by coloring process. To get inputs to these MUXes we should refer to the connection graph in Fig. 1.18,b. Let us discuss MUX1 with outputs *Adr1, Adr0* and *PC*. We go along the last but one column *targets* in this figure and search for target *Adr1*. The first time *Adr1* appears as a target with source *Ext_Adr* and the target weight equal *2* (last column). So we include *ext_adr* as an input with input weight equal to *2* (Fig. 1.22,a). Then we continue to descend and find *Adr1* with source *IR2*, its target weight is equal to *2* as well. We put the second input to MUX1. Going down with source *Adr1* we find the third input *x"ffff"* (weight = *1*).

Recall that the weight of microoperation *Adr := Ext_Adr* is equal to the number of appearances of this microoperation in the combined functional ASM. This means that *Ext_Adr* is used twice as the source for target *Adr*, or which is the same, *Adr* is twice the target for *Ext_Adr*. Thus, when we talk about a pair (*source, target*), their weights are equal. On the other hand, the weight in the left column of Fig. 1.18,b is the total weight of the source for all targets. For example, weight 4 on the left of *IR2* is the sum of corresponding weights of source *IR2* with all targets written on the right side. The zero weight there means that the direct connection is used for the corresponding transfer.

Now we should repeat the same for target *Adr0*. The first appearance of *Adr0* in column *targets* is with input *Ext_Adr*, target weight 2. Since we already have such input in MUX1, we add the new weight *2* to the old weight 2 (the weight of input *Ext_Adr* became equal to *4*) and write *Adr0* over the arrow for *Ext_Adr* near *Adr1* to show that this source *Ext_Adr* sends information to *Adr1* and *Adr0* using MUX1 (Fig. 1.22,b). Coming down with target *Adr0*, we insert one more input *PC* with weight 2. Execution the same procedure for target PC gives us the final picture of MUX1 in Fig. 1.22,c.

Continuing in the same way, we constructed MUX2 in Fig. 1.22,d. Note, that the same input can appear in the different MUXes if such input has several targets distributed between several MUXes. For example, *ir2* sends information to *Adr1* and *PC* through MUX1, and to *BoR* – through MUX2. Inputs *PC* and *BoR2* occur in both MUXes as well. We will talk about the input encoding later.

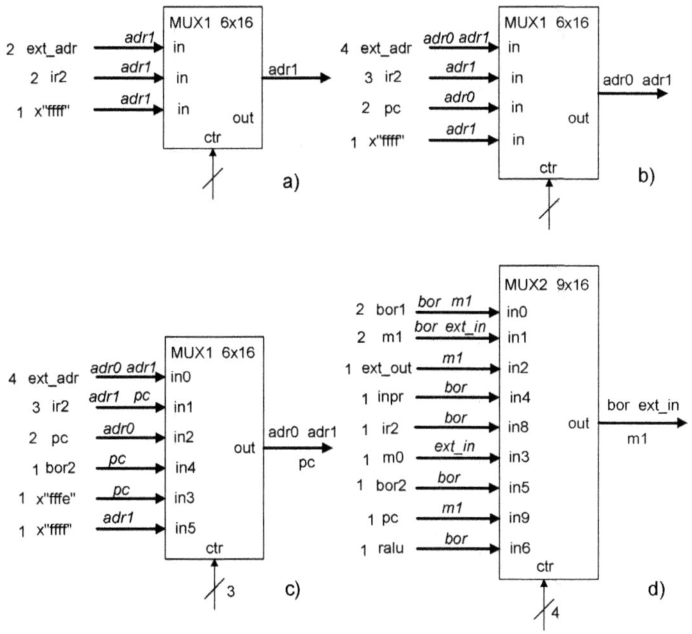

Fig. 1.22 Constructing MUXes for our Processor

Fig 1.23 presents the connection graph for 4-bit transfers. The corresponding graph of incompatibility contains only one vertex *AdrW*. The list of Parallel microoperations prior to considering direct connections is shown in Fig. 1.24.

```
weight : sources       targets : weight
     1 : IR1(12-15)     AdrR2  : 0
                        AdrW   : 1
     5 : IR1(8-11)      AdrR1  : 0
                        AdrW   : 5
```

Fig. 1.23 Connection graph for 4-bit transfers as a list

```
Y2      : y3     AdrR1:=IR1(8-11)    d
          y4     AdrR2:=IR1(12-15)   d
Y6      : y1     AdrW:=IR1(8-11)
          y4     AdrR2:=IR1(12-15)   d
Y8      : y3     AdrR1:=IR1(8-11)    d
          y17    AdrW:=IR1(12-15)
```

Fig. 1.24 Parallel microoperations for 4-bit transfers

After considering direct connections, we will find that there are no parallel microoperations for 4-bit transfers. Thus, we get one *MUX3* presented in Fig. 1.25.

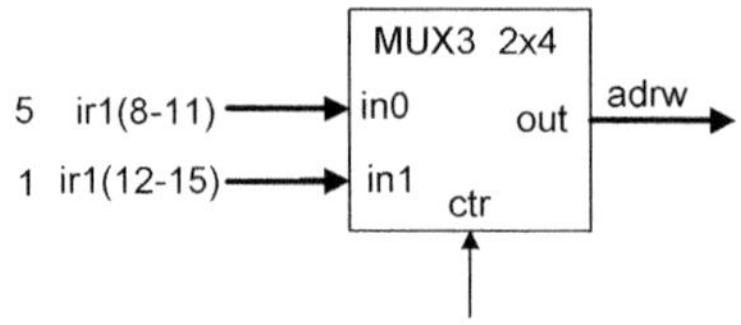

Fig. 1.25 Main MUX for 4-bit transfers

Data Path for our example is presented in Fig. 1.26. The dotted lines correspond to direct connections. To clear this picture up we removed wires for two signals – reset asynchronous (*rst*) and clock (*clk*). Thus, we finished the system design for Data Path. What have we got? We have got a "naked Data Path" – it means that our Data Path doesn't contain "cloud circuits".

1.4 Control Unit Design

1.4.1 Transformation of Functional ASM into Structural ASM

Our next step is design of the Control unit for our Processor. For this, let us return to the process table (Table 1.2) and implement each functional microoperation from the third column by structural microoperation (or by the set of structural microoperations) in the fifth column of this table. Once again, we will postpone consideration of input encoding for MUXes and will use codes from Fig. 1.22,cd and Fig. 1.25.

Let we have a functional microoperation *A* := *B*. To implement this assignment we must *read* information from *B* and *write* it to *A*. Now we will discuss reading and writing information in our design.

Read1. When source *B* is a combinational circuit or a register, we should not supply special signals to read information from such devices. This information is always at its output. Examples:

RALU := *ALU; zf* := *z;* (reading from combinational circuits).
Adr1 := *IR2; BoR[AdrW]* := *InpR;* (reading from registers).

Read2. Source *B* is *BoR* (Block of Registers). BoR has two outputs *BoR[AdrR1] (bor1)* and *BoR[AdrR2] (bor2)*. To choose the register from BoR, the Control unit should give address to *AdrR1* or *AdrR2* (or to both of them) and send information from the output(s) to the target. Examples:

$$AdrR2 := IR1(12\text{-}15);\ ALU2 := BoR[AdrR2];$$
$$AdrR1 := IR1(8\text{-}11);\ OutR := BoR[AdrR1].$$

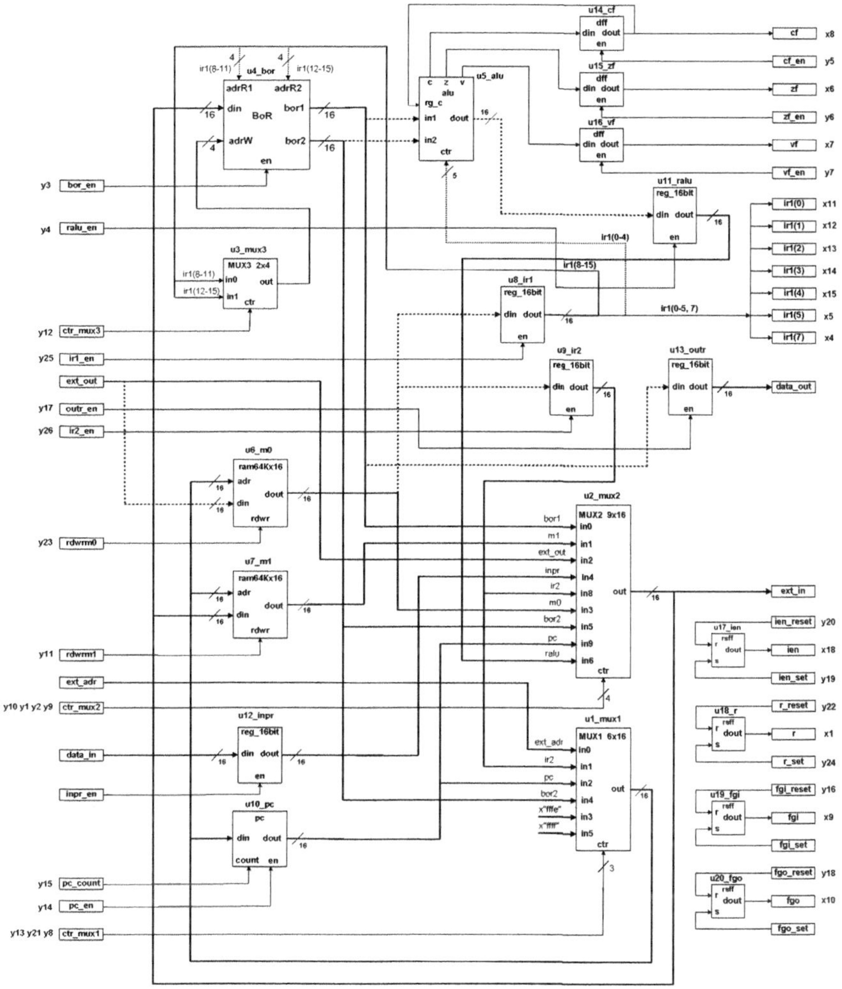

Fig. 1.26 Data Path

Read3. Source *B* is a memory (*M0* or *M1*). As *B* is a memory, Processor should send an address to the address bus *Adr0* for *M0* or *Adr1* for *M1* and *rdwr0 := 0* for *M0 or rdwr1 := 0* for *M1*. Then the corresponding word of the memory *M0[Adr0]* or *M1[Adr1]* will appear at the output of *M0* or *M1*. Examples:

$$Adr0 := PC;\ IR1 := M0[Adr0];$$
$$Adr1 := Ext\_Adr;\ Ext\_in := M1[Adr1].$$

Write1. Target *A* is a register. To write information from the source to the register we should supply signal *reg_en := 1* for direct connection or pass this information from the source to the input of the register through the MUX and supply the same signal *reg_en := 1*. Examples:

OutR := BoR[AdrR1]; PC := IR2.

Write2. Target *A* is *BoR*. In our Processor we write source *B* into *BoR* by using *AdrW* (see Fig. 1.2 (*BoR*)) and signal *bor_en := 1*. Examples:

AdrW:=IR1(8-11); BoR[AdrW]:=IR2;
AdrW:=IR1(8-11); BoR[AdrW]:=M1[Adr1].

Write3. Target *A* is a memory. To write information to the memory from some source we must provide an address to the address bus *Adr0* for *M0* or *Adr1* for *M1* and *rdwr0 := 1* for *M0 or rdwr1 := 1* for *M1*. The information from the input of the memory will be written into the memory word with a given address. *Examples:*

Adr0:=Ext_Adr; M0[Adr0]:=Ext_Out;
Adr1:=x"FFFF"; M1[Adr1]:=PC.

Let us discuss several examples of transformation of functional microinstructions into structural ones in Table 1.2.. Really, it is very similar to the transformation in our first structure – see Fig. 1.14 and Fig. 1.15.

1. *PC := IR2*. Case Read1 – Write1 (row Y_{11}, column 3).
To pass information from *IR2* to the input of *PC* through *MUX1* (see Fig. 1.22,c or Fig. 1.26) we must send signal *ctr_mux1 := 001*, because input *ir2* of *MUX1* has code *001*. The signal *pc_en := 1* will write information from the output of *MUX1* into *PC*. Finally we use the following microoperations at the structure level:

ctr_mux1 := 001; pc_en := 1.

2. *AdrW := IR1(8-11); BoR[AdrW] := InpR*. Case Read1 – Write2 (row Y_{13}).
To pass information from *IR1(8-11)* to the input *AdrW* of *BoR* through *MUX3* we must supply signal *ctr_mux3 := 0*, because input *IR1(8-11)* of *MUX3* has code *0* (Fig. 1.25 or Fig. 1.26). To pass information from *InpR* to the input of *BoR* through *MUX2* we must supply signal *ctr_mux2 := 0100*, because input *inpr* of *MUX2* has code *0100* (Fig. 1.22,d or Fig. 1.26). The signal *bor_en := 1* will write information from the output of *MUX2* into the register of *BoR* with address *AdrW*. Finally we use the following microoperations at the structure level:

ctr_mux3 := 0; ctr_mux2 := 0100; bor_en := 1.

3. *Adr1 := x"FFFF"; M1[Adr1] := PC*. Case Read1 – Write3 (row Y_{18}).
The content of *PC* should be written to the word of memory *M1* with address *x"FFFF"*. To pass information from the constant *x"FFFF"* to the address bus *Adr1* of the *M1* through *MUX1* we must supply signal *ctr_mux1 := 101*, because input *x"ffff"* of *MUX1* has code *101* (Fig. 1.22,c or Fig. 1.26). To pass information from *PC* to the input of memory *M1* through *MUX2* we must supply signal

ctr_mux2 := 1001, because input *pc* of *MUX2* has code *1001*. The signal *rdwrM1 := 1* will write information from the output of *PC* into the cell of the *M1* with address *X"FFFF"*. Finally we use the following microoperations at the structure level:

ctr_mux1 := 101; ctr_mux2 := 1001; rdwrM1 := 1.

4. *AdrR2 := IR1(12-15); PC:=BoR[AdrR2]*. Case Read2 – Write1 (row Y_{10}).

To pass information from *BoR2 (BoR[AdrR2])* to *PC* through *MUX1* we must give signal *ctr_mux1 := 100*, because input *bor2* of *MUX1* has code *100*. *IR1(12-15)* is connected directly with *AdrR2* so we do not need special signal for the first assignment. The signal *pc_en := 1* will write information from the output of *MUX1* into *PC*. Finally we use the following microoperations at the structure level:

ctr_mux1 := 100; pc_en := 1.

5. *AdrR1 := IR1(8-11); AdrW := IR1(12-15); BoR[AdrW] := BoR[AdrR1]*. Case Read2 – Write2 (row Y_8).

To pass information from *BoR1 (BoR[AdrR1])* to *BoR* through *MUX2* we must give signal *ctr_mux2 := 0000*. To write information to the register with the second address *(IR1(12-15))* we should send this address to the input *AdrW* of *BoR* through *MUX3* (*ctr_mux3 := 1*). The signal *bor_en := 1* will write information from the output of *MUX2* into the register of *BoR* with the second address. Finally we use the following microoperations at the structure level:

ctr_mux3 := 1; ctr_mux2 := 0000; bor_en := 1.

6. *Adr0 := PC; IR2 := M0[Adr0]*. Case Read3 – Write1 (row Y_{25}).

The content of the word in the memory *M0* with the address in *PC* should be written into *IR2*. To pass information from *PC* to the address bus *Adr0* of memory *M0* through *MUX1* we must supply signal *ctr_mux1 := 010*, because input *pc* of *MUX1* has code *010*. The signal *rdwrM0 := 0* will read information from the cell of *M0* with address *Adr0* equal to *PC*. Because memory *M0* is connected directly with the input of *IR2* (see Fig. 1.18,b and Fig. 1.26), no *MUX* is used to pass information from *M0* to *IR2*. To write information into *IR2* it is sufficient to supply signal *ir2_en := 1*. Finally, we use the following microoperations at the structure level:

ctr_mux1 := 010; rdwrM0 := 0; ir2_en := 1.

7. *Adr1 := IR2; AdrW := IR1(8-11); BoR[AdrW] := M1[Adr1]*. Case Read3 – Write2 (row Y_4).

The content of the word in memory *M1* with the address in *IR2* should be written into the register of *BoR* with address in *IR1(8-11)*. For this we should send this address to the input *AdrW* of *BoR* through *MUX3* (*ctr_mux3 := 0*). To pass information from *IR2* to the address bus *Adr1* of *M1* through *MUX1* we must give signal *ctr_mux1 := 001*, because input *ir2* of *MUX1* has code *001*. The signal *rdwrM1 := 0* will read information from the word of memory *M1* with address *Adr1* equal to *IR2*. To pass information from the memory *M1* to the input of *BoR* through *MUX2* wc must supply signal *ctr_mux2 := 0001*, because input *m1* of *MUX2* has code *0001*. The signal *bor_en := 1* will write information from the output of *MUX2* into the register of the *BoR* with the first address. Finally we use the following microoperations at the structure level:

ctr_mux1 := 001; ctr_mux3 := 0; ctr_mux2 := 0001; rdwrM1 := 0; bor_en := 1.

In this manner, we have filled the whole fifth column *"Structural Microoperations"* of Table 1.2.

1.4.2 Synthesis the Finite State Machine (FSM) from ASM

We use Algorithmic state machines to describe the behavior of digital systems, mainly of their control units. But if we must construct a logic circuit of the control unit we should use a Finite state machine (FSM). We will shortly consider a method of synthesis of FSM Mealy implementing a given ASM. As an example we will use ASM G_1 in Fig. 1.27. A Mealy FSM for a given ASM may be constructed in two stages [7]:

Stage1. Construction of a marked ASM;

Stage 2. Construction of an FSM transition table.

At the first stage, the inputs of vertices following operator vertices are marked by symbols $a_1, a_2, \ldots, a_M$ as follows:

1. Symbol a_1 marks the input of the vertex following the initial vertex *"Begin"* and the input of the final vertex *"End"*;
2. Symbols $a_2, \ldots, a_M$ mark the inputs of all vertices following operator vertices;
3. Vertex inputs are marked only once;
4. Inputs of different vertices, except the final one, are marked by different symbols.

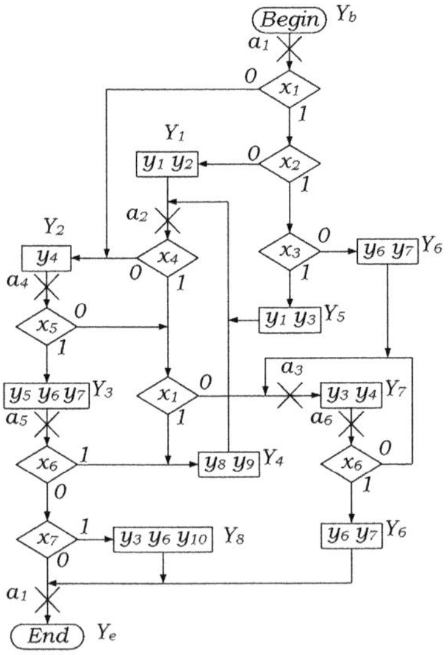

Fig. 1.27 ASM G_1 marked for the Mealy FSM synthesis

Marked ASM G_1 in Fig. 1.27 is a result of the first stage. At the second stage, we will consider the following transition paths in the marked ASM:

$$a_m \tilde{x}_{m1} \ldots \tilde{x}_{mR_m} Y_g a_s \qquad (P1)$$

$$a_m \tilde{x}_{m1} \ldots \tilde{x}_{mR_m} a_1 \qquad (P2)$$

Next we construct an FSM Mealy with states (marks) $a_1, \ldots, a_M$, obtained at the first stage. We have six such states $a_1, \ldots, a_6$ in our example. FSM has a transition from state a_m to state a_s with input $X(a_m, a_s)$ and output Y_g if, in ASM, there is transition path *P1*

$$a_m \tilde{x}_{m1} \ldots \tilde{x}_{mR_m} Y_g a_s .$$

Here $X(a_m, a_s)$ is the product of logical conditions written in this path:

$$X(a_m, a_s) = \tilde{x}_{m1} \ldots \tilde{x}_{mR_m} .$$

For the second transition path *P2*, FSM transits from state a_m to the initial state a_1 with input $X(a_m, a_1)$ and output Y_0 which is the operator containing an empty set of microoperations.

Fig. 1.28 present the transition table of FSM, constructed from ASM in Fig. 1.27. Each row of this table corresponds to one transition path *P1* or *P2*. We remind you that if some microinstruction, for example, $Y_5 = \{y_1, y_3\}$ is written in the operator vertex, it means that $y_1 = y_3 = 1$ and other microoperations are equal to zero. Our understanding of output signals in FSM is just like this. If y_1 and y_3 are written in the column for output signals (see row 1, column 4 in Fig. 1.28), only these signals are equal to one at the transition from *a1* to *a2* with the input signal *x1*x2*x3*, but other output signals are equal to zero.

a1	a2	x1*x2*x3	y1y3	1
a1	a3	x1*x2*~x3	y6y7	2
a1	a2	x1*~x2	y1y2	3
a1	a4	~x1	y4	4
a2	a2	x4*x1	y8y9	5
a2	a6	x4*~x1	y3y4	6
a2	a4	~x4	y4	7
a3	a6	1	y3y4	8
a4	a5	x5	y5y6y7	9
a4	a2	~x5*x1	y8y9	10
a4	a6	~x5*~x1	y3y4	11
a5	a2	x6	y8y9	12
a5	a1	~x6*x7	y3y6y10	13
a5	a1	~x6*~x7	--	14
a6	a1	x6	y6y7	15
a6	a6	~x6	y3y4	16

Fig. 1.28 FSM constructed from ASM in Fig. 1.27

In consideration of this, let us continue to fill in Table 1.2. In the sixth column of this table, we write only assignments, which assign "*ones*" to the signals in the fifth column of Table 1.2. Doing this we present each vector signal (control signal of MUX) as a set of separate binary components and we write assignments only for components equal to one. Look, for example, at microinstruction *Y4* in Table 1.2. In the fifth column, the following structural microoperations are written:

ctr_mux1 := 001; ctr_mux3 := 0; ctr_mux2 := 0001; rdwrM1 := 0; bor_en := 1.

We write in the column 6:

ctr_mux1(2) := 1; ctr_mux2(3) := 1; bor_en := 1.

In this column, we do not write ctr_mux1(0) := 0, ctr_mux1(1) := 0, ctr_mux3 : = 0, ctr_mux2(0) := 0, ctr_mux2(1) := 0, ctr_mux2(2) := 0 and rdwrM1 := 0, because zeroes are assigned in these microoperations.

1.4.3 Synthesis of Control Unit (FSM) for Processor

The combined structural ASM is presented in Fig. 1.29.

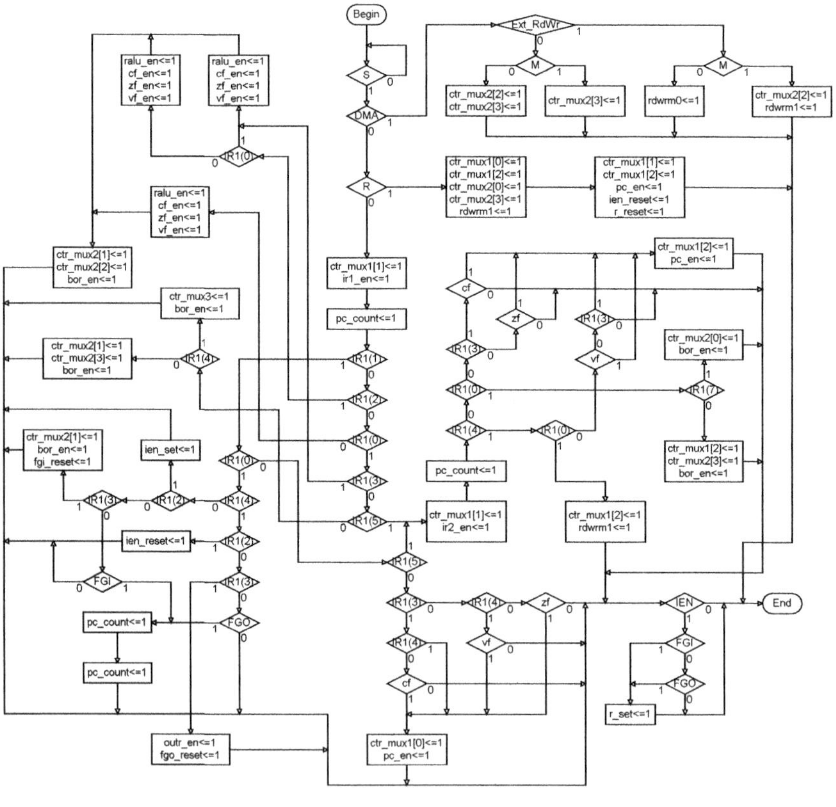

Fig. 1.29 Combined structural ASM

This ASM was constructed from functional ASM (Fig. 1.13) by replacing the functional microoperations in operator vertices, written in column 3 of Table 1.2 by structural microoperations from the column last but one in this table. As graphs, these two ASMs are absolutely identical. They have the same conditional and operator vertices and the same arcs (connections between these vertices), only the contents of operator vertices were changed in the structural ASM.

In the last column of Table 1.2, the microoperations from the previous column are numbered by *y1, y2, …, y26*. From the structural ASM at Fig. 1.29 we constructed ASM in Fig. 1.30. To do this, we replace each microoperation in operator vertex by its number from the last column of Table 1.2. The list of logical condition in this ASM is shown in Fig. 1.31. This list is the same for the functional and structural ASMs.

We use ASM in Fig. 1.30 to construct FSM Mealy in Fig. 1.32. Note, that this FSM has only 9 states whereas ASM in Fig. 55 has vertex *Begin*, vertex *End* and 28 operator vertices.

Thus, we constructed Data path and Control unit. Our next step is to combine two components – Control unit and Data path in one final block. The top level of our design is presented in Fig. 1.33.

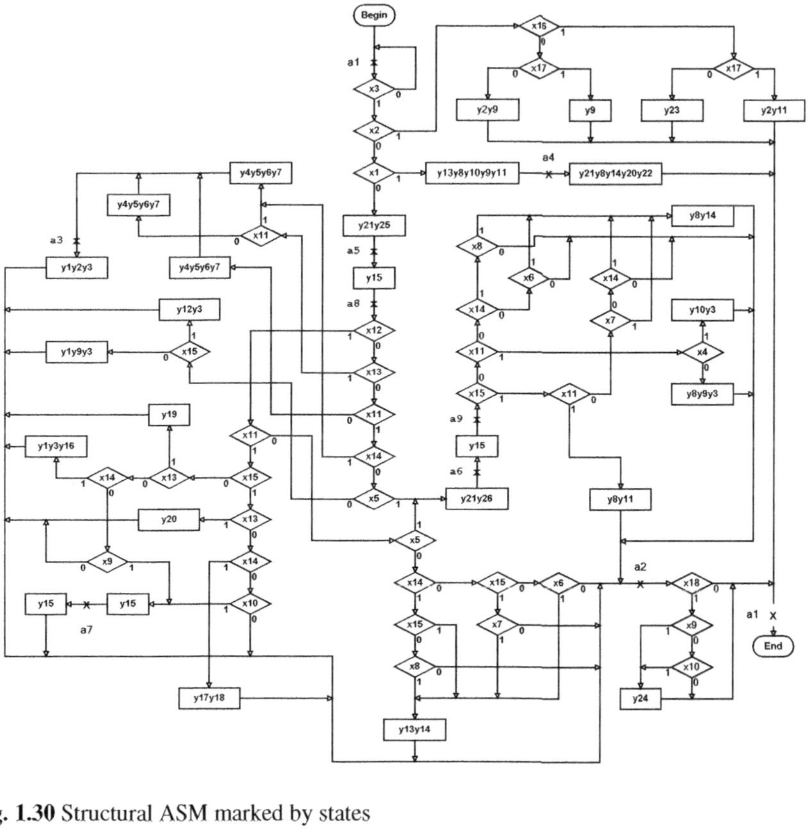

Fig. 1.30 Structural ASM marked by states

x1	**: R**	**x10**	**: FGO**
x2	**: DMA**	**x11**	**: IR1(0)**
x3	**: S**	**x12**	**: IR1(1)**
x4	**: IR1(7)**	**x13**	**: IR1(2)**
x5	**: IR1(5)**	**x14**	**: IR1(3)**
x6	**: zf**	**x15**	**: IR1(4)**
x7	**: vf**	**x16**	**: Ext_RdWr**
x8	**: cf**	**x17**	**: M**
x9	**: FGI**	**x18**	**: IEN**

Fig. 1.31 Logical conditions

a1	a1	x3*x2*x16*x17	y2y11	1
a1	a1	x3*x2*x16*~x17	y23	2
a1	a1	x3*x2*~x16*x17	y9	3
a1	a1	x3*x2*~x16*~x17	y2y9	4
a1	a4	x3*~x2*x1	y8y9y10y11y13	5
a1	a5	x3*~x2*~x1	y21y25	6
a1	a1	~x3	--	7
a2	a1	x18*x9	y24	8
a2	a1	x18*~x9*x10	y24	9
a2	a1	x18*~x9*~x10	--	10
a2	a1	~x18	--	11
a3	a2	1	y1y2y3	12
a4	a1	1	y8y14y20y21y22	13
a5	a8	1	y15	14
a6	a9	1	y15	15
a7	a2	1	y15	16
a8	a2	x12*x11*x15*x13	y20	17
a8	a2	x12*x11*x15*~x13*x14	y17y18	18
a8	a7	x12*x11*x15*~x13*~x14*x10	y15	19
a8	a1	x12*x11*x15*~x13*~x14*~x10*x18*x9	y24	20
a8	a1	x12*x11*x15*~x13*~x14*~x10*x18*~x9	--	21
a8	a1	x12*x11*x15*~x13*~x14*~x10*~x18	--	22
a8	a2	x12*x11*~x15*x13	y19	23
a8	a2	x12*x11*~x15*~x13*x14	y1y3y16	24
a8	a7	x12*x11*~x15*~x13*~x14*x9	y15	25
a8	a1	x12*x11*~x15*~x13*~x14*~x9*x18*x10	y24	26
a8	a1	x12*x11*~x15*~x13*~x14*~x9*x18*~x10	--	27
a8	a1	x12*x11*~x15*~x13*~x14*~x9*~x18	--	28
a8	a6	x12*~x11*x5	y21y26	29
a8	a2	x12*~x11*~x5*x14*x15	y13y14	30
a8	a2	x12*~x11*~x5*x14*~x15*x8	y13y14	31
a8	a1	x12*~x11*~x5*x14*~x15*~x8*x18*x9	y24	32
a8	a1	x12*~x11*~x5*x14*~x15*~x8*x18*~x9*x10	y24	33
a8	a1	x12*~x11*~x5*x14*~x15*~x8*x18*~x9*~x10	--	34
a8	a1	x12*~x11*~x5*x14*~x15*~x8*~x18	--	35
a8	a2	x12*~x11*~x5*~x14*x15*x7	y13y14	36
a8	a1	x12*~x11*~x5*~x14*x15*~x7*x18*x9	y24	37
a8	a1	x12*~x11*~x5*~x14*x15*~x7*x18*~x9*x10	y24	38
a8	a1	x12*~x11*~x5*~x14*x15*~x7*x18*~x9*~x10	--	39
a8	a1	x12*~x11*~x5*~x14*x15*~x7*~x18	--	40
a8	a2	x12*~x11*~x5*~x14*~x15*x6	y13y14	41

Fig. 1.32 Control Unit as FSM

```
a8 a1   x12*~x11*~x5*~x14*~x15*~x6*x18*x9         y24          42
a8 a1   x12*~x11*~x5*~x14*~x15*~x6*x18*~x9*x10    y24          43
a8 a1   x12*~x11*~x5*~x14*~x15*~x6*x18*~x9*~x10   --           44
a8 a1   x12*~x11*~x5*~x14*~x15*~x6*~x18           --           45
a8 a3   ~x12*x13*x11                              y4y5y6y7     46
a8 a3   ~x12*x13*~x11                             y4y5y6y7     47
a8 a3   ~x12*~x13*x11*x14                         y4y5y6y7     48
a8 a6   ~x12*~x13*x11*~x14*x5                     y21y26       49
a8 a2   ~x12*~x13*x11*~x14*~x5*x15                y3y12        50
a8 a2   ~x12*~x13*x11*~x14*~x5*~x15               y1y3y9       51
a8 a3   ~x12*~x13*~x11                            y4y5y6y7     52
a9 a2   x15*x11                                   y8y11        53
a9 a2   x15*~x11*x7                               y8y14        54
a9 a2   x15*~x11*~x7*x14                          y8y14        55
a9 a1   x15*~x11*~x7*~x14*x18*x9                  y24          56
a9 a1   x15*~x11*~x7*~x14*x18*~x9*x10             y24          57
a9 a1   x15*~x11*~x7*~x14*x18*~x9*~x10            --           58
a9 a1   x15*~x11*~x7*~x14*~x18                    --           59
a9 a2   ~x15*x11*x4                               y3y10        60
a9 a2   ~x15*x11*~x4                              y3y8y9       61
a9 a2   ~x15*~x11*x14*x8                          y8y14        62
a9 a1   ~x15*~x11*x14*~x8*x18*x9                  y24          63
a9 a1   ~x15*~x11*x14*~x8*x18*~x9*x10             y24          64
a9 a1   ~x15*~x11*x14*~x8*x18*~x9*~x10            --           65
a9 a1   ~x15*~x11*x14*~x8*~x18                    --           66
a9 a2   ~x15*~x11*~x14*x6                         y8y14        67
a9 a1   ~x15*~x11*~x14*~x6*x18*x9                 y24          68
a9 a1   ~x15*~x11*~x14*~x6*x18*~x9*x10            y24          69
a9 a1   ~x15*~x11*~x14*~x6*x18*~x9*~x10           --           70
a9 a1   ~x15*~x11*~x14*~x6*~x18                   --           71
```

Fig. 1.32 (*continued*)

Table 1.4 The process table with poor encoding for input *bor1* of *MUX2*

Y7	y12	Adr1:=IR2	16	ctr_mux1 := 001	ctr_mux1[2] := 1
	y3	AdrR1:=IR1(8-11)	4	ctr_mux2 := 0000	rdwrm1:= 1
	y16	M1[Adr1]:=BoR[AdrR1]	16	rdwrM1 := 1	ctr_mux2[0] := 1
					ctr_mux2[1] := 1
					ctr_mux2[2] := 1
					ctr_mux2[3] := 1
Y8	y3	AdrR1:=IR1(8-11)	4	ctr_mux3 := 1	ctr_mux3 := 1
	y17	AdrW:=IR1(12-15)	4	ctr_mux2 := 0000	bor_en := 1
	y18	BoR[AdrW]:=BoR[AdrR1]	16	bor_en := 1	ctr_mux2[0] := 1
					ctr_mux2[1] := 1
					ctr_mux2[2] := 1
					ctr_mux2[3] := 1

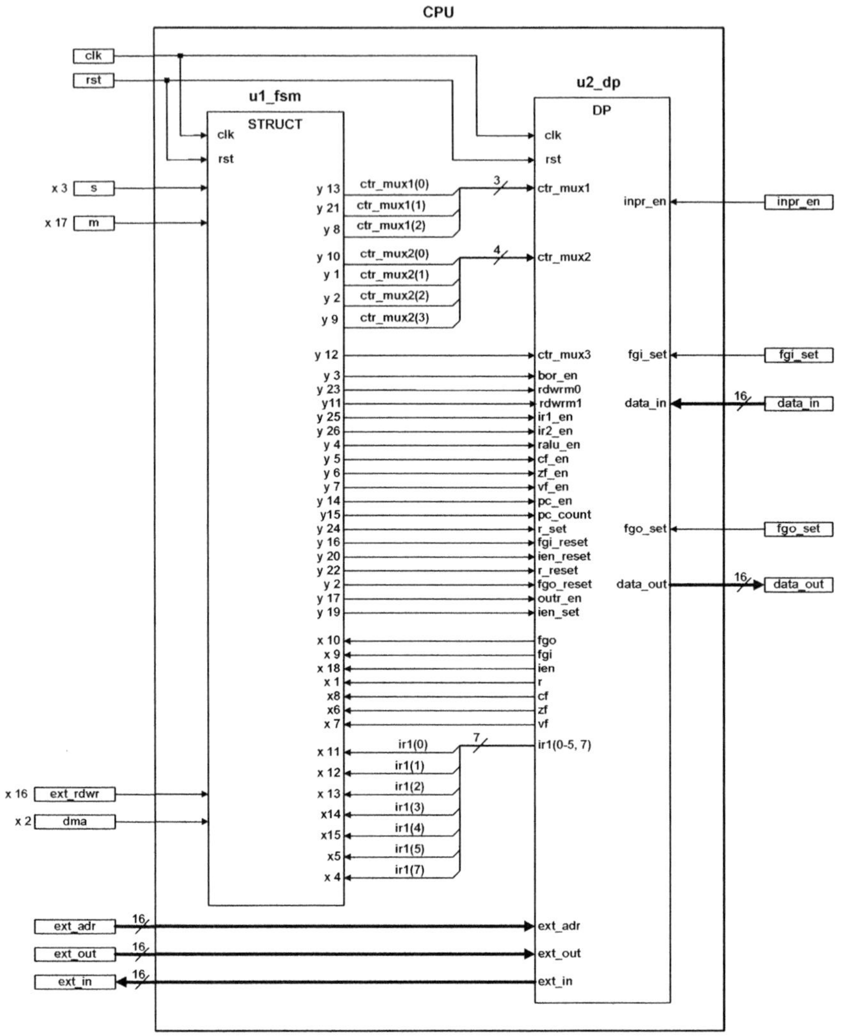

Fig. 1.33 Top level of Processor

1.4.4 Encoding of Inputs of MUXes

Our goal in this process is to minimize the number of outputs $y_1, \ldots, y_N$ in FSM table or, which is the same, in the Control unit. To explain this, we appeal to the rows with microinstructions *Y7* and *Y8* in Table 1.2. We used code *0000* for input *bor1 (BoR[AdrR1])* in *MUX2* and therefore the components of vector *ctr_mux2* didn't appear in the sixth column of Table 1.2 when we turned from the structural microinstructions (column 5) to the minimized structural microinstructions (column 6). If, for example, we used code *1111* for the same input *bor1* of *MUX2*, we would get *ctr_mux2 := 1111* instead of *ctr_mux2 := 0000* in the same rows (see Table 1.4). This leads to appearance of the eight additional rows in the sixth column of Table 1.4 (*ctr_mux2(0) := 1, ctr_mux2(1) := 1, ctr_mux2(2) := 1, ctr_mux2(3) := 1*), twice for all of them. It entails eight additional outputs in the transition table of the finite state machine.

To encode the inputs of *MUX1* we constructed Table 1.5 where *p(in)* is the weight of each input of *MUX1* in Fig. 1.22,c. The algorithm for the input encoding is absolutely the same as in the state assignment of FSM. First, we used the zero code for input *ext_adr* with max *p(ext_adr) = 4*. Then codes with one *'1'* are used for inputs *ir2, pc* and *bor2* with the next max input weights and, finally, codes with two 'ones' are used for the left inputs *x"fffe"* and *x"ffff"*. In the same manner, we encoded inputs of *MUX2* and *MUX3* (see Fig.1.22,d and Fig. 1.25).

Table 1.5 Optimal input encoding for MUX1

input	p(in)	code
ext_adr	4	0 0 0
pc	2	0 1 0
ir2	3	0 0 1
bor2	1	1 0 0
x"fffe"	1	0 1 1
x"ffff"	1	1 0 1

1.5 Conclusions

This Chapter presents a new methodology for high level design of complicated digital systems. This methodology is based on Algorithmic State Machine (ASM) transformations (composition, minimization, extraction, etc.), special algorithms for Data Path and Control Unit design and a very fast optimizing synthesis of FSMs as well as combinational circuits with hardly any constraints on their size, i.e., the number of inputs, outputs and states. Design tools supporting this methodology allow us to implement, check and estimate many possible design versions very fast, to find an optimized decision of a design problem and to simplify the verification problem for digital systems.

Problem orientation regarding the design system is nonessential – it can be a processor, a robot, a controller, etc. If the system is rather complicated, it is possible to pick out some subbehaviors (modes) in its behavior. For a processor it can be an instruction or a set of instructions that can be described together; for a mobile robot – its different modes (cruise, follow, avoid, escape etc.). We also suppose that any digital system is usually regarded as a composition of a *Control unit* and an *Operational unit* (*Data path*). In a processor, for example, a data path contains such regular blocks as memory, registers, ALU, counters, coders, encoders, multiplexers, demultiplexers, etc. A control unit produces a sequence of control signals that force an implementation of microoperations in a data path. The design process contains the following stages.

Stage 1 Design ASMs $G_1, \ldots, G_M$ for each separate mode. We can present these ASMs in VHDL or use the special tool *ASM Creator* from the EDA tool *Abelite* supported by the described design methodology. In Abelite, it is very simple to draw an ASM and compile it in VHDL and/or several other representations. It is really important that an ASM may contain any number of *generalized operators*. Each of such operators is an ASM itself and it will be automatically inserted in the combined ASM at the fourth stage. Moreover, there are no restrictions on the number of such generalized operators in an ASM and on the number of included levels – each of such operators can contain any number of generalized operators itself.

Stage 2 Combine separate ASMs into one combined functional ASM. After constructing separate ASMs we combine them into one combined functional ASM still containing generalized operators. At this stage each microoperation is presented at the functional level. Really, we do not have the real architecture for our project, we only know some units of our future Data path. Thus, microoperations at this level are similar to assignments of variables in some programming language and are not connected with any specific Data path. During ASM combining we minimize the number of operator vertices in the combined ASM. If several ASMs contain the same operator vertex, there will be only one such operator vertex in the combined ASM.

Stage 3 Minimize the combined functional ASM. At this stage, the number of conditional vertices in the combined ASM is minimized. Such minimization allows us to reduce dramatically the number of vertices in the ASM (sometimes for two or three times) and to reduce the complexity of logic circuits at the stage of logic design.

Stage 4 Include generalized operators. At this stage, generalized operators constructed at the first stage are included into the minimized ASM constructed at the previous stage. It is the last stage of the functional ASM design.

Stage 5 Data path synthesis. First, we construct a Connection graph from the functional ASM designed on Stage 4. Such a graph contains a list of sources and targets for each component of an operational unit and some metrics that will be used in the optimization of the Data path. Next we construct an optimized List of parallel microoperations to increase the speed of the design system. Then we design the Graph of incompatibility from the Connection graph and the List of parallel microoperations. On the final step of Data path synthesis we construct Muxes (by coloring the Graph of incompatibility) and the List of direct connections from the Connection graph.

Stage 6 Control unit design. Using the functional ASM (stage 4) and the Muxes and the List of direct connections (stage 5) we immediately construct the structural ASM. This ASM describes the behavior of the Control unit corresponding to the Data Path. On the last step of this stage we construct the Finite state machine and its multilevel logic circuit.

Stage 7 VHDL code design. The Data path constructed according to our design methodology does not contain any "cloud" (irregular) circuits. It makes it possible to simplify considerably VHDL or Verilog code for the Data path using the structure style of VHDL to combine VHDL or Verilog codes of units. VHDL code for the Control unit can be constructed automatically. VHDL code for the top level of the system is the result of combining VHDL codes of the Data path and the Control unit.

References

[1] Ellervee, P.: High-level synthesis of control and memory intensive applications. Royal Institute of Technology, Stockholm (2000)
[2] Gajski, D., Dutt, N., Wu, A., Lin, S.: High-level synthesis: Introduction to chip and system design. Kluwer Academic Publishers, Boston (1993)
[3] De Micheli, G.: Synthesis and optimization of digital circuits. McGraw-Hill, Inc., New Jersey (1994)
[4] Eles, P., Kuchcinski, K., Peng, Z.: System synthesis with VHDL. Kluwer Academic Publishers, Boston (1998)
[5] Lin, Y.: Recent development in high-level synthesis. ACM Transactions on Design Automation of Electronic Systems 2(1), 2–21 (1997)
[6] Adamski, M., Barkalov, A.: Architectural and sequential synthesis of digital devices. UZG Press, Zielona Góra (2006)
[7] Baranov, S.: Synthese des Automates Microprogrammes. MIR, Moscow (1983)
[8] Baranov, S.: Logic synthesis for control automata. Kluwer Academic Publishers, Dordrecht (1994)

2 Rectangular Function Π(x) and Its Application for Description of Some Logical Devices Operation

Edward Hrynkiewicz

Silesian University of Technology, Institute of Electronics,
ul. Akademicka 16, 44-100 Gliwice, Poland
e-mail: ehrynkiewicz@polsl.pl

Abstract. A carrying out of the logic operations on pulses and pulse series described by means of rectangular function Π(x) is presented in the section. The logic sum, logic product, logic negation and Ex-OR operations were investigated. The utilizing of these operations for mathematical description of frequency multiplying is shown as the example of application of Π(x) function. Moreover section deals with the problem of rectangular function Π(x) utilization for description of operation of such logical devices as digital sine wave generators and for nonlinear distortions analyzing in such generators.

2.1 Introduction

For description of logic operations on pulses or pulse trains and for analysis of same digital devices in time domain one can use several rectangular functions. One of them is rectangular function Π(x) which is defined as [2,4,7]:

$$\begin{aligned} \prod(x) &= 1 \; for \; 0 \le x \le 1 \\ &= 0 \; for \; x < 0 and x > 1 \end{aligned} \tag{2.1}$$

Rectangular pulse which duration time is equal to τ and which has unit amplitude may be expressed as:

$$\mathrm{I(t)} = \prod\left(\frac{t}{\tau}\right) \tag{2.2}$$

and a pulse train which period is equal to T:

$$\mathrm{F(t)} = \sum_{m=0}^{\infty} \prod\left(\frac{t-mT}{\tau}\right) \tag{2.3}$$

In some cases rectangular functions Π(x) are more convenient for description of a system operation or for system designing then for example Haar functions or very

M. Adamski et al. (Eds.): Design of Digital Systems and Devices, LNEE 79, pp. 43–57.
springerlink.com

popular Walsh functions. Below there are examples of utilisation of rectangular function Π(x) for mathematical description of several logical devices.

2.2 Logic Operations on Rectangular Functions

Let us take into account two rectangular pulses appearing in time period T (Fig.2.1):

$$I_1(t) = \prod\left(\frac{t - p_1T}{s_1T}\right) \tag{2.4}$$

$$I_2(t) = \prod\left(\frac{t - p_2T}{s_2T}\right) \tag{2.5}$$

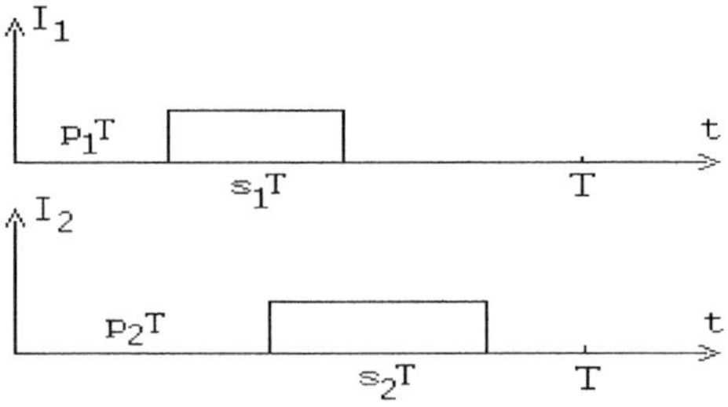

Fig. 2.1 The rectangular pulses $I_1(t)$ and $I_2(t)$

A. Logic Sum

Logic sum of such pulses may be written as:

$$I_1(t) + I_2(t) = \prod\left(\frac{t - p_1T}{s_1T}\right) + \prod\left(\frac{t - p_2T}{s_2T}\right) \tag{2.6}$$

For $p_2 < p_1 + s_1$

$$I_1(t) + I_2(t) = \prod\left(\frac{t - p_1T}{(p_2 + s_2)T - p_1T}\right) = \prod\left(\frac{t - p_1T}{(p_2 - p_1 + s_2)T}\right) \tag{2.7}$$

In particular case $p_1 = p_2 = 0$

and $$I_1(t) + I_2(t) = \prod\left(\frac{t}{\max(s_1, s_2)T}\right) \tag{2.8}$$

B. Logic Product

If $p_2 < p_1 + s_1$

$$I_1(t)\cdot I_2(t)=\prod\left(\frac{t-p_1T}{s_1T}\right)\cdot\prod\left(\frac{t-p_2T}{s_2T}\right)$$
$$=\prod\left(\frac{t-p_2T}{(p_1+s_1)T-p_2T}\right)=\prod\left(\frac{t-p_2T}{(p_1-p_2+s_1)T}\right) \tag{2.9}$$

if $p_1 = p_2 = 0$

$$I_1(t)\cdot I_1(t)=\prod\left(\frac{t}{\min(s_1,s_2)T}\right) \tag{2.10}$$

and for $p_2 \geq p_1 + s_1$ $\qquad I_1(t) \cdot I_2(t) = 0$

C. Negation

$$\overline{I_1(t)}=\overline{\prod\left(\frac{t-p_1T}{s_1T}\right)}=\prod\left(\frac{t}{p_1T}\right)+\prod\left(\frac{t-(p_1+s_1)T}{T-(p_1+s_1)T}\right) \tag{2.11}$$

D. EX-OR (logic inequality)

Logic inequality (Ex-OR function) may be written as:

$$I_1(t)\oplus I_2(t)=\prod\left(\frac{t-p_1T}{s_1T}\right)\oplus\prod\left(\frac{t-p_2T}{s_2T}\right)=$$

$$=\prod\left(\frac{t-p_1T}{s_1T}\right)\cdot\left[\prod\left(\frac{t}{p_2T}\right)+\prod\left(\frac{t-(p_2+s_2)T}{T-(p_2+s_2)T}\right)\right]+$$

$$+\left[\prod\left(\frac{t}{p_1T}\right)+\prod\left(\frac{t-(p_1+s_1)T}{T-(p_1+s_1)T}\right)\right]\cdot\prod\left(\frac{t-p_2T}{s_2T}\right)= \tag{2.12}$$

$$=\prod\left(\frac{t-p_1T}{(p_2-p_1)T}\right)+\prod\left(\frac{t-(p_1+s_1)T}{(p_2+s_2)T-(p_1+s_1)T}\right)$$

Let us assume that: $p_1 = 0,\quad p_2 = \frac{1}{4},\quad s_1 = s_2 = \frac{1}{2}$ than

$$I_1(t)\oplus I_2(t)=\prod\left(\frac{t}{\frac{T}{4}}\right)+\prod\left(\frac{t-\frac{1}{2}T}{\frac{T}{4}}\right)=\sum_{m=0}^{1}\prod\left(\frac{t-m\cdot\frac{T}{2}}{\frac{T}{4}}\right) \tag{2.13}$$

E. Binary Counters

Let us consider of counting of the pulses represented by the pulse train shown in Fig. 2.2.

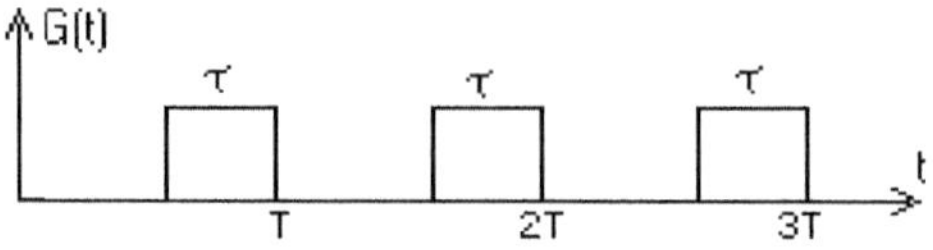

Fig. 2.2 The pulse train

Such pulse train may be expressed using function Π(x) as:

$$G(t) = \sum_{m=0}^{\infty} \Pi\left(\frac{t+\tau-(m+1)T_i}{\tau}\right) \tag{2.14}$$

where: τ - pulse duration; T_i - pulse train period

The state of the binary up-counter outputs may be written as:

$$Q_0 = \sum_{m=0}^{\infty} \Pi\left(\frac{t-2^0 T_i - \mathrm{int}(\frac{m}{2})2T_i}{2^0 T_i}\right)$$

$$Q_1 = \sum_{m=0}^{\infty} \Pi\left(\frac{t-2^1 T_i - \mathrm{int}(\frac{m}{2^2})2^2 T_i}{2^1 T_i}\right) \tag{2.15}$$

$$Q_2 = \sum_{m=0}^{\infty} \Pi\left(\frac{t-2^2 T_i - \mathrm{int}(\frac{m}{2^3})2^3 T_i}{2^2 T_i}\right)$$

and $$Q_k = \sum_{m=0}^{\infty} \Pi\left(\frac{t-2^k T_i - \mathrm{int}(\frac{m}{2^{k+1}})2^{k+1} T_i}{2^k T_i}\right)$$

For the same input pulse train the states of a ring counter flip-flops are expressed by:

$$\mathrm{Q_k} = \sum_{m=0}^{\infty} \Pi\left(\frac{t-kT_i - \mathrm{int}(\frac{m}{n})nT_i}{T_i}\right) \tag{2.16}$$

where: n - number of counter bits

2.3 Utilization of the Rectangular Functions Π(x) for Analysis of Pulse or Frequency Multiplying

Let it will be given pulse train described by the following expression:

$$F(t)=\sum_{m=0}^{\infty}\Pi\left(\frac{t-mT}{sT}\right) \tag{2.17}$$

m - integer number, $0< s < 1$
For sT < T/K pulse train which frequency is K times greater may be written as [3]:

$$F_K(t)=\sum_{p=0}^{\infty}\Pi\left(\frac{t-p\frac{T}{K}}{sT}\right) \tag{2.18}$$

Putting p = mK + r
where for m = 0,1,......., ∞; r = 0,1,......., K-1
one obtains the following formula:

$$F_K(t)=\sum_{r=0}^{K-1}\sum_{m=0}^{\infty}\Pi\left(\frac{t-(mK+r)\frac{T}{K}}{sT}\right) \tag{2.19}$$

which represent the sum of K trains of pulses delayed each other about T/K. The basic diagram of such frequency multiplier is shown in Fig. 2.3 and its second version with one delay block in Fig. 2.4.

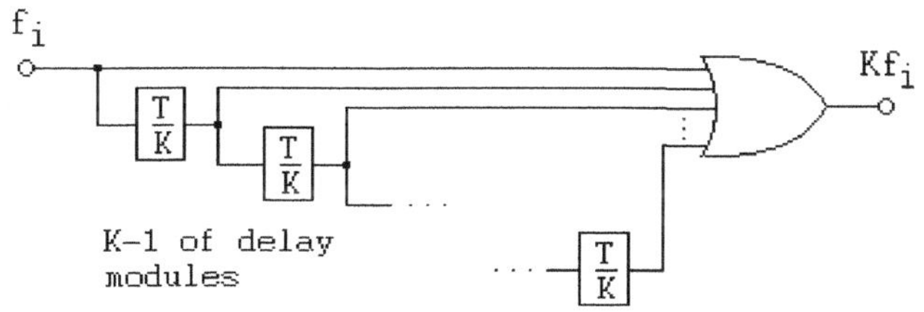

Fig. 2.3 Frequency multiplying circuit for sT < T/K (I version)

Pulse train which period is equal to T and duty cycle is equal to 1/2 may be expressed by:

$$F(t)=\sum_{m=0}^{\infty}\Pi\left(\frac{t-mT}{\frac{T}{2}}\right) \tag{2.20}$$

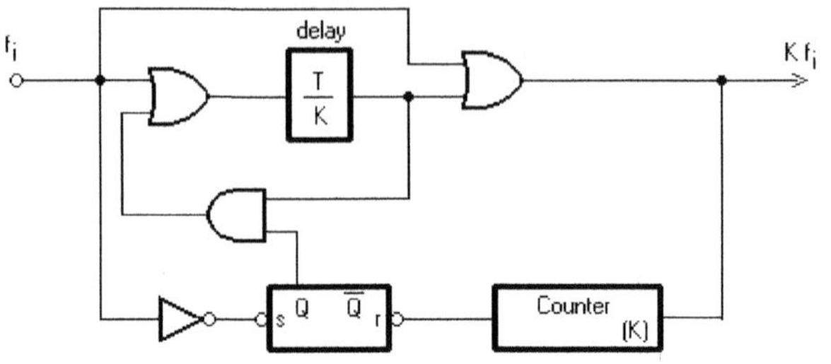

Fig. 2.4 Frequency multiplying circuit for sT < T/K (II version)

A pulse train which frequency is K times greater and duty cycle equal to 1/2 may be written in the following form:

$$F_K(t)=\sum_{p=0}^{\infty}\prod\left(\frac{t-p\frac{T}{K}}{\frac{T}{2K}}\right) \tag{2.21}$$

Putting p = mK + r   m = 0,1,......., ∞
r = 0,1,......., K-1

we obtain

$$F_K(t)=\sum_{r=0}^{K-1}\sum_{m=0}^{\infty}\prod\left(\frac{t-(mK+r)\frac{T}{K}}{\frac{T}{2K}}\right) \tag{2.22}$$

Taking into account (2.13) we can rewrite the above formulae in the following form:

$$F_K(t)=\sum_{r=0}^{K-1}\oplus\sum_{m=0}^{\infty}\prod\left(\frac{t-\left(mK+\frac{r}{2}\right)\frac{T}{K}}{\frac{T}{2}}\right) \tag{2.23}$$

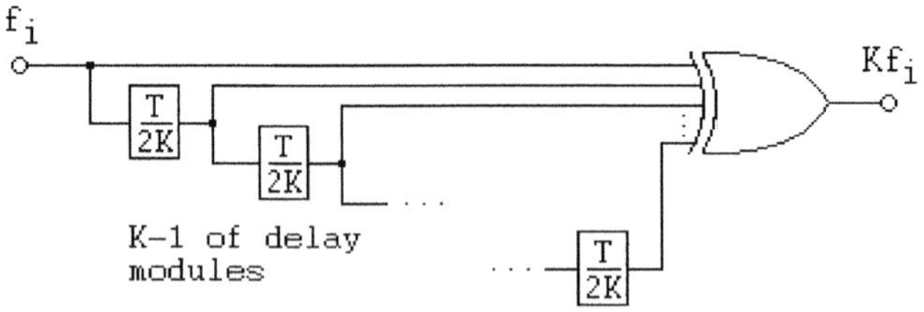

Fig. 2.5 Frequency multiplying circuit for duty cycle of a pulse train equal to ½ [8]

As it was shown for a pulse train which duty cycle is equal to 1/2, frequency multiplication by K one can obtain summing modulo K pulse trains delayed each other about T/2K (Fig. 2.5).

2.4 Utilizing the Function Π(x) for Harmonic Analysis of Digital Sine Wave Generator

2.4.1 *Digital Sine Wave Generator Based on Digital Integrators*

The generator of sinusoidal wave may be formed with digital integrators. Let us assume the integrator consisted of binary rate multiplier and counter. The block diagram of such a generator is shown in Fig. 2.6.

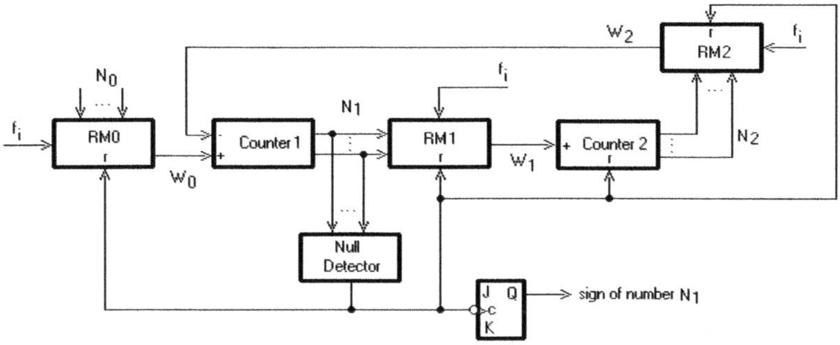

Fig. 2.6 The block diagram of sine wave generator based on digital integrators

Assuming big numbers N_1 and N_2, the system may be described by approximate relationships [2,5,6] given below.

$$\left(\frac{N_0}{2^n s} f_i - \frac{N_2(s)}{2^n}\right)\frac{1}{s} = N_1(s) \quad \text{where: n- number of BRM stages} \quad (2.24)$$

$$\frac{N_1(s)}{2^n} \cdot f_i \cdot \frac{1}{s} = N_2(s) \quad \text{s - Laplace operator}$$

hence

$$N_1 = N_0 \sin \omega t \quad \text{where: } \omega = \frac{f_i}{2^n} \quad (2.25)$$

$$N_2 = N_0(1 - \cos \omega t)$$

Because maximum value of number N_2 is equal to $2N_0$ and maximum value of number N_0 must satisfy the following formula:

$$N_0 \leq 2^{n-1} - 1 \qquad \text{where } n \text{ - number of BRM stages.}$$

Due to the fact that reversible counter does not operate normally when at its inputs "up" and "down", pulses will appear simultaneously it is necessary to introduce anti-coincidence circuit at the counter 1 inputs the task of which is to block the pulses appearing simultaneously in both counting lines.

The determination of nonlinear distortion of the output wave from the generator is not possible by means of approximate description given by the equations (2.24) and (2.25). It may be done using rectangular function Π(x) in the way presented below.

The presence of the k+1 clock pulse at the output of a RM may be expressed as [2]:

$$W_m(k+1) = \bigcup_{s=0}^{n-1} [\overline{Q}_{ms}(k) N_{m(n-1-s)} \cdot \bigcap_{r=0}^{s-1} Q_{mr}(k)] \qquad m = 0,1,2$$
$$\bigcap_{r=0}^{-1} Q_{mr}(k) \equiv 1 \tag{2.26}$$

where: $W_m(k+1) = 1$ - means that (k+1) clock pulse is present at the RM output
$N_{m(n-1)}..N_{m0}$ - binary expression of a number N_m
$Q_{m(n-1)}..Q_{m0}$ - binary expression of a counter state of the m-th RM
and a new state of a RM counter may be calculated in the following way:

$$Q_{ms}(k+1) = Q_{ms}(k) \oplus \bigcap_{r=0}^{s-1} Q_{mr}(k) \quad s = 0,1,2,\cdots,n-1$$
$$\bigcap_{r=0}^{-1} Q_{mr}(k) \equiv 1 \tag{2.27}$$

The state of the reversible counter we can write as:

$$N_1(k+1) = N_1(k) + W_0(k+1) - W_2(k+1) \tag{2.28}$$

On the base of the above formulas it is easy, using computer, to calculate the shape of $N_1(t)$ wave. In real system the number N_1 is converted to the output voltage using D/A converter. If the conversion factor is equal to 1V the output wave may be shown in form presented in Fig. 2.7.

Due to resetting of the system every half period, the wave in each half period has the same shape and opposite sign. As it was shown in Fig. 2.7 output wave can be regarded as the sum of rectangular waves φ_k (t):

$$N_1(t) = \sum_{k=1}^{k_p} \varphi_k(t)$$

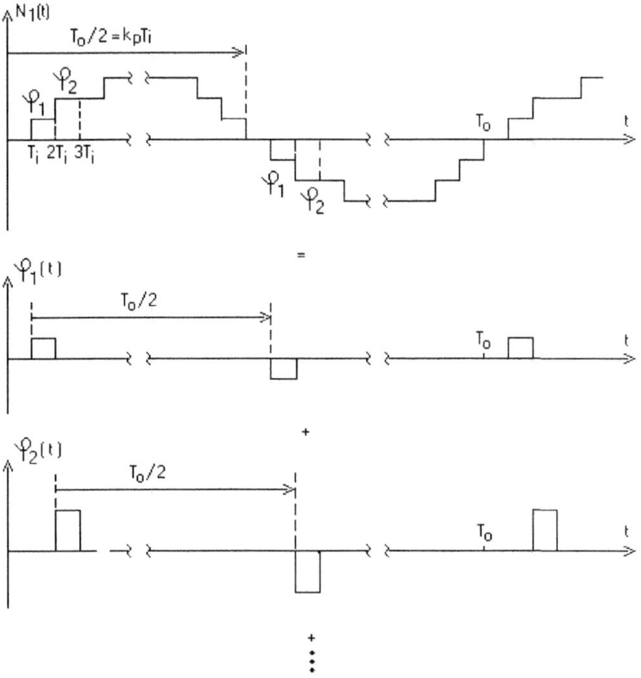

Fig. 2.7 The output wave from the generator and the component waves

Using the definition of the rectangular function Π(x) the wave φ_k (t) can be written as follows:

$$\varphi_k(t)=\sum_{m=-\infty}^{\infty} N_1(k)\left[\Pi\left(\frac{t-mT_o-kT_i}{T_i}\right)-\Pi\left(\frac{t-(m+\frac{1}{2})T_o-kT_i}{T_i}\right)\right] \tag{2.29}$$

Then

$$N_1(t)=\sum_{k=1}^{k_p}\sum_{m=-\infty}^{\infty} N_1(k)\left[\Pi\left(\frac{t-mT_o-kT_i}{T_i}\right)-\Pi\left(\frac{t-(m+\frac{1}{2})T_o-kT_i}{T_i}\right)\right] \tag{2.30}$$

Decomposing the wave $\varphi_k(t)$ into Fourier series and considering the fact that the complex amplitude of the nth harmonic of a sum of k_p waves may be calculated as:

$$\hat{C}_n=\sum_{k=1}^{k_p}\hat{C}_{nk}$$

for the wave $N_1(t)$ we obtain:

$$\hat{C}_{2n}=2\frac{T_i}{T_o}\sin c(n\omega_o\frac{T_i}{2})e^{-jn\omega_o\frac{T_i}{2}}\sum_{k=1}^{k_p}N_1(k)e^{-jn\omega_o kT_i} \quad \text{for - odd } n \qquad (2.31)$$

$$\hat{C}_{2n} = 0 \qquad \text{for - even } n$$

$$\text{where}: \sin c n\omega_o\frac{T_i}{2}=\frac{\sin n\omega_o\frac{T_i}{2}}{n\omega_o\frac{T_i}{2}}$$

$$T_o=2k_pT_i;\omega_o=\frac{\pi}{k_pT_i}$$

$$\text{and} \quad C_{2n}=2\left|\hat{C}_{2n}\right|$$

hence:

$$C_{2n}=\frac{2}{k_p}\left|\sin c\frac{n\pi}{2k_p}\right|\cdot\left|\sum_{k=1}^{k_p}N_1(k)e^{\frac{-jnk\pi}{k_p}}\right| \quad \text{for odd } n$$

$$C_{2n} = 0 \qquad \text{for even } n \qquad (2.32)$$

So by using the values $N_1(k)$ obtained from formulae (2.28), it is possible to calculate real amplitudes of harmonics present in the wave $N_1(t)$ and the value of non-linear distortion factor h (Table 2.1):

$$h=\frac{\sqrt{C_3^2+C_5^2+\ldots\ldots}}{C_1} \qquad (2.33)$$

Table 2.1 Values of N_1 and non-linear distortion factor h

N	N_0	$2k_p$	N_1	h
8	126	1014	125	0.0100
9	254	3200	249	0.0051
10	510	6245	510	0.0021

2.4.2 Digital Sine Wave Generator Based on ROM

The generator of sinusoidal wave which is a part of function generator or which is used in low and infra-low frequency range may be formed as a logical device. Let us take into account a generator which block diagram is shown in Fig.2.8.

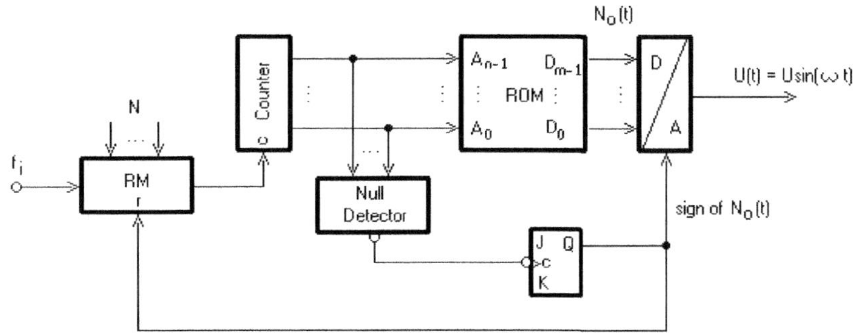

Fig. 2.8 Digital sine wave generator

The one half-period of a sine wave have been tabularised in the ROM memory in the form of K_p digitalized samples (a value of the K-th sample is equal to the nearest whole number to $N_{omax}\sin(K\pi/K_p)$). The number of the output pulses from the binary rate multiplier (RM) [5,6] may be calculated as:

$$K = \sum_{i=0}^{r-1} \mathrm{int}(k2^{i-r} + \frac{1}{2})N_i \tag{2.34}$$

where: r - number of RM stages; k - number of input (clock) pulses to the RM; N_i - i-th bit of binary representation of N number.

This number represent additionally the state of the address counter then it implicates the read-out from the memory of the sample which on the output of D/A converter may be denoted as U(K). Because K depends on k then it means that this way one can obtain the samples of U(k). It may happens - due to the rate multiplier operation - that for few or many successive values of number k the number K does not change so it means that successive U(k) does not change too. (see Fig.2.9).

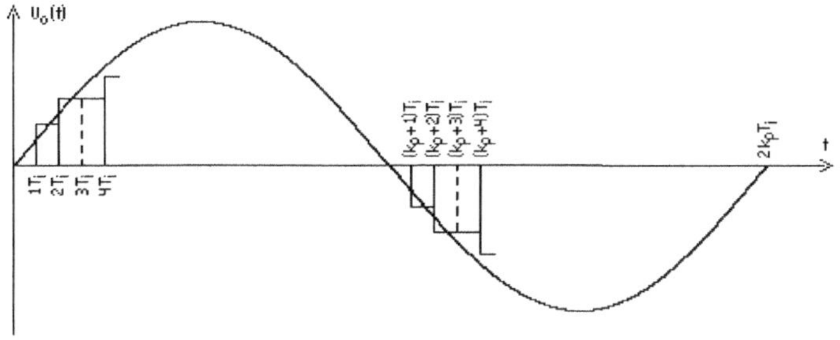

Fig. 2.9 The shape of the output wave from generator shown in Fig.2.8

Using the definition of the rectangular function **Π**(x) the output wave from the generator may be written as follows ($T_i = 1/f_i$, $T_o = 2k_pT_i$; k_p - number of clock pulses per one half period of output wave):

$$U(t)=\sum_{k=1}^{k_p}\sum_{m=-\infty}^{\infty}U(k)\left[\prod\left(\frac{t-mT_o-kT_i}{T_i}\right)-\prod\left(\frac{t-(m+\frac{1}{2})T_o-kT_i}{T_i}\right)\right] \qquad (2.35)$$

Decomposing the wave U(t) into Fourier series we obtain [2,5]:

$$C_n=\frac{2}{k_p}\left|\sin c\frac{n\pi}{2k_p}\right|\cdot\left|\sum_{k=1}^{k_p}U(k)e^{\frac{-jnk\pi}{k_p}}\right| \quad \text{for odd } n \qquad (2.36)$$

$$C_n = 0 \qquad \text{for even } n$$

$$\text{where: } \sin c\frac{n\pi}{2k_p}=\frac{\sin\frac{n\pi}{2k_p}}{\frac{n\pi}{2k_p}}$$

So by using the values U(k) it is possible to calculate real amplitudes of harmonics present in the wave U(t) and the value of nonlinear distortion factor.

In particular case, when the frequency does not have to be controlled digitally the digital generator of sine wave may have the form presented in the Fig. 2.10.

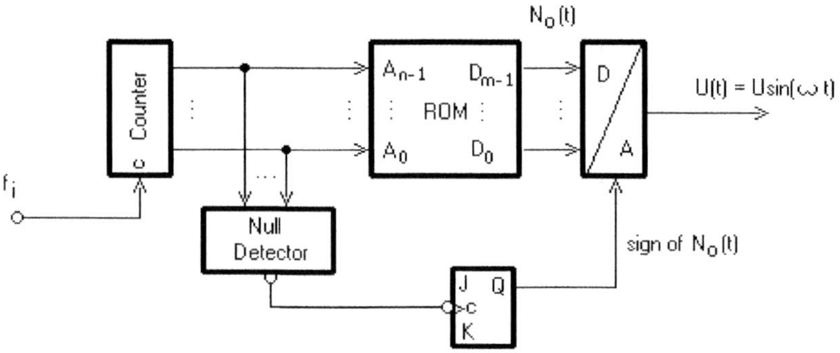

Fig. 2.10 The sine wave generator with uniformly sampled output wave

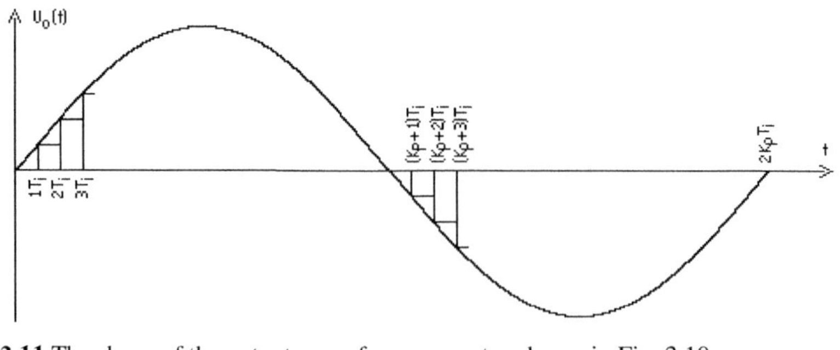

Fig. 2.11 The shape of the output wave from generator shown in Fig. 2.10

An output wave from such generator has a form of uniformly sampled sine wave and next reproduced in a 0-rank extrapolator (see Fig.2.11).

In this case $T_o = 2K_pT_i$, K_p - number of clock pulses per one half period of output wave and $U(k)=U_m \sin(k\pi/K_p)$. The output wave may expressed as:

$$U_0(t)=\sum_{k=0}^{K_p-1}\sum_{m=-\infty}^{\infty}U(k)\left[\Pi\left(\frac{t-mT_o-kT_i}{T_i}\right)-\Pi\left(\frac{t-\left(m+\frac{1}{2}\right)T_o-kT_i}{T_i}\right)\right] \tag{2.37}$$

and spectrum may be calculated as:

$$C_n=4U_m\frac{T_i}{T_o}\left|\sin cn\omega_o\frac{T_i}{2}\right|\cdot\left|\sum_{k=0}^{K_p-1}\sin\omega_o kT_i\cdot e^{-jn\omega_o kT_i}\right| \quad for\,odd\,n \tag{2.38}$$

$C_n = 0$ for even n where: $\omega_o = \frac{2\pi}{T_o} = \frac{\pi}{K_pT_i}$

Because $\sin\omega_o kT_i = \frac{e^{j\omega_o kT_i} - e^{-j\omega_o kT_i}}{2j}$

then for odd n

$$C_n=2U_m\frac{T_i}{T_o}\left|\sin cn\omega_o\frac{T_i}{2}\right|\left|\sum_{k=0}^{K_p-1}\left(e^{-j\omega_o kT_i(n-1)}-e^{-j\omega_o kT_i(n+1)}\right)\right|$$

Taking into account that:

$$\sum_{k=0}^{K_p-1}e^{-ak}=\frac{e^{-aK_p}-1}{e^{-a}-1}$$

then $$C_n=2U_m\frac{T_i}{T_o}\left|\sin cn\omega_o\frac{T_i}{2}\right|\cdot\left|\frac{e^{-j\omega_o K_pT_i(n-1)}-1}{e^{-j\omega_o T_i(n-1)}-1}-\frac{e^{-j\omega_o kT_i(n+1)}-1}{e^{-j\omega_o T_i(n+1)}-1}\right|$$

Because $\omega_o K_p T_i = \pi$ we can the above equation rewrite as:

$$C_n=\frac{U_m}{K_p}\left|\sin cn\frac{\pi}{2K_p}\right|\cdot\left|\frac{\sin\frac{\pi}{2}(n-1)}{\sin\frac{\pi}{2}\frac{n-1}{K_p}}\cdot e^{-j\frac{\pi}{2}\frac{(n-1)(K_p-1)}{K_p}}-\frac{\sin\frac{\pi}{2}(n+1)}{\sin\frac{\pi}{2}\frac{n+1}{K_p}}\cdot e^{-j\frac{\pi}{2}\frac{(n+1)(K_p-1)}{K_p}}\right|$$

Because for odd n

$\sin\frac{\pi}{2}(n-1)=0$ oraz $\sin\frac{\pi}{2}(n+1)=0$ then $C_n = 0.$

But for such n for which $$\frac{\sin\frac{\pi}{2}(n\pm1)}{\sin\frac{\pi}{2}\frac{n\pm1}{K_p}}=\frac{0}{0},$$

Cn may be different then zero. It means that $n \pm 1=2pK_p$ and $n = 2pK_p \pm 1$. Then

$$C_{2pK_p\pm1}=\frac{U_m}{K_p}\left|\sin c\left(2pK_p\pm1\right)\frac{\pi}{2K_p}\right|\cdot\left|\lim_{(n\pm1)\to 2pK_p}\frac{\sin\frac{\pi}{2}(n\pm1)}{\sin\frac{\pi}{2}\frac{(n\pm1)}{K_p}}\right| \tag{2.39}$$

$$=U_m\sin c[(2pK_p\pm1)\frac{\pi}{2K_p}]=U_m\sin c(\frac{\pi}{2K_p})\cdot\frac{1}{2pK_p\pm1}$$

This result was first publicized by Fulford [1] but the above proof is new. It is easier and shorter and it does not require proofing auxiliary theorem.

Using the result (2.39) one can calculate a distortions factor on the base well known formula [1]:

$$h=\frac{\sqrt{\sum_{p=1}^{\infty}U_m^2\sin c^2\left(\frac{\pi}{2K_p}\right)\left(\frac{1}{2pK_p-1}\right)^2+\sum_{p=1}^{\infty}U_m^2\sin c^2\left(\frac{\pi}{2K_p}\right)\left(\frac{1}{2pK_p+1}\right)^2}}{U_m\sin\frac{\pi}{2K_p}} \tag{2.40}$$

Or after transformations

$$h=\sqrt{\sum_{p=1}^{\infty}\left[\left(\frac{1}{2pK_p-1}\right)^2+\left(\frac{1}{2pK_p+1}\right)^2\right]} \tag{2.41}$$

The calculated values of factor h are presented in the Table 2.2.

Table 2.2 Values of factor h

K_p	10	20	50	100	200	500	1000
h	0.20	0.089	0.035	0.018	0.008	0.0035	0.002
number of hamonics	100	the first and two successive non equal to zero					

2.5 Conclusions

Rectangular function $\Pi(x)$ can be well used for mathematical description of frequency multiplication process carried-out by means of logic gates. The delay modules used in such circuits create a certain problem. They limits of using of these frequency multipliers to the cases in which input frequency does not change or the changes are small. It is easy to notice that in the same way one can describe and analyse, for example, pulse multiplication, pulse counting and other operations on pulses or pulse trains.

Rectangular function $\Pi(x)$ can be also used for mathematical description such devices as digital generators and digital frequency multipliers. Apart from this function $\Pi(x)$ may be used for carrying-out logic operations on pulses and pulse trains, for counting process description and so on. This function gives the possibility to carry out this description in time domain what may be convenient, in some practical cases.

References

[1] Fulford, J.F.: Generation of waveforms at very low frequencies using the sampling technique. In: Proceedings of the Institution of Electrical Engineers, vol. 111(12), pp. 1993–2001 (1964), doi:10.1049/piee.1964.0325

[2] Hrynkiewicz, E.: An Application of the Binary Rate Multiplier in Sine Wave Generator. In: VI National Conference on Circuit Theory and Electronic Circuits, Kozubnik k/Bielska-Białej, Poland (1983) (in Polish)

[3] Hrynkiewicz, E.: Multiplying of a Square Wave Frequency in Devices which Contain a Delay Circuits. Rzeszów-Myczkowce (1989)

[4] Hrynkiewicz, E.: Digital Frequency Multipliers of Square Wave. D.Sc. Thesis, ZN Pol. Sl., ser. Automatyka, Gliwice (1992) (in Polish)

[5] Hrynkiewicz, E.: Rectangular Function $\Pi(X)$ and Its Application for Digital Circuits Design. In: Proceedings of 4th International Scientific-Technical Conference on Actual Problems of Electronics Instrument Engineering, APEIE 1998, Novosibirsk, Russia (1998)

[6] Lancaster, D.J.: Matrix Representation of the Multiplying Properties of Binary Rate Multipliers. Transactions on Industrial Electronics and Control Instrumentation IECI-23(1), 70–75 (1976), doi:10.1109/TIECI.1976.351351

[7] Sobkowski, J.: Signal Analysis in Frequency Domain. MON, Warszawa (1975) (in Polish)

[8] Zagajewski, T.: Logic Operations on Walsh Functions and Some of Their Applications. Bull. Acad. Pol. Sci., Ser. Sci. Techn. XXVI, 8–9 (1978)

3 Design and Application of the PLD-Based Reconfigurable Devices

Alexander V. Palagin and Vladimir M. Opanasenko

Department of Microprocessor Devices, Institute of Cybernetics,
Ukraine, Kiev
e-mail: {Palagin_a,vlopanas}@ukr.net

Abstract. Theoretical bases of construction and designing of the PLD–based reconfigurable devices, including the new formalized design techniques of construction and dynamic reconfiguration of architecture and structure of digital devices with a high degree of reconfiguration, corresponding with properties of performing algorithms, constructive and technological features PLD, and also tool means of their designing, are presented. Bases of the theory of adaptive logic networks, intended for the solution of a wide class of tasks by direct structural realization of algorithms of processing and direct representation of input data to output data by functional and structural customization for universal components of a network, are developed. Synthesis algorithms of adaptive logic networks on the classes of tasks set are developed. Design techniques of the computer systems with using of the standard CAD PLD (ISE Foundation) are developed. The structure of the reconfigurable computer system with the open library of configuration files for basic parametrical blocks, including the threshold device, Hemming adder, sorting devices, median filters, matrix multipliers etc. are designed.

3.1 Introduction

The level of development and manufacture of products of high technologies among which one of leading places is occupied by tools of computer engineering (CE), appreciably defines technological progress of many industries. Now scientific researches and practical development in the field of CE on perspective element base, i.e. microprocessors, microprocessor complete sets and systems on the chip in a combination to the LSI circuit of memory and Programmable Logic Devices (PLD) are called to satisfy requirements of the broad audience of users are carried out and to put in pawn bases of development of new effective means of computer engineering.

Development of batch production VLSI demands greater expenses both for development, and for the equipment for their manufacturing. In the schemes realized by a method of printed circuit, change to bring difficultly enough, and in the schemes executed in the form of the LSI and VLSI, any, in advance not stipulated

M. Adamski et al. (Eds.): Design of Digital Systems and Devices, LNEE 79, pp. 59–91.
springerlink.com

changes, are impossible in general. It not only limits opportunities of their specialization for concrete applications, but also prospects of modernization, expansion with addition of new functions, modification in algorithm of functioning. Therefore one of actual requirements to modern digital systems and devices is increase of their adaptability (flexibility). The basic direction of increase of an adaptability of devices (systems) now is specialization of digital devices by programming their structure.

The first theoretical researches, devoted to synthesis reconfigurable devices, concern to the beginning of 60th years. Base work [1] on the organization reconfigurable computer, presented it as two basic parts: a constant – a computer with fixed structure, and a variable – in the form of a set of computers which can reconstruct the structure by means of the program. It promoted occurrence of a new direction in computer facilities on designing reconfigurable devices with virtual (programmed) architecture on the basis of PLD – Reconfigurable Computing (RC). The term "Reconfigurable Computing" generally designates two-uniform concept: as reconfigurable structure of a computer (hardware), and the process of data processing which is performed by a computer. The significant contribution to development of the given problematic was brought also with works [2–6].

Also subject domains in which reconfigurable computer systems (RCS) have found the "lawful" niches were defined and continue to develop intensively. It is first of all:

- The hardware systems guaranteeing safety of control by especially important objects;
- Complex physical experiments with modeling and management in real time;
- Effective digital processing of high-frequency signals;
- Acceleration of tool means of the automated designing of objects of new techniques and technologies;
- Emulation and designing of wireless communication systems, etc.

Importance and perspectives of the specified scopes testify to urgency of direction Reconfigurable Computing and the problems connected with development of technology RC.

Application of PLD gives an opportunity to realize structures of devices with dynamic reconfiguration and by that to solve problems of effective adjustment for the set algorithm, survivability and reliability. *Reconfigurability* – property of system to redefine set of hardware and connections between them in conformity algorithm of functioning. New physical principles and technical opportunities of microelectronic components are, in turn, a source of new principles of construction and new architecture of modern means CE.

Most a wide circulation has received PLD two types: *CPLD* – Complex Programmable Logic Devices; *FPGA* – Field Programmable Gate Array [7].

CPLD consists of set of PAL–like functional blocks (36V18) which contain macrocells and incorporate a matrix of switching to blocks of input-output. Use of FastFLASH-technology allows realizing intrasystem programming with non-volatile storage of configuration data. Feature CPLD is predictability of delays of the signals.

Architecture FPGA generally represents a matrix of logic cells – configurable logic blocks (CLB), surrounded by peripheral cells – Input/Output Blocks (IOB). Connections between cells are carried out by means of programmed matrixes of interconnections. Everyone CLB contains the combinational logic part, remembering element and internal blocks of management and trace. A basis of combinational part CLB is high-speed static CMOS memory and for realization any Boolean functions the technology of Look-Up Table (LUT) is used and the delay of distribution of a signal through the combinational block is independent of generated function. Programmed interconnections provide all communications inside of a crystal. IOB provide the interface between contacts of a crystal and its internal components.

Under existing forecasts, crystals Virtex series by present time should reach logic capacity up to 100 million logic gates (at initial 50 thousand). If first crystals FPGA were manufactured on technology 0,34 microns, now – 65 nanometers.

FPGA series of type Spartan and Virtex [8] possess similar architecture. PLD the considered series except for the logic sold in logic cells (LC), contain block memory (BR) which unlike the distributed memory sold on logic cells, is built in and does not borrow logic resources. Memory BR is organized in the form of blocks, each of which represents the two-port synchronous device of various capacities depending on type FPGA. To modules of a general purpose, except for BR, multiplier units, and built in receiver-transmitter blocks are entered into series Virtex-II Pro with speed of the transfer reaching some Gbit/sec on the channel in a duplex mode, and also RISC-processors blocks of PowerPC type.

3.2 Evolution of Computer Systems

The RC grows out evolution of computer systems (CS). One of the important stages of evolution is creation of emulating computer systems, possessing ability of modification and full change of internal language. The concept of flexibility of architecture CS has been formulated.

In particular, has been developed and has received practical approbation logic–information method (LIM) designing of the microprocessor systems, uniting in itself theoretical concepts of the theory of digital automatic devices and theories of the information. Essence of LIM is illustrated by the scheme:

$$\begin{array}{l} \forall_{i}(\exists\Omega : A_N \Rightarrow \Lambda_N \Rightarrow R_N \\ \qquad\qquad \vdots \\ \qquad\qquad \downarrow \\ \qquad\qquad A_i \Rightarrow \Lambda_i \Rightarrow R_i \\ \qquad\qquad\qquad \vdots \\ \qquad\qquad\qquad \downarrow \\ \qquad\qquad\qquad A_0 \Rightarrow \Lambda_0 \Rightarrow R_0(\Theta = \Theta_{extr}(Q,t)) \end{array} \tag{3.1}$$

where: A_i, Λ_i, R_i $(i = \overline{N,0})$ – accordingly sets of algorithms, operators and their information-code representations at i -th level of programming, Θ – a set of the generalized characteristics (hardware resources (Q), time (t), etc.).

Modern PLD have defined *the new stage* of evolution connected with creation high–efficiency CS. For the formalized representation of model reconfigurable devices updating of method LIM which is illustrated by the following scheme is offered:

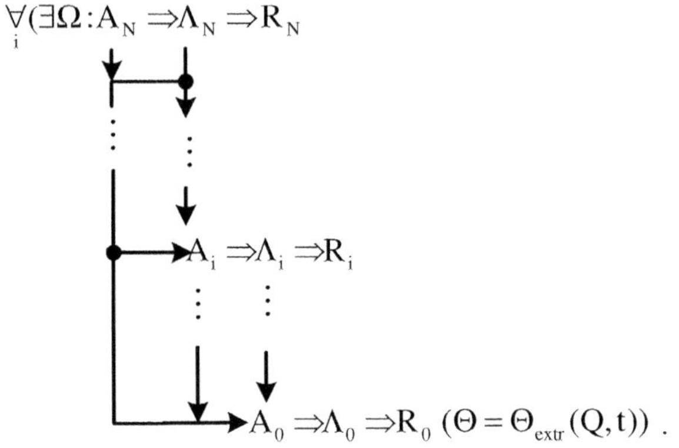

(3.2)

In the scheme (3.2) for classical architecture following levels of programming are used: τ_0 – physical or "zero"; τ_1 – microprogrammed; τ_2 – programmed; τ_3 – algorithmic. Programming at a "zero" level defines physical structure of the device which finally realizes the set algorithm of functioning, i.e. carries out programming structure of the device. In difference from the scheme (3.1), updating (3.2) realizes not microprogrammed, but hardware realization of algorithms on gate level. In it difference reconfigurable devices with programmable structure from modern computers consists. Thus the logic structure reconfigurable devices can dynamically vary both by preparation for the decision of a problem, and during computing process.

The most widespread requirement shown to facility CS, high speed is. The given problem, in particular, arises at use of means CS for problems of management and modeling. At use of computers for control of moving objects, technological processes, fighting operations, etc., they should work with anticipation of real processes in operated object or, and generally speaking, in real time. There is a class of problems in which it is necessary to operate quickly the changing processes proceeding in short time intervals, and highly dynamical, quickly functioning objects. Thus simultaneously with high speed maintenance of high accuracy of management is required. Therefore the computer for maintenance of high speed and accuracy of management should possess ultrahigh speed to provide simultaneously set accuracy and work in real time. The similar problem arises and at use of facility CS for modeling complex dynamic objects, and also quickly proceeding processes and the phenomena.

With the advent of modern crystals FPGA began possible to use the results received earlier for construction reconfigurable devices and systems of the raised complexity on the basis of PLD with completely programmed architecture. One of approaches for increase of productivity of facility CS is the combination conveyorization and parallelism.

If frequent change of carried out functions down to new function takes place at each new execution, the conveyor with a dynamic configuration takes place. The given approach is realized in technology RC. As an example of such realization hypercomputer HAL (Star Bridge System Corp.) which is constructed on crystals FPGA can serve. During functioning crystals change the structure and functions is continuous at the decision of numerous computing problems in a mode of real time.

Opportunities RC are introduced in supercomputers Cray XD1 with the purpose of increase of productivity for target appendices by use of a subsystem of acceleration of the appendices, based on crystals of type FPGA (Virtex–4) firm Xilinx which can be programmed on acceleration of key algorithms, such as search, sorting, digital processing of signals, etc. the Given subsystem functions as the coprocessor in relation to base processor AMD Opteron.

In Berkeley Wireless Research Center supercomputer system High-End Reconfigurable Computing System (HERC) on crystals FPGA [9] is developed. Basis of HERC is system prototype BEE (Berkeley Emulation Engine) with two modules, intended for designing, construction and programming HERC for of some applied areas. In opinion of developers, use BEE the system based on processors DSP with similar power consumption and cost, and more than on two orders in comparison with the systems realized on the basis of standard microprocessors provides on the order greater productivity, than. The main components of architecture – computing blocks and the programmed communication environment. The computing block is structurally presented in the form of the printed-circuit-board, which contains four crystals FPGA – processing modules with memory (up to 4 GB everyone) and one for management (the operating module).

3.3 Architecture and Structure of PLD-Based Computer Systems

The typical reconfigurable *computer system* (CS) consists, as a rule, of 2 parts: constant (or "fixed") part F – a Host-computer and a variable part V – reconfigurable subsystem (RSS) which can be united in various configurations. The architecture of reconfigurable systems depends on capacities of sets of algorithms: (N_F), carried out on the equipment F, and (N_v), carried out on the equipment V. The parity of these sizes defines offered classification of reconfigurable computing systems:

a) The computing systems focused on a Host-computer in which the basic computing capacities are concentrated, and reconfigurable computer provides increase of productivity only for a narrow class of problems ($N_F \to N, N_V \to 0, N_F >> N_V$);

b) The computing systems focused on RSS in which the Host-computer is used, basically, for performance of auxiliary functions (service, input-output), and all algorithms are carried out mainly in RSS which can have own field of external devices (through payments of expansion) or the general field of external devices with a Host-computer to which RSS has direct access;
c) Reconfigurable computing systems in which a Host-computer and RSS have approximately identical complexity, thus RSS it is focused on the decision of labour-consuming problems, and the Host-computer provides strong support regarding translation, input-output, service, etc.;
d) RSS is the independent device in case of $N_F = 0, N_V = N$, and the Host-computer is absent.

RSS connects to a Host-computer through one of the standard trunks, the variants of connection most widespread today are realized through trunks PCI and PCI–Express. RSS have functional processing field (FPF) the set dimension which is configured for performance of the set algorithm or its part, providing, thus, optimum realization of this algorithm both under time characteristics, and on hardware expenses.

Introduction in practice of crystals PLD and HDL–technology (Hardware Description Language) for performance of projects in this element basis intensified development of a wide spectrum of the digital modules representing ready technical decisions, essentially reducing time of designing and an output for the market of new products. Such opportunities of HDL–technology as hierarchical designing, bearableness of libraries, platform- independence, allow using available soft cores as macrocells for development of new technical decisions. The architecture of modern crystals FPGA is optimized for use both hard and soft cores, and allows integrating them into projects easily. For example, crystals of Virtex type have in advance built in multipliers and PowerPC processors as hard cores, and also other functional blocks.

In RSS, or devices with programmable architecture the functional field of the set dimension configured specially for performance of certain set algorithm or its part is fixed, providing, thus, realization of this algorithm optimum, by the set criteria, by way. Adjustment of structure for performance of demanded algorithm and its realization in a crystal on a gated level allow increasing speed of the device by some orders in comparison with universal decisions.

The algorithm can be broken into the fragments which are carried out consistently in this connection, structures corresponding these fragments also are loaded into a crystal consistently (by way of their performance), that leads to essential economy of resources. Complexity of fragments of algorithm thus is defined by only logic capacity of a crystal, i.e. dimension of a processing field.

Thus, reconfigurable data processing represents to a certain extent change of the central paradigm of designing of modern means of computer facilities.

The model of the projected computing system is offered:

$$S = <M, A, B, P>,$$

where: M – set of mathematical methods, characteristic for a subject domain, reflecting functioning of system; A – set of algorithms of realization of a method;

$B = \{b\}$ – the components of alphabet from which the structure is synthesized; P – procedure of the description of the project (the description of object). Thus, process of designing consists in the decision of a problem of synthesis of structure on the basis of components $\{b\}$ the alphabet B for performance of the certain algorithm A realizing a method M, underlying functioning of structure, according to requirements of specifications. Result of procedure P is the description of the project by means language CAD.

Synthesis of structural realization of sequence of algorithms is offered, when the method/problem (M) is represented sequence of algorithms ($A_i, \forall i = \overline{1 \div n}$):

$$M = \bigcup_i A_i .$$

In RSS, the base (zero) architecture realized on chip of PLD in the form of a functional processing field of fixed dimension, the controller of the trunk of a host-computer, a field of memory, and also well structured library of configurations files (LCF) structural realizations of the methods (algorithms) which are carrying out display of algorithm in structural realization ($F: A_i \Rightarrow B_i$) is initially set. Each algorithm has display $F: A_i \Rightarrow B_i$ in structural realization (B_i) which represents a configuration file for a crystal PLD. Generally there are some variants of realization of algorithm (for example, consecutive, series-parallel and parallel):

$$B_i = \bigcup_z B_{iz}, (z = 1 \div k) .$$

Each variant is characterized by parameters of speed (time of performance – t_{iz}) and hardware expenses (q_{iz}). And we assume, that capacity of set B is sufficient for realization of a wide set of algorithms. In the event that demanded realization of *i*-th algorithm in library is absent ($B_i = \varnothing$), it is necessary to create by means of CAD PLD it and to include as a standard element in library. Thus, the problem of optimization is reduced to the ordered purpose to each *i*-th top the column of sold algorithm (B_{iz})-th element of library with the purpose of reception of extreme value of some criterion of quality. I.e. any operator is displayed only by one element from library. The structure realizing set columns is as a result defined. Then the decision of a problem can be received by methods of integer mathematical programming and, depending on demanded criterion of quality, it is possible to define following variants of statement of a problem of optimization.

The problem of optimization consists in definition of a minimum of criterion function, and criteria of quality are total hardware expenses for realization of all algorithms:

$$\alpha \sum_i \sum_z t_{iz} x_{iz} + \beta \sum_i \sum_z q_{iz} x_{iz} = min, \quad (\forall i = \overline{1 \div n}, \forall z = \overline{1 \div k}),$$

under conditions of restrictions $\sum_{z=1}^{k} x_{iz} = 1$, $\sum_{i}\sum_{z} q_{iz} x_{iz} \le Q_0$ $\sum_{i}\sum_{z} t_{iz} x_{iz} \le T_0$,

where α, β – weight coefficients which can be certain, for example, a method of expert estimations; x_{iz} – z-th realization of i-th algorithm A_i; T_0 – admissible time of performance of all algorithms; Q_0 – admissible hardware expenses.

Methods of the decision of such problems are well enough developed and allow receiving for admissible time the comprehensible decision.

Presented approaches are put in a basis of the generalized algorithm of designing PLD-based reconfigurable devices which represents system of the interconnected algorithms, the part from which is formalized and shown to statement and the decision of a problem of synthesis and a choice of optimum structural realizations from set, the others use heurism. Each algorithm is a separate fragment of designing to which the certain section of the dissertation where it is presented in the form of the formalized technique of designing with a theoretical substantiation of its basic positions and the description of methods of the decision of concrete applied problems of the analysis, synthesis and optimization of separate structural realizations is devoted.

The algorithm of designing of the structural realizations RSS representing a Basic board (for the coprocessors connected to the standard Bus of a Host-computer) or Carrier board (for independent devices) with a set of expansion boards and expansion modules, or crystal PLD for realization SoC (System–on–Chip) is developed. The algorithm represents sequence of stages (Fig. 3.1).

The analysis of problem area statement of a problem a choice of suitable algorithm (in case of absence of a configuration file for realization of corresponding algorithm its synthesis with the subsequent record in LCF) from LCF imaging at a level of the general architecture (function chart) preparation of the formalized technical project programming of structure on the basis of a configuration file a programming of algorithm the decision of a problem an estimation of characteristics of parameters (structure is carried out, process of the decision) check of parameters on conformity to the established criteria (if necessary following iteration) commissioning. The block diagram of algorithm (Fig. 3.1) provides also correction of criteria. It is analyzed features of designing of digital devices on the basis of PLD with use of HDL-technology and CAD PLD.

The developed technique of designing, leaning on the given system of algorithms and logical-information model RSS laying in its basis, allows to decide – to formalize the main task of designing process of search of optimum pair «algorithm-structural realization». The technique intends for designing: the task-oriented coprocessors and the independent devices working with the algorithms set; reconfigurable processors with conveyor data processing; parametrical IP-Core for realization of the algorithms set which are represented by elements of library of configuration files; SoC. It can be modified depending on the initial task, a class of problems, element -technological base, etc.

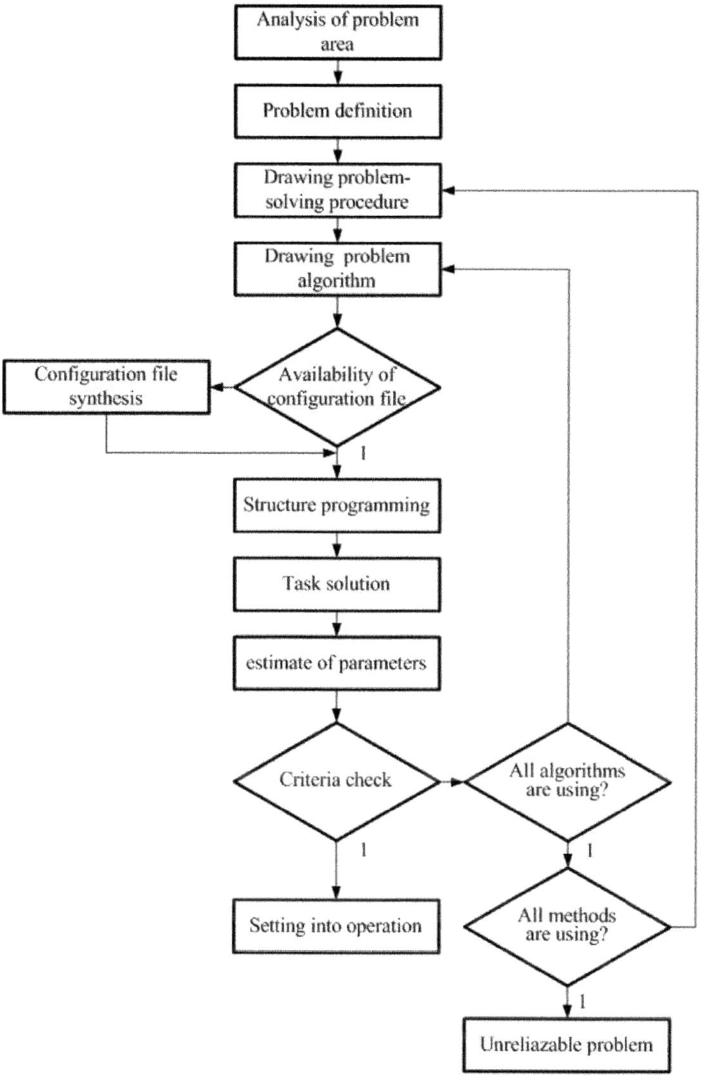

Fig. 3.1 Algorithm of designing of the structural realizations RSS

3.4 Adaptive Logical Network (ALN)

The adaptive logic network is a discrete converter of codes of type of the asynchronous combinational automatic device, set directed graph which tops are logic functions, and edges – communications between them type "output–input".

From the point of view of topology of system ALN represents a matrix of universal logic elements (LE) which are grouped into functional units (FU) and blocks (FB) which site is fixed, thus change of their functioning occurs depending on a class of problems and from their purpose.

Universal LE we shall name the combinational automatic device: $L = \langle n, F \rangle$, where: n – quantity of binary inputs or dimension of entrance variables LE; $F = \{f_\rho\}$, $\rho = [1 \div 2^{2^n}]$ – the totality set of Boolean functions. Universality LE consists in an opportunity of its adjustment for realization any Boolean functions.

Structure ALN can be described by following system:

$$A = \langle n, h, F, S, L, m, D, X, Y \rangle,$$

where: *n* – word length of input binary vectors (dimensionality of ALN on an input); *h* – target word length ($h = \overline{1 \div n}$), dimensionality of ALN on an output; $F = \{F_{ij}\}$ – set of logic functions of system; S – structure of communications between LEs; $L = \{L_{ij}\}$ – set LE (i –a serial number of element LE; j – number of a level of processing); m – quantity of levels of processing; $D = \{d\}$ – set of *n*-dimensional binary vectors (training sample); X – full set of input binary vectors; $Y = \{Y_{ij}\}$ – the generalized function of system, $Y_{ij} = f_{ij}(Y_{v,(j-1)}, Y_{w,(j-1)})$ – value of the function f_{ij} sold by an element L_{ij}, $Y \in \{0,1\}$ which structure is resulted on Fig. 3.2 (*v*, *w* – value of an index *i* for inputs LE).

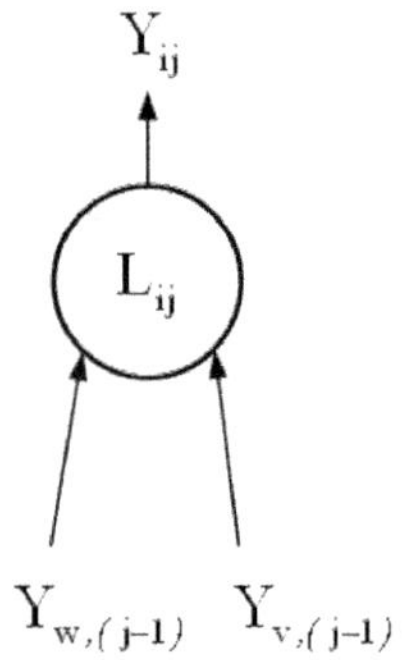

Fig. 3.2 Structure of universal logic element

Each level ALN represents a ruler of LE (ρ inputs for all), each of which can be adjusted on performance of any of a full set (2^{2^ρ}) Boolean functions of its input variables and realizes imaging of l -dimensional $(l \le n)$ binary vectors into a u -dimensional $(l \ge u)$ binary vector. Matrix of LEs or FU represents the combinational automatic device without the memory, a having l -digit input, a u -digit output and m – quantity of lines of a matrix. Within the limits of one level the type of function can be set for everyone LE separately (stepwise adjustment) or for all LEs (by the level adjustment).

The functional block represents a network of consistently included automatic devices (hierarchical assembly of l ($l=\overline{1 \div m}$) functional units). In the further we shall be limited to consideration of three types FB distinguished to a topological attribute: «rectangular matrix» (RM) – ($h=n$); «triangular matrix» (TM) – ($h=1$); «trapeziform matrix» (TrM) – ($h=\overline{2 \div (n-1)}$). Depending on structure of communications following TM types are offered: with logarithmic structure of communications (LSC); with cellular structure of communications (CSC); with asymmetric structure of communications (ASC). The offered structures of communication differ on capacity sold Boolean functions and to hardware resources.

Problems of the structural organization and synthesis of multilevel structure ALN of type TM consists in definition of types of logic functions f_{ij} for all LEs network. For definition of set of logic functions $F=\{f_{ij}\}$ the approach based on the description of a Boolean network by polynomials which factors are set, in particular, by means of Adamaar matrixes is used [10].

At coding values Boolean functions and its arguments transition to coding with use of values (1) and (–1) is carried out. Thus, the set of variables $X=\{x_1, x_2, ..., x_n\}$ for Boolean from n variables will be represented function f by set $E=\{e_1, e_2, ..., e_n\}$, where $e_i=(-1)^{x_i}$, and set of values $Y=\{y_1, y_2, ..., y_{2^n-1}\}$, where $y_j \in \{0,1\}$; set $V=\{v_0, v_1, ..., v_{2^n-1}\}$, where $v_j=(-1)^{y_j}$. For any Boolean functions f from n the variables accepting values from set $\{1,-1\}$, there is an equivalent polynomial $P_{f(n)}$ with factors from set of real numbers– $f(X)=P_{f(n)}(X)$. Factors of a polynomial for function f enter the name by means of Adamaar matrix ($A=\frac{1}{2^n} H_n V_n$,), where $A=\{a_0, a_1, ..., a_{2^n-1}\}$ – set of factors of a polynomial, H_n – Adamaar matrix dimension 2^n, V_n – set of values Boolean functions.

By way of illustration applications ALN of type TM a number of functional devices of the average complexity focused, mainly on problems of recognition of images is synthesized.

So, on the basis of the scheme of transformations using as base bit operations of addition and multiplication on the module 2 (logic operations XOR and AND) the problem of synthesis of adder Hemming of any word length is solved [11]. The offered synthesis algorithm of adder Hemming of any word length carries out imaging: $\Im:(g,d) \Rightarrow T$, where: $g \in G, d \in D, T=\sum_{l=1}^{n}(g_l \oplus d_l)=\sum_{\lambda} 2^{(\lambda-1)}\tau_\lambda$; τ_λ – a component of the vector, containing value λ–th bit of binary representation T of a mismatch of vectors g and d; $\lambda=1 \div (\mathrm{Ent}\{log_2(n+1)\})$; $2^{(\lambda-1)}$ – weight λ-th bit of binary representation of a mismatch T.

The problem of synthesis of the threshold device of the any word length realizing threshold operation from set $\Psi = \{\psi_\xi\}$ is solved also, where $\xi = 1 \div 4$ (ψ_1 –operation $\leq$, ψ_2 – operation $\geq$, ψ_3 – operation>, ψ_4 – operation <), types of logic functions for each level of structure TM (type ASC) depending on value of a threshold are as a result defined.

The synthesis algorithm of the threshold device is based on the bit-by-bit analysis of value of a threshold vector Θ according to binary data presentation: $\Theta = \sum_{\lambda} 2^{\lambda-1} \theta_\lambda$, where $\tau_\lambda, \theta_\lambda$ – the components of vectors containing value λ-th bit of binary representation T and Θ accordingly, and $2^{\lambda-1}$ – weight of λ-th bit ($\lambda = 1 \div n$, n – dimension (length) of a binary vector).

The synthesis algorithm of the symmetric threshold device of the any word length realizing symmetric threshold operation with the top and bottom borders, symmetric concerning the center of a numerical axis on a piece $[0 \div (2^n - 1)]$ is developed also. As the set operation concerns to symmetric functions at the first level of structure TM logic function XOR is used, and for other levels types of logic functions F_s^ξ are defined to algorithm similarly considered above for the threshold device.

3.5 Problem-Oriented Structures of Digital Devices

The technique is developed and process of designing the typical reconfigurable problem-focused devices with hardware PLD-based realization in the form of base library parametrical functional blocks by means of their VHDL language description and the Schematic editor is considered. Library of functional devices of the wide application providing use by the broad audience of developers at designing of digital devices by the task of corresponding parameters and a choice of optimum structure (by criteria speed-complexity of realization) are developed.

The following developed functional blocks are included in structure of library of files of configurations: Hemming adders which are carrying out calculation of distance Hemming for 4, of 8 and 16-digit numbers; two variants of realization of algorithm of sorting (the linear sorter and the memory-based sorter); multiplier square matrixes of the order $m = 10$ for the whole 16-digit numbers; the median filters using consecutive, it is serial–parallel and parallel computing models; arithmetic devices of multiplication with a floating point of unary accuracy (compatible to standard IEEE–754). The developed functional blocks are verified at real stands and the reconfigurable device (board ADS–XLX–SP3–EVL400) that proves their functioning.

3.5.1 Functional Blocks with a Floating Point

During designing of mathematical coprocessors, the DSP processors, the in-built arithmetic coprocessors wide application is found with floating point functional blocks. Many vendors (for example, Nallatech Corp.) are developed own soft cores for realization of such arithmetic operations, has developed Core for processing operands with a floating point (standard IEEE-754) under Virtex series.

The problem of designing of arithmetic devices and algorithms for processing operands in a floating point format is actual and now. Standard IEEE–754 gives most the general representation for numbers with a floating point in modern computers, including Intel PC, Macintosh and majority Unix platforms.

Let's consider development of the devices which are carrying out the floating point operations in conformity with standard IEEE-754. The generalized structure of functional blocks with a floating point (Fig. 3.3) and contains of three compound modules: the module input arguments checking module (IAC); the functional module (FM) and the result creation module (RCM). The description of modules is executed by means VHDL language, by development synthesizer FPGA Compiler II from Synopsys is used, system CORE Generator System is applied to formation of IP-Core blocks. The developed modules are verified by a modeling method with definition of time and hardware parameters. Modules represent the finished typical technical decisions and can be used in other projects as soft cores.

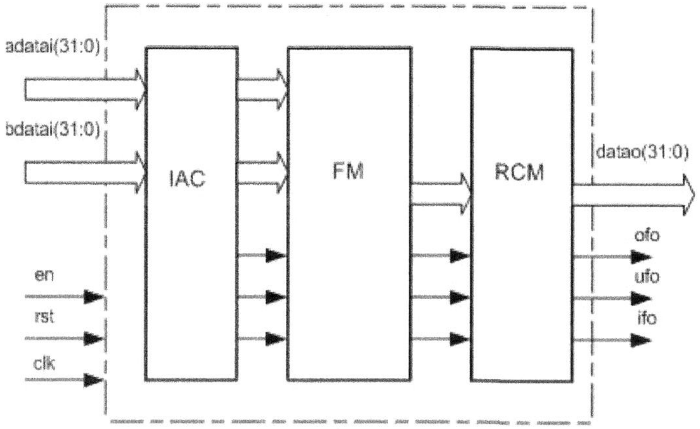

Fig. 3.3 Structure of the functional block with a floating point

IAC will transform input data, analyzes them on conformity to standard IEEE-754 with formation of corresponding attributes. Corresponding numbers and the information concerning classes of input data gives out as results to the functional module with a floating point.

FM carries out the set operation from a floating point with formation of corresponding attributes.

RCM carries out the conformance of a format result data with standard IEEE-754 and final setting of flags.

Inputs and outputs of the floating point block are not adhered to the fixed input-output contacts of a specific FPGA, because using of any chips therefore is supposed. Assignment of inputs and outputs of the block is shown on Fig. 3.3: clk – a global signal clock; rst – a global signal reset; en – Enable signal; adatai (31:0) – the input data bus A; bdatai (31:0) – the input data bus B; datao (31:0) – the output data bus; ofo – a flag "Overflow"; ufo – a flag "Underflow"; ifo – a flag «Inadmissible operation».

For the agreement of the obtained result of transformation with standard IEEE–754 it is necessary to present numbers as normalized form. Therefore it is required to define a high-order digit «1» and to realize shift aside to the high-order digit on demanded number of bits with simultaneous subtraction of this value from the resulting exponent part. Presence of powerful logic resources in crystals Virtex series allows accelerating this procedure by fast definition of number of shifts. Then, unlike realization of serial shift with the simultaneous analysis of the high-order bit, is carried out parallel shift on demanded number of position for normalization of a mantissa.

Floating point Addition a includes strictly serial five operations: comparison of exponents, shift to the right mantissas of smaller number, summation of mantissas, search of left unit of a mantissa of result, normalization of a resulting mantissa.

For realization of operation of search of left unit using priority coder is offered. Let is available ($n = 24$) meaning bits of a mantissa. It is required to define number of the high-order "nonzero" position and to carry out of normalization of a mantissa $F = \{ f_{23}, f_{22}, ..., f_i, ..., f_0 \}$.

Priority coder represents the combinational scheme, having n inputs and ($\mathrm{Ent}\{\log_2 n\}$) outputs which consists of two sequentially connected schemes – the first allocates high-order unit, and the second its number (number of demanded shifts) in an operand.

The first scheme has n inputs and n outputs, realizing following system of the logic equations:

$$a_i = f_i \left(\bigcap_{i=i+1}^{(n-1)} \overline{(f_i)} \right) \forall i = 0 \div (n-1) \tag{3.3}$$

The second scheme has n inputs and ($Ent\{\log_2 n\}$) outputs, realizing following system of the logic equations:

$$y_j = \bigcup_{k=1}^{N=\mathrm{Ent}\{(n-1)/(2^{(j+1)})\}} \left[\bigcup_{i=2^j+(k-1)\times 2^{(j+1)}}^{2^j+(2^j-1)+(k-1)\times 2^{(j+1)}} a_i \right], \tag{3.4}$$

$$(\forall j = 0 \div (\mathrm{Ent}\{\log_2 n\} - 1)),$$

Thus, mathematical expressions (3.3) and (3.4) allow to synthesize priority coder for any word length, the representing parametrical module which can be used by development of new projects by other users.

By development of typical modules as well as by development of usual projects, use already well fulfilled accessible IP–Core is expedient.

Let's consider an example of designing of the 32–bit floating point block of multiplication. The block consists of three elements, first two of which, according to Fig. 3.3 enter into functional module FM, and the third – in functional module MFR.

The first element forms 24-digit operands for the block of multiplication ("1" in the high-order – 23-rd position and 23 digits of fraction of a mantissa), summarize exponents of multiplied numbers (8 bits) and defines a sign on result.

The second element carries out operation of multiplication and is formed by means of the Core Generator (Xilinx Corp.).

The third element carries out check of conditions and formation of result. Following conditions are checked: if the sum of exponents of numbers is equal or more than 255, the signal "Overflow" is formed; if 24-th bit of product of numbers is equal "1", then shift of product of numbers on one position aside low-order digits and increase in the exponent per unit is made; if, after increase in the exponent per unit, value the exponent becomes equal 255, then the signal "Overflow" is formed.

Using of an element of multiplication of combinational type the result of multiplication is formed on a step following a step of registration of operands. When it is necessary to multiply arrays of the numbers acting synchronously with any clock sequence CLK, using of an conveyor-based element of multiplication is preferable. In the developed module elements of multiplication, both with the multiplier of combinational type, and with 4–levels (LUT–based realization are used) or the 2–level (in–built blocks of multiplication 18x18 are used) conveyor that allows to reduce essentially due to increase in clock speed time of multiplication of arrays of numbers. For the timing agreement four or two series registers in this case are entered into the first element of the module for conveyor transfer on an output of the exponent and a sign of product of numbers. The delay (Latency) between registration of the first operands and registration of the first product of the module of multiplication is equal to 5-th or 3-th periods CLK accordingly at use a 4-level or 2-level conveyor-based element of multiplication.

In Fig. 3.4 the diagram of work of the module of control IAC, executed by means of editor State Editor is represented. On the first step at presence of signal EN=1 the block passes in status STATE1, on which (digits 0–31) from input operand A formed signals EXP_F (exponent – digits 23–30) and FRAC (fraction of a mantissa – digits 0–22).

Further check of conditions is made, at performance of one of which block passes on the second step in one of statuses (STATE2 – STATE6) with formation of a corresponding flag:

- If value of the exponent to equally zero, and fraction of a mantissa nonzero the input operand is nonnormalized number;
- If values of the exponent and to fraction of a mantissa are equal to zero the input operand is zero;
- If value of the exponent is more than zero and less than 255 the input operand is the normalized number;
- If value of the exponent equally 255 and fraction of a mantissa zero, the input operand is infinity ($\pm\infty$);
- If values of the exponent equally 255 and fraction of a mantissa nonzero the input operand is not real number *(NAN).*

Transition of the block in an initial status is made on a signal of reset (RESET) or setting of signal EN in a zero status.

In a chip of series Spartan–II (XC2S50–5) the block borrows 46 Slices and operates on clock speed 103MHz.

Advantage of the offered realizations in comparison with known is reached due to optimum distribution of descriptions of constituent modules in different modes, and also original priority coder which allows to define number of high-order "1" for the subsequent performance of operation of normalization of a mantissa for one timing step.

Comparative estimations of hardware resources and are presented to productivity of the developed modules of the multiplication realized with using Core (Xilinx Corp.), with similar modules of Digital Core Design.

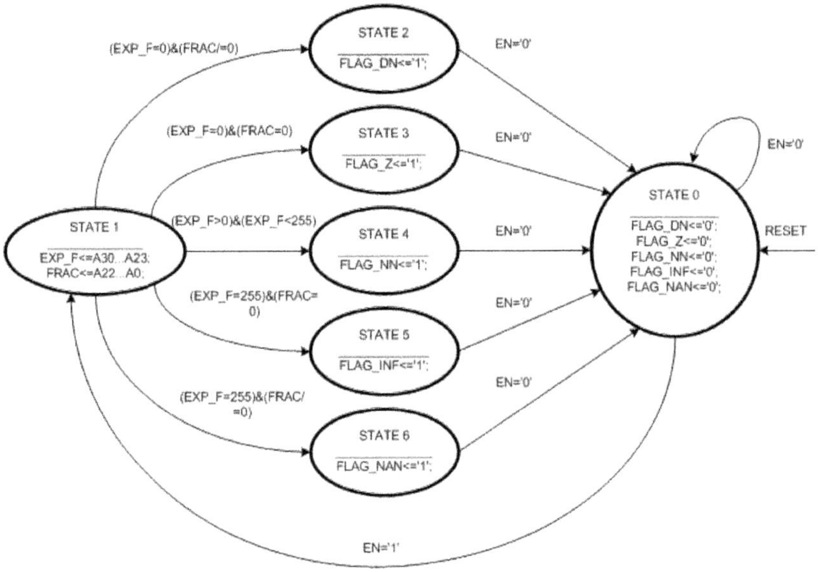

Fig. 3.4 The diagram of work of module IAC

Resources are estimated by quantity Slices. Productivity is estimated by frequency CLK. Hardware resources we shall estimate concerning known realizations:

$$\Delta Q_i = Q_0 / Q_\mu ,$$

where: Q_0 – hardware resources of the module of Digital Core Design; Q_μ – hardware resources of the offered module; Q_1 – hardware resources of the module without the conveyor; Q_2 – hardware expenses of the conveyor-based module.

For a FPGA of type 2S200–6: $\Delta T_1 = T_0/T_1 = 2{,}54$; $\Delta T_2 = T_0/T_2 = 2{,}33$; $\Delta Q_1 = Q_0/Q_1 = 2{,}5$; $\Delta Q_2 = Q_0/Q_2 = 2{,}3$.

For a FPGA of type V300–6: $\Delta T_1 = T_0/T_1 = 2{,}5$; $\Delta T_2 = T_0/T_2 = 2{,}27$; $\Delta Q_1 = Q_0/Q_1 = 3{,}5$; $\Delta Q_2 = Q_0/Q_2 = 2{,}26$.

For a FPGA of type 2V250–5: $\Delta T_1 = T_0/T_1 = 6{,}15$; $\Delta T_2 = T_0/T_2 = 9{,}67$; $\Delta Q_1 = Q_0/Q_1 = 6{,}15$; $\Delta Q_2 = Q_0/Q_2 = 3{,}98$.

The variant of realization of the module on FPGA 2V250–5 with using LUT is absent in offers of Digital Core Design, however regarding hardware resources it we shall compare with offered realizations on FPGA V300–6, but allows to work (approximately on third) with greater clock time.

The synthesized the functional floating point blocks (compatible to standard IEEE–754), can be used as a library element by development of complex computers.

3.5.2 Functional Blocks for Multiplication of Matrixes

One of the basic features of programmable logic is the opportunity of using a principle of parallel data processing at the solving of the wide problems. The increasing of resources of modern programmable logic allows to raise essentially speed of developed devices and to realize by hardware the algorithms working in real time. Multisequencing of calculations or logic operations it can be carried out both at a level of digits of representation of the information, and at a level of the blocks which are carrying out corresponding algorithms of mathematical model. An example of such successful realization is the principle of Parallel Distributed Arithmetic used in digital signals processing.

Let's consider realization of multiplication algorithm of a matrix $A = \| a_{ij} \|$ of the size $m \times n$ on a matrix $B = \| b_{jk} \|$ of the size $n \times r$. The resulting matrix $C = \| c_{ik} \|$ in the size $m \times r$ is formed as follows:

$$C = AB = \| a_{ij} \| \times \| b_{jk} \| = \| c_{ik} \|,$$

Where

$$c_{ik} = \sum_{i=1}^{n} a_{ij} b_{jk} \tag{3.5}$$

Thus, according to (3.5), each *j*-th element of *i*-th line of a matrix A is consistently multiplied by corresponding *j*-th element of a column of a matrix B and the received products are added.

For definition of each element of resulting matrixes are used operations of multiplication and summation of partial products. Summation can be carried out by two ways: accumulation of partial products at their serial receipt on an input of the accumulator from an output of the multiplier and parallel summation of partial products. The first way assumes presence of the block which is carrying out multiplication and summation (accumulation) of received partial products. The second way uses a set of multipliers and the multiport adder for reception of an element of resulting matrixes.

These ways are realized by several variants:

- Serial (SL), when the processing field consists of one block consistently calculating the sum of pair products in (3.5);
- Parallel-serial (PS1), when the processing field contains set of which quantity correspond to quantity (i) lines of a matrix A, by means of which the sums of pair products for elements c_{ij} are simultaneously calculated, and results in (3.5) further are consistently formed;
- Parallel-serial (PS2), when the processing field contains such quantity of blocks, in which quantity of multipliers correspond to quantity (i) lines of a matrix A, in parallel realizing, thus, calculation of one element c_{ij} of a matrix C, and further other elements c_{ij} are consistently calculated.

Let's consider realization of the device which are carrying out multiplication of square matrixes of the order $m = 10$ for the whole 16-bit numbers, realized in a crystal Virtex-E series. The quantity of using Slices includes input, output and intermediate registers for realization of conveyor-based calculations. Execution time of operation of multiplication of two 16-bit numbers with accumulation of the 32-bit sum (summation of result of multiplication with the number which is being the accumulator) for specified type of a crystal is 6,424 nanoseconds. In Table 3.1 hardware and time estimations for the considered variants of realization are resulted.

Table 3.1 Results for different implementations

Variant of realization of algorithm of multiplication of matrixes	Quantity of multipliers / adders	Speed (full time of multiplication of matrixes), nanosecond	Hardware expenses (quantity Slices)
SL	1/1	6424	181
PS1	10/10	642,4	1810
PS2	10/1	890	1665

3.5.3 Designing and Realization of Median Filters

Digital methods of processing of images now play a significant role in scientific researches, the industries, medicine, space researches and information-telecommunication systems. One of methods of digital processing the images applied to elimination of defects of the image, caused by handicapes and noise, is the median filtration. Median filters (MF) differed robustness and are convenient for smoothing the information in cases when noise characteristics are unknown. Stepped changes of a signal pass through the median filter without distortion. This feature is used, for example, in the image filtering where data should be smoothed, but distortion of the form of fronts of a signal is inadmissible.

Let's consider realization of the PLD-based median filter, it is using serial, serial–parallel and parallel computing models.

Generally, the median can be defined as magnitude x_{med}, for which at any values z fairly expression:

$$\sum_{i=1}^{n} | x_{med} - x_i | \le \sum_{i=1}^{n} | z - x_i |.$$

Median filtration realizes a choice x_{med} for odd n, thus is unequivocal (for even value n there is an infinite number of possible values x_{med}). So, for the two-dimensional window containing 3x3 of elements of the image (pixels), the median filter with nine vectors describing brightness for halftone image or color for the color image, chooses a vector with average value which then is appropriated to central pixel of windows. Median filtration can be carried out also for a window of any other form, for example, crosswise with number of pixels, equal 5 or 9, etc. Irrespective of the form of a window the filter realizes the same algorithm and is characterized by number (n) and word length (m) of processing pixels.

Thus, a median x_{med} of discrete sequence of binary vectors $x_i(i = 1 \div n)$ for odd n is that its element for which exists $(n-1)/2$ elements, smaller or equal to it on size, and $(n-1)/2$ elements, greater or equal to it on size. Let for $\forall x_i \in X$ input set of binary vectors $X = \{x_i\}$ $(i = 1 \div n)$ it is necessary to define a median x_{med}. With the purpose of increase of speed of the scheme it is offered to use algorithm of definition of the median, allowing manipulating not input data that is inherent in some algorithms of sorting, and results of comparison of input codes among themselves. The algorithm of definition of a median in this case represents sorting data with the subsequent choice of the code having number $(n-1)/2$ from sorted sequence which numbering begins with zero.

The square matrix is formed:

$$Y = \|y_{ij}\| \; (i, j = 1 \div n),$$

where $y_{ij} \in \{0,1\}$ – an element of a matrix which is defined by a rule:

$$\begin{cases} y_{ij} = 1, \text{if } x_i \ge x_j; \\ y_{ij} = 0, \text{if } x_i < x_j. \end{cases} \tag{3.6}$$

Elements y_{ij} of the main diagonal of a matrix $(i = j)$ accept zero value. And if $(\forall i, j)\, y_{ij} = 1$, then $y_{ji} = 0$ and on the contrary. Therefore values of elements y_{ij} with the indexes $i > j$, laying above the main diagonal (the quantity of these elements is defined by size $(n^2 - n)/2)$ are defined. Values of the elements y_{ij} $(i < j)$ laying below the main diagonal are defined as follows:

$$\text{if } (\forall i, j)\, y_{ij} = 1, \text{ then } y_{ji} = 0;$$

$$\text{if } (\forall i, j)\, y_{ij} = 0, \text{ then } y_{ji} = 1.$$

For every line received matrix $Y = \|y_{ij}\|$ the arithmetic sum of values of elements y_{ij} is calculated:

$$s_i = \sum_j y_{ij} \tag{3.7}$$

Depending on numerical value s_i which unequivocally corresponds to input vector x_i, we receive result of sorting of set of vectors X:

$$\begin{cases} if\ s_i = 0, \text{then } x_i = min\{X\}; \\ if\ s_i = (n-1), \text{then } x_i = max\{X\}; \\ if\ s_i = (n-1)/2, \text{then } x_i = \text{med}\{X\}. \end{cases} \tag{3.8}$$

The offered algorithm is realized by various ways, depending on quantity of simultaneously formed elements of a matrix $Y = \|y_{ij}\|$.

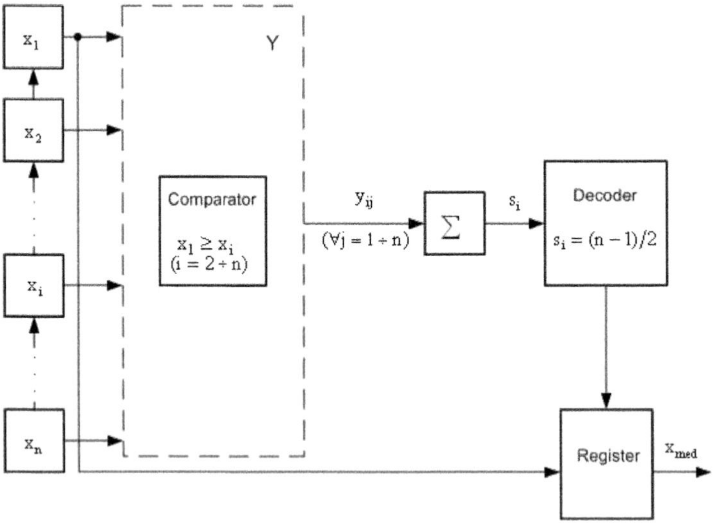

Fig. 3.5 Functional scheme of the median filter SL

At serial realization (SL) it is consecutive on each step of algorithm one comparison is carried out only and one element of a matrix Y is formed. The maximum quantity of steps necessary for realization of algorithm, is equal to quantity of operators of comparison (comparators) plus one step for record of initial data. The quantity of comparators is defined by size $(n^2 - n)/2$. Thus function scheme is presented on Fig. 3.5.

In the given realization it is carried out step-by-step formation of a matrix $Y = \|y_{ij}\|$ by comparison of an input vector $x_i\ (i = 1 \div (n-1))$ with vectors $x_{i+1}, x_{i+2}, \ldots, x_n$ according to (3.7) and (3.8). The received elements $y_{ij}\ (\forall j = 1 \div n)$ line-by-line are summarized in conformity with (3.9) and the result is compared to a constant equal $(n-1)/2$. In case of equality $s_i = (n-1)/2$ the corresponding vector x_i gets out as a median x_{med} and performance of algorithm stops.

In serial-parallel realizations (SP) for one step it is carried out in parallel from one before $(n-1)$ comparisons, and lines of a matrix Y are consistently formed. The maximum quantity of steps necessary in this case for realization of algorithm, in view of a step of record of initial data equally $(n+1)$. In this realization consecutive formation of a matrix $Y = \|y_{ij}\|$ by parallel comparison of input vector $x_i\ (i = 1 \div (n-1))$ with vectors $x_{i+1}, x_{i+2}, \ldots, x_n$ according to (3.7) and (3.8) is carried out. The number of comparisons on everyone $(i+1)$–th step of formation of a matrix Y in relation to i –th step decreases on unit. On penultimate $(n-1)$–th step is carried out only one comparison (x_{n-1}, x_n). Elements $y_{ij}\ (\forall j = 1 \div n)$ line-by-line are summarized according to (3.9) and the result is compared to a constant equal $(n-1)/2$. In case of equality $s_i = (n-1)/2$ the corresponding vector x_i gets out as a median x_{med}.

Further we shall consider a variant of construction MF, in which the median is defined for one step.

In parallel realization (PR) all comparisons are carried out simultaneously and elements of a matrix Y are formed in parallel. In the given variant the square matrix $Y = \|y_{ij}\|$ is formed of set of input vectors $x_i\ (\forall i = 1 \div n)$ in conformity with (3.7) and (3.8). Elements y_{ij} are simultaneously summarized line-by-line, according to (3.9), and the received values in parallel are compared to a constant $(n-1)/2$. Equality $s_i = (n-1)/2$ unequivocally defines a choice x_i as a median.

PLD-based realization of the median filter.

Synthesis of structure MF can be executed by means of Schematic Editor and libraries of components Project Libraries of system of designing Foundation Series. Except for opportunities of the description of the scheme in the specified way the system of designing puts at disposal of the designer more progressive means of the description of project HDL Editor and State Editor. First of editors serves for the description of the equipment on one of languages (VHDL or Verilog), the second – for the description of work of automatic devices by means of diagrams of statuses which further are automatically translate into the HDL description.

Essential advantage of the HDL description of the project is the opportunity of the description, both architecture of the projected device, and its behaviour. Besides such description in comparison with using of the Schematic editor is easily modified. Presence of modern the synthesis programs which is carrying out transformation of construction of a HDL-code in the scheme of logic elements, allows to carry out complex projects to similarly development of programs in language of a high level.

Example of serial-parallel realization of algorithm of median definition.
The description of the project is spent with using of construction of VHDL-language for the most simple in the description of variant MF for five 8–bit pixels.

Let's present MF as a "black" box on which input acts five 8–bit vectors: *a [7:0], b [7:0], c [7:0], d [7:0], e [7:0]*, signals *init* (initial installation), *ld* (loading), *clk*, and from an output are removed values of a median *out_m [7:0]* and a signal of interruption *int*. Further vectors, i.e. multidigit signals and variables will be designated, as well as in the text of the VHDL-description after the announcement of ports and signals, without the directive of quantity of digits in square brackets.

Process of designing consists in the description of functioning of a "black" box. We shall designate registers in which input data will enter, accordingly: *xa [7:0], xb [7:0], xc [7:0], xd [7:0], xe [7:0]*.

Loading of input data is made at initialization of a signal *ld* and can be described by expression:

if ld = ' 1 ' then xa <=a; xb <=b; xc <=c; xd <=d; xe <=e;
end if;

(If the signal *ld* is equal to *'1'* then signals *xa, xb, xc, xd* and *xe* values *a, b, c, d, e* are appropriated, respectively).

After record of initial data into registers it is possible to spend comparison of each entrance signal with other signals. The maximal number of steps in this case will be equal six: on the first step (we shall designate it as status S1) is made record of input data, on other five steps (S2, S3, S4, S5 and S6) – comparison of signals among themselves, summation of results of comparison, comparison of the received sums with a constant and formation of a signal of the enable of record of a code of a median into the output register. With the purpose of reduction of the hardware resources demanded for connection of compared signals to comparators, on each step parallel shift of data in registers is made, i.e. data from the register *xb* correspond in the register *xa*, from *xc* in *xb*, from *xd* in *xc*, from *xe* in *xd*:

a <=xb; xb <=xc; xc <=xd; xd <=xe;

Shift, as well as other assignment operations, is made at switching clocked signal *clk* from a status *' 0 '* in a status *' 1 '*.

For fixing of comparison results with the purpose of their further processing we shall enter variables of integer type *Integer* with area of values from 0 up to 1: *ab, ac, ad* and *ae* which are valid in a current status, and signals *ba, ca, da, ea, cb, db, eb, dc, ec* and *ed* the same type, valid for all time of process. Then the comparison operations which are carried out in various statuses can be presented in the form of:

S2 – comparison of the code containing in the register *xa* with codes, containing in registers *xb, xc, xd* and *xe* (comparison of a vector *a* with vectors *b, c, d* and *e*):

```
if xa> =xb then ab: = 1; ba <=0; else ab: = 0; ba <=1;
end if;
if xa> =xc then ac: = 1; ca <=0; else ac: = 0; ca <=1;
end if;
if xa> =xd then ad: = 1; da <=0; else ad: = 0; da <=1;
end if;
if xa> =xe then ae: = 1; ea <=0; else ae: = 0; ea <=1;
end if;
```

(if the vector *a* more or is equal *b, c, d, e* then variables *ab, ac, ad, ae* value *1*, and to signals *ba, ca, da, ea – 0* is appropriated, otherwise *ab, ac, ad, ae* is appropriated *0*, and *ba, ca, da, ea – 1*).

S3 – comparison of the code containing in the register *xa* with codes, containing in registers *xb, xd* and *xd* (comparison of a vector *b* with vectors *c, d* and *e*):

```
if xa> =xb then ab: = 1; cb <=0; else ab: = 0; cb <=1;
end if;
if xa> =xc then ac: = 1; db <=0; else ac: = 0; db <=1;
end if;
if xa> =xd then ad: = 1; eb <=0; else ad: = 0; eb <=1;
end if;
```

S4 – comparison of the code containing in the register *xa* with codes, containing in registers *xb* and *xc* (comparison of a vector *c* with vectors *d* and *e*):

```
if xa> =xb then ab: = 1; dc <=0; else ab: = 0; dc <=1;
end if;
if xa> =xc then ac: = 1; ec <=0; else ac: = 0; ec <=1;
end if;
```

S5 – comparison of the code containing in the register *xa* with a code, containing in the register *xb* (comparison of a vector *d* with a vector *e*):

```
if xa> =xb then ab: = 1; ed <=0; else ab: = 0; ed <=1;
end if;
```

After comparison of codes summation of results of comparison is made, thus for record of result of summation the variable of integer type *yz* with area of values from 0 up to 4 is entered:

S2: yz: = (ab+ac) + (ad+ae);
S3: yz: = (ab+ac) + (ad+ba);
S4: yz: = (ab+ac) + (ca+cb);
S5: yz: = (ab+da) + (db+dc);
S1: yz: = (ea+eb) + (ec+ed);

Final operation is comparison of result of summation with a constant (in this case 2), record, in case of the equality, current value of the code which is being the register *xa*, into output register, transition in status S1 and formation of a signal of interruption *int*:

```
if yz=2 then out_m <=xa;
else xa <=xb; xb <=xc; xc <=xd; xd <=xe;
end if; int <= ' 1 ';
```

At an inequality of result of summation to a constant shift of data in registers is made and transition to a following status, and a signal *int* is appropriated value *'0'*. The full description of the project can be executed by means of State Editor as flowgraph.

Example of parallel realization of algorithm.
For realization of operations of comparison and calculation of results variables of integer type with area of values from 0 up to 4 are entered: *xa*, *xb*, *xc*, *xd* and *xe*. Each of variables defines a place of input code in sorted sequence and is formed as follows:

```
if (a> =d) then xa: = xa+1; else xd: = xd+1; end if;
if (a> =e) then xa: = xa+1; else xe: = xe+1; end if;
if (b> =c) then xb: = xb+1; else xc: = xc+1; end if;
if (b> =d) then xb: = xb+1; else xd: = xd+1; end if;
if (b> =e) then xb: = xb+1; else xe: = xe+1; end if;
if (c> =d) then xc: = xc+1; else xd: = xd+1; end if;
if (c> =e) then xc: = xc+1; else xe: = xe+1; end if;
if (d> =e) then xd: = xd+1; else xe: = xe+1; end if;
```

At presence of a signal *clk* and a condition of equality any from variables to a constant (number 2), record in the output register of the input code corresponding the given variable is made:

```
if (clk'event and clk = ' 1 ') then
if (xa=2) then out_m <=a;
elsif (xb=2) then out_m <=b;
elsif (xc=2) then out_m <=c;
elsif (xd=2) then out_m <=d;
elsif (xe=2) then out_m <=e;
end if; end if;
if (a> =b) then xa: = xa+1; else xb: = xb+1; end if;
if (a> =c) then xa: = xa+1; else xc: = xc+1; end if;
```

Description MF for nine 8-digit pixels is similarly carried out.

On examples descriptions of serial-parallel and parallel realizations of algorithms in VHDL language are presented. Results of synthesis and realization of projects in crystal XCV50CS144–6, received with use of tool means FPGA Express and programs of placement and routing of system Xilinx Foundation Series, are presented in Table 3.2.

In realization of variant SL the maximal time of definition of a median T_1 to equally product tN_1, where t – the minimal period clk, $N_1 = [(n^2 - n)/2 + 1]$ – the number of statuses equal (for $n = 5$) in view of loading eleven. If a median is the input code x_1 it will be certain on the fifth step (the first step corresponds to loading), x_2 – on the eighth step, x_3 – on the tenth step, x_4 – on the eleventh and x_5 – on the first step of a following cycle corresponding loading of a new portion of input data. Thus, the size T_1 will be within the range of from $(n-1)t$ up to $[(n^2 - n)/2 + 1]t$.

Table 3.2 Results of synthesis

Type of realization of algorithm	Quantity of 8-digit pixels	Hardware resources (quantity of Slices)	Min. period [ns] / max. freq. [MHz]	Max time of definition of a median [ns]
SL	5	73	12 / 87	132
SP	5	62	13 / 77	65
PR	5	87	15 / 67	15
PR	9	317	40 / 25	40

Maximal time of definition of a median T_2 for realization of a variant of software to equally product tN_2, where $N_2 = n$ – the number of statuses equal (for $n = 5$) in view of loading 5. Definition of a median occurs on a step which number corresponds to number of a code in entrance sequence plus 1. If a median is the input code x_1 it will be certain on the second step (the first step corresponds to loading), x_2 – on the third step, etc. Time T_2 will be within the limits of from t up to nt. At fixing the fact of definition of a median return to an initial status – S1 is carried out and the signal of interruption of process is formed.

At realization of variant PR time of definition of a median T_3 is size of a constant (t) and is equal 15ns / 40ns accordingly for five / nine pixels.

Thus, variant PR (for $n = 5$) uses approximately in 1,4 (in comparison from software) and accordingly in 1,2 times (in comparison with SL) more slices in a crystal (an estimation of hardware expenses), possessing thus approximately in 4,5 (in comparison from software) and accordingly in 9 (in comparison with SL) time greater speed.

3.5.4 Hemming Adder Realization

Let's consider following variants of realization of Hemming adder:

HA1 is realization of the multilevel combinational scheme on the basis of logic elements AND, XOR by means of Schematic editor.

HA2 realizes the adder, using a tree chart of the adder on which top level in pairs weighed sums two components are formed, and further on the basis of standard schemes of adders - result of the weighed sum. All elements HA2 are created by means of system Core Generator as functionally completed blocks and, eventually by means of the schematic editor the resulting scheme is formed.

HA3 is realized by the behavioral description by VHDL language which is resulted below.

```
library IEEE;
use IEEE.std_logic_1164.all;
use IEEE.std_logic_unsigned.all;
entity hamm_v is
          generic (N:Integer: = 32);
               port (
               a: in STD_LOGIC_VECTOR (N-1 downto 0);
               y: out INTEGER range 0 to N
               );
end hamm_v;
architecture Behavioral of hamm_v is
begin
process (a)
variable x: integer range 0 to N;
begin
                    x: = 0;
                    for I in 0 to N-1 loop
                    if a (I) = ' 1 ' then
       x: = x+1;
                    end if;
                    end loop;
                    y <=x;
       end process;
end Behavioral;
```

Synthesis HA3 is executed by means of FPGA Express (Synopsys).

In tab. 3.3 comparative characteristics of devices (HA1, HA2, HA3), carrying out calculation of Hemming distance for 4–, 8– and 16–bit numbers realized in a crystal Virtex series are resulted.

The offered realizations of algorithms differ hardware resources (hardware resources are understood as dimension of a processing field or the logic capacity of a crystal defined by quantity Slices) and speed.

Table 3.3 Comparative characteristics of devices HA1, HA2, HA3

Variant of realization	Speed (the period [ns] / frequency [MHz])			Hardware resources (quantity Slices)		
	n = 4	n = 8	n = 16	n = 4	n = 8	n = 16
HA1	4,4 / 227,2	8,1 / 123,4	11,9 / 84	4	10	38
HA2	7,8 / 128,2	10,2 / 98	13,4 / 74,6	4	10	22
HA3	3,6 / 277,8	5,5 / 181,8	12,7 / 78,8	2	5	22

On the basis of the received estimations it is possible to draw following conclusions. Variant HA3 has the best parameters of estimations on hardware resources for any word length (for $n = 16$ an resources coincide with variant HA2), on speed slightly conceding only to variant HA1 for $n = 16$. HA2 has no advantages before other realizations, confirming known regulations about volume, that the complex system from optimum components not necessarily is optimum in aggregate. However the basic advantage of variant HA3 is that the presented behavioral description, is the parametrical description of Hemming adder, i.e. universal (for any word length). The task of parameter *(N)* in the description *(generic)* defines word length of the synthesized adder.

3.6 Verification of Projects by Means of Stands

Let's consider the description process project on an example of the median filter which block–diagram (Fig. 3.6), contains the block from five 8-bit registers, outputs of each of which are connected to inputs of other register and inputs of the device for definition of a median.

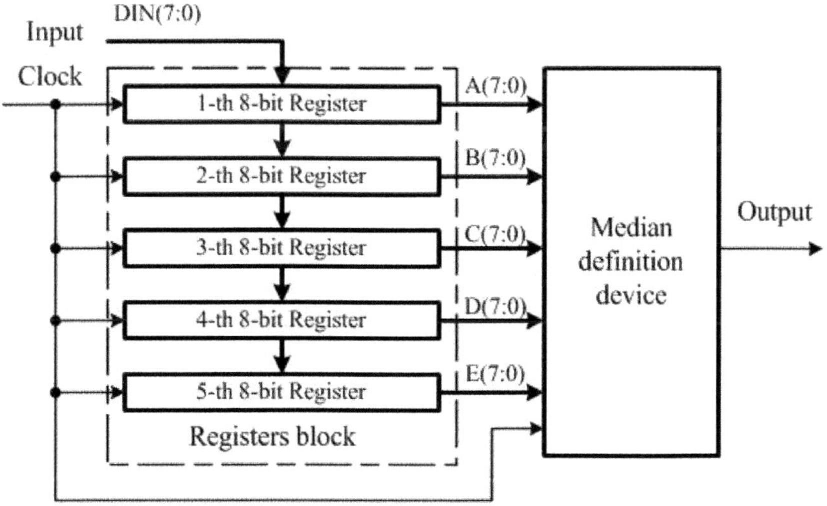

Fig. 3.6 Block–diagram of the median filter

On an input of the filter (n = 5, m = 8) the file of 8–bit codes, for example from output ADC consistently acts. With signal Clock codes move from 1–st register in 2–nd, from 2–nd in 3–rd, etc.

Simultaneously with everyone Clock transfer of codes from all 5 registers on inputs of the device for a median definition is carried out. Definition of a median is carried out on 5 pixels (codes) according to algorithm and the VHDL–description, and modified due to inclusion in the device for median definition a of four registers blocks for realization of a conveyor mode. The project of the median filter contains, thus, the scheme, being top level of hierarchy of the project, and two modules: the registers block and the device for the median definition, executed as VHDL–descriptions.

In Fig. 3.7 the scheme of the filter synthesized by means of the schematic editor, registers block is presented in the form of the text of VHDL–description, is resulted. Description of the registers block as text is much more compact and easier, than the description of this module by means of the schematic editor.

In Fig. 3.7 this module is presented in the form of an environment of a macrocell from the user library. VHDL description of the median filter:

```
library IEEE; -- library declaration
use IEEE.STD_LOGIC_1164. ALL; -- using declaration
use IEEE.STD_LOGIC_ARITH.ALL;
use IEEE.STD_LOGIC_UNSIGNED.ALL;
entity reg_s is -- object declaration
  Port (din: in std_logic_vector (7 downto 0); -- port declaration
      clk: in std_logic;
      dout_a: out std_logic_vector (7 downto 0);
      dout_b: out std_logic_vector (7 downto 0);
      dout_c: out std_logic_vector (7 downto 0);
      dout_d: out std_logic_vector (7 downto 0);
      dout_e: out std_logic_vector (7 downto 0));
end reg_s;
architecture Behavioral of reg_s is – architecture declaration
signal va, vb, vc, vd, ve: std_logic_vector (7 downto 0); -- signal declaration
begin
```

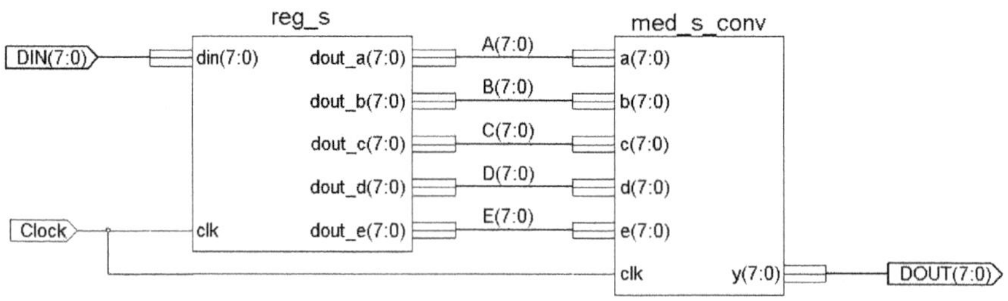

Fig. 3.7 Functional schema of the median filter

At the description of this module by means of the Schematic editor it would be necessary to execute the scheme containing five 8-bit registers FD8CE where register FD8CE would be presented by the scheme of lower level of the hierarchy consisting of eight triggers FDCE of system library.

The considered example of the median filter has been synthesized, placed and routing in crystal XC2S150–5PQ208 (Spartan–II series).

At realization the project has borrowed 9 % of resources of a crystal, i.e. 169 of 1728 Slices, one global buffer, 16 pinouts (IOB). The maximal clock frequency is equal 95 MHz.

Project Verification

The basic tool of verification of the project is the modeling system of Model Technology ModelSim, Xilinx Edition (MXE II). One of laborious processes of verification is a file processing of input influences on model-based object. For simple projects can be used HDL Bencher – the graphic interface for creation of input influences in the form of sequences of the impulses set by the user (Waveform). For projects where input influences are a product of complex logic transformations or a codes file, creation of the virtual stand or stands for verification of the project is expedient. Such stand can be executed in the form of the subproject included in a separate branch of a tree of hierarchy of the developed project. Further the test (HDL Test Bench) is described in the form of structure where the stand and the project are presented in the form of the interconnected components.

Let's consider as an example development of the stand for functional check of the median filter which description has been resulted above (see Fig. 3.6).

Let the filter makes "clearing" the signal consisting of a "useful" signal of the sine wave form and formed "noise". Thus, the stand can consist of the shaper of a signal of the sine wave form, the generator of random numbers and the multiplexer which is carrying out transfer of a code from an output of the shaper or the generator on an input of the filter (Fig. 3.8).

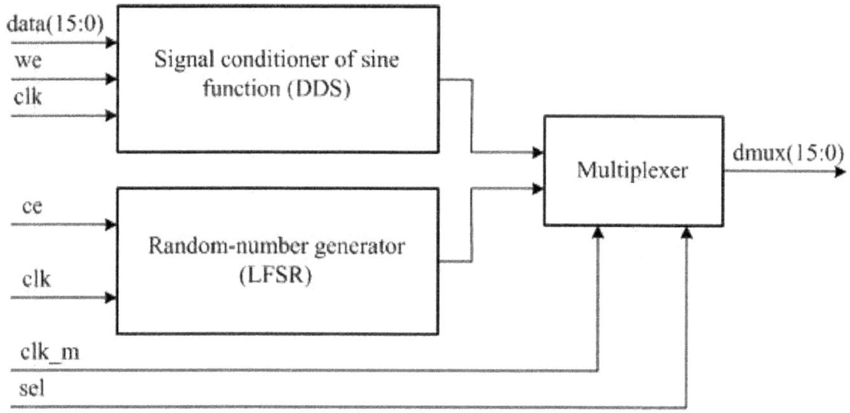

Fig. 3.8 Block–diagram of the stand for functional check of the median filter

As the shaper of a signal of the sine wave form it is used IP–Core – Direct Digital Synthesizer (DDS), as the generator of random numbers – the Linear Feedback Shift Register (LFSR).

In Fig. 3.9 the timing diagram of the filtration process, received is resulted at functional modeling of the median filter with using of the specified stand.

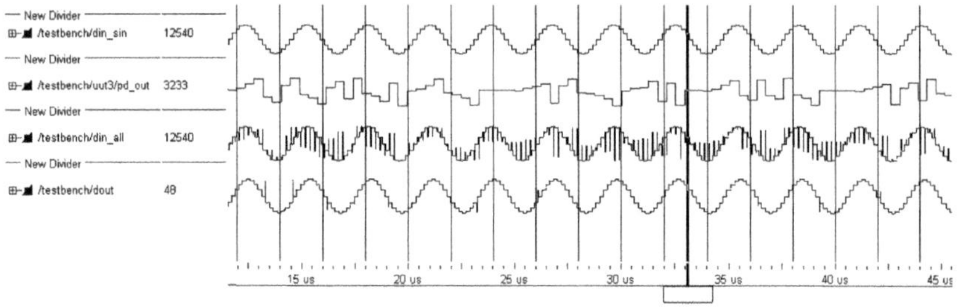

Fig. 3.9 The timing diagram of a filtration process:
a – signal on an output of the shaper of the sine wave form;
b – signal on an output of the generator of random numbers;
c – signal on an input of the median filter;
d – "cleared" signal on an output of the median filter.

3.7 Reconfigurable Processors

The typical structure of reconfigurable processor (RP) allows the developer (user) to realize any algorithm, i.e. to change structure depending on a carried out problem (the set algorithm). Last can be broken into the fragments which are carried out consistently on fixed hardware that leads to the general economy of hardware, thus complexity of fragments of algorithm is defined only by logic capacity of a crystal FPGA. Presence of set of functional processing fields (FPF) allows is hardware to realize parallel data processing, and set of configuration files – conveyor programming of the structure realizing fragments of algorithm.

Reconfigurable processors have FPF the set dimension which is configured for performance of the set algorithm or its part, providing, thus, optimum realization of this algorithm, both under time characteristics, and on hardware expenses. At the conveyor mechanism of realization of algorithm additional matrixes are entered into structure RP. Structure RP is presented in Fig. 3.10 and contains S matrixes FPF, the channel of input-output (CIO) for connection to the standard Bus of a Host-computer, a memory of a configuration files (MCF), the data RAM, the controller (CT), data bus (DB) and control bus (CB). The conveyor mechanism assumes loading a configuration file in the next matrix in parallel with data processing in a current matrix.

The format of a configuration file is standard for FPGA and contains the information about of a configuration matrix, i.e. forms the corresponding basic electric scheme realizing set algorithm. Matrix FPF represents a matrix of universal elements which under control of a configuration file F_γ direct function is appointed and the structure of communications between them is formed. Configuration files F_γ enter the name in matrix FPF from a memory of configuration files under control of CT.

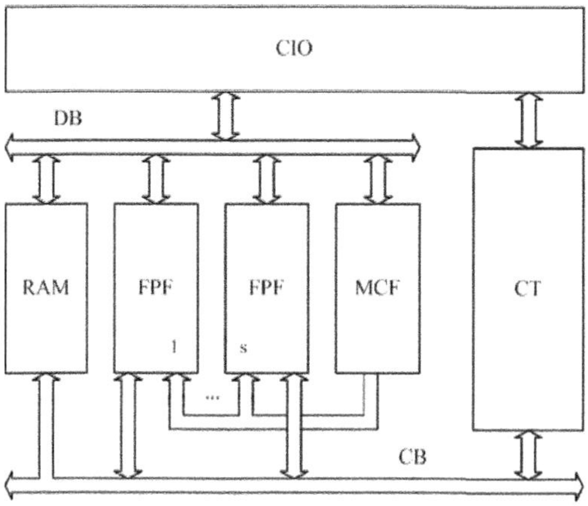

Fig. 3.10 Structure of reconfigurable processor

In matrix FPF on data bus details from the RAM or external entrance data through CIO can act. Results of processing from matrix FPF can be transferred in channel CIO as external target data or in the RAM as intermediate results. The set of files of a configuration $F = \{F_\gamma\}$ enters the name in the MCF through channel CIO under control of CT.

Initialization of system consists of three stages: record of set of files of a configuration F in the MCF; loading of files of a configuration F_γ in FPF from the memory of configuration files; functioning of system – realization of algorithm.

Procedure of data processing is carried out as follows. In a command number (α) matrixes FPF is underlined, and CT forms a signal initializing corresponding matrix FPF. Then loading of a corresponding file of a configuration in next matrix FPF is carried out. After end of data processing by α-th matrix results of data processing enter the name in the RAM and serve as intermediate (initial) data for ($\alpha+1$)-th matrix. Upon termination of work of algorithm with given FPF (the termination of the microprogramma) is formed interruption which acts on operating input CT where its processing is carried out.

3.8 Conclusions

The received results have allowed to raise efficiency display of initial problems and algorithms to architecture and structure of projected PLD-based devices and systems by criteria « speed – complexity of realization » on the basis of the developed formalized techniques of construction and dynamic reorganization of their architecture and structure, proceeding from properties of sold algorithms, and also logic, constructive and technological features PLD, and tool means of their designing.

As a result of the executed analysis of evolution, tendencies of development and technology of realization of a new class of components computer engineering – programmable logic devices it is certain, that PLD opportunities of construction on their basis of devices and the systems possessing properties of the reconfigurability, providing give adaptation to a wide spectrum of problems and reception of high characteristics of projected devices and systems.

Principles of construction and functioning of a new class of computers and systems with reconfigurable the architecture are developed, differing from traditional (von Neumann type) properties of high dynamic reorganization, multilevel and parallelism of data processing that functional means of computer engineering for any algorithms allow the developer (user) to create, providing thus an opportunity of structural adaptation, including in real time, according to a solved problem (algorithm) and also to duplicate them for a wide range of developers, reducing process of designing of digital devices to a choice from library of optimum structure by criteria «speed – complexity of realization» with adjustment of corresponding parameters.

The known logical-information method of designing reconfigurable devices and systems which basic difference became orientation to functionalities PLD is modified. In the offered kind it allows to operate with any quantity of levels of programming, to define optimum quantity of such levels and to synthesize the optimum structure of the device represented by multilevel hierarchical system with unlimited number of levels on a class of criteria « speed – complexity of realization ».

The new class of computing structures – adaptive logic networks (ALN), principles and techniques of their construction and functioning are offered. It is shown, how for base set of structures ALN and training samples, the binary vectors set by set, using polynomial representation, which factors are set by means of Adamar matrix, it is possible to receive analytically set of logic functions (functional adjustment) components ALN at the functional restrictions preliminary certain also analytical by that allows, passing process of direct synthesis to execute predesign estimations of a realizability of developed devices. Process of designing consists in correct display of entrance set of data in target set of data and is reduced to is formal-analytical procedure of decomposition with use of preliminary received functional restrictions. The offered device effectively supports process of adaptation ALN on classes of problems which are reduced to procedure of classification, including problems of natural classification.

A number of structural realizations ALN is offered: in the form of "triangular", "trapezoidal" and "rectangular" matrixes which covers a wide class of problems. Process of adjustment of matrixes is reduced to definition of types of logic functions elementary a component and structures of communication from the limited set is set. The offered structures differ on capacity sold Boolean functions and to hardware expenses, are accompanied received analytical by asymptotic by estimations of complexity (depending on word length of entrance binary vectors) and capacities of target set of binary vectors.

The open library of functional devices which structure can extend and be oriented to problem is developed. In particular, it is devices: definitions of a median with time step by step conveyor processing of input data; memory-based sorting of data; adders Hemming (for any word length); multiplication of matrixes; multipliers with a floating point (standard IEEE-754), etc.

The base structure of reconfigurable processor with set of functional fields which allows to focus functionally it on an any class of problems (algorithms) is developed, supporting, in particular, parallel, conveyor and in parallel-conveyor data processing. The developed processor is a basis for construction of a lot of computing systems of high complexity, productivity and survivability.

References

[1] Estrin, G., Turn, R.: Parallel processing in a restructurable computer. IEEE Transaction on Electronic Computers EC 12(6), 747–755 (1963)

[2] Athanas, P.M., Schewel, J., McHenry, J.T., James–Roxby, P.B.: Reconfigurable Technology: FPGAs and Reconfigurable Processors for Computing and Communication. In: SPIE International Society for Optical Engineering, p. 174 (2001)

[3] Villasenor, J., Mangione–Smith, W.H.: Configurable Computing, http://www.vcc.com

[4] Lord, E., Cantle, A.J., Dr Devlin, M., Shand, D.: COTS Platform for the Development of Re-configurable Processing in Aerospace Systems. White paper, http://www.nallatech.com

[5] Schewel, J.: Hardware / Software Co–Design System using Configurable Computing Technology, http://www.vcc.com

[6] Palagin, A.V., Opanasenko, V.N.: Reconfigurable computing technology. Cybernetics and Systems Analysis 43(5), 675–686 (2007)

[7] Palagin, A.V., Opanasenko, V.N., Sakharin, V.G.: Features of Digital Devices Design of Modern PLD of the Xilinx Incorporation. Journal of Automation and Information Sciences 33(3), 80–89 (2001)

[8] Cosoroaba, A., Rivoallon, F.: Achieving Higher System Performance with the Virtex–5 Family of FPGAs. Xilinx Inc. White Paper WP245 (v1.1) May 17 (2006), http://www.xilinx.com

[9] Chang, C., Wawrzynek, J., Brodersen, R.W.: BEE2: A High–End Reconfigurable Computing System. IEEE Design and Test of Computers 22(2), 114–125 (2005)

[10] Bruck, J., Blaum, M.: Neural networks, error–correcting codes, and polynomials over the binary n–cube. IEEE Transactions on information theory 35(5), 976–987 (1989)

[11] Palagin, A.V., Opanasenko, V.N., Chigirik, L.G.: Synthesis of a Hamming network on a basis of programmable logic integrated circuits. Engineering Simulation 13, 651–666 (1996)

[12] Astola, J., Haavisto, P., Neuvo, Y.: Vector median filters. Proc. of the IEEE 78(4), 678–689 (1990)

4 Application of Multilevel Design on the Base of UML for Digital System Developing

Raisa Malcheva

Donetsk National Technical University, Department of Computers, Apt. 41, 204a, Artema str., 83122 Donetsk, Ukraine
e-mail: raisa@cs.dgtu.donetsk.ua

Abstract. In this chapter the features of image generation and performing systems' design are analyzed. Estimation of complexity of the standard rendering pipeline is done. The architectural decisions and algorithm approaches for the real-time rendering systems' creation are discussed. Adaptation of a multilevel designing method of the built-in systems with realization of separate modules on reconfigurable devices, based on application of architecture operated by models and the unified modeling language, is offered. The graphical application of the modified method is shown.

4.1 Introduction

Until the early 1980s, computer graphics was a small, specialized field, largely because the hardware was expensive. The typical application was the image generators for trainer development and simulators. These systems must generate high quality images in real time. The concept of a "desktop" now became a popular metaphor for organizing screen space. Modern small personal devices, such as Nokia N810 Internet Tablet [1] and Nokia M810 WiMAX Edition, are in want of real-time computer graphics which concerned with animation, video processing, 2D and 3D performing. Computer graphics systems' developing, as well as alteridem, typically is performed years in advance of subsystem development and integration. In this process, models of functions and possible solutions for the physical architecture must be defined and matched to evaluate quality and select the most effective algorithm and the best possible hardware platform. For small personal systems designers the primary architectural/design issue the partitioning of system functionality across both hardware and software. Separate specifications for hardware and software, often written in non-formal languages, are delivered with functionality a priori, because changes to the partition may necessitate extensive redesign. Because software rework is viewed as easier than hardware redesign often drawbacks' corrections have a heavy software decision.

M. Adamski et al. (Eds.): Design of Digital Systems and Devices, LNEE 79, pp. 93–117.
springerlink.com

4.2 Features of Digital Systems for Real-Time Image Generation

Real-Time Image Generator can be represented as a rendering pipeline - a logical model for computations needed in a raster-display system. It is not necessarily a physical mode, since the stages of the pipeline can be implemented in either software or hardware. Fig. 4.1 shows a version of the rendering pipeline that is typical for systems using conventional primitives (lines and polygons) and conventional shading techniques (constant, Gouraud, or Phong).

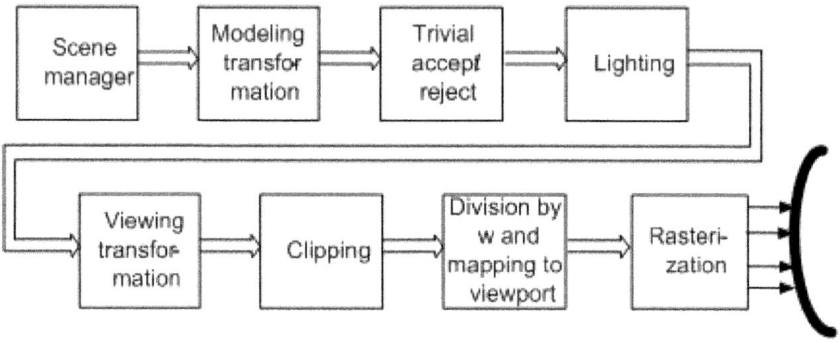

Fig. 4.1 Standard graphics pipeline

4.2.1 Estimation of the Complexity of the Standard Rendering Pipeline

Scene Manager

The first stage of the pipeline is traversal of the display model or scene manager. This is necessary because the image may change by an arbitrary amount between successive frames. All the primitives in the database must be fed into the remainder of the display pipeline, along with context information, such as colors and current-transformation matrices. Newman [2] described the two types of traversal: immediate mode and retained mode. Both methods have advantages and disadvantages, and the choice between them depends on the characteristics of the application and of the particular hardware architecture used.

Immediate mode offers flexibility, since the display model does not need to conform to any particular display-list structure and the application has the luxury of recreating the model differently for every frame. The main CPU must perform immediate-mode traversal, however, expending cycles it could use in other ways. Retained mode, on the other hand, can be handled by a display processor if the structure database is stored in its local memory. Retained-mode structure traversal can be accelerated by optimizing the database storage and access routines or by using a dedicated hardware traverser. Furthermore, since the main CPU only edits the database each frame, rather than rebuilding it from scratch, a low-bandwidth channel between the main CPU and the display processor is sufficient. Of course, relatively few changes can be made to the structure database between frames, or astern performance will suffer.

The choice between traversal modes is a controversial matter for system designers. Many argue that retained mode offers efficiency and high performance. Others believe that immediate mode supports a wider range of applications and does not necessarily lead to reduced performance if the system has a sufficiently powerful CPU.

Unfortunately, it is difficult to estimate the processing requirements for display traversal, since they depend on the traversal method used and on the characteristics of the Particular display model. At the very least, a read operation and a write operation must be performed for each word of data to be displayed. The processing requirements may be much greater if the structure hierarchy is deep or if it contains many modeling transformations.

Modeling Transformation

In this stage of the pipeline, graphics primitives are transformed from the object-coordinate system to the world-coordinate system. This is done by transforming the vertices of each polygon with a single transformation matrix that is the concatenation of the individual modeling transformation matrices. In addition, one or more surface-normal vectors may need to be transformed, depending on the shading method to be applied.

Constant shading requires world-space surface-normal vectors for each polygon. We compute these by multiplying object-space surface *normals* by the transpose of the inverse modeling transformation matrix. Gouraud and Phong shading require world-space *normals* for each vertex, rather than for each polygon, so each vertex-normal vector must be multiplied by the transpose inverse transformation matrix.

Let us compute the number of floating-point calculations required to transform a single vertex if Gouraud shading is to be applied. Multiplying a homogeneous point by a 4×4 matrix requires 16 multiplications and 12 additions. Multiplying each vertex normal by the inverse transformation matrix requires 9 multiplications and 6 additions (only the upper-left 3x3 portion of the matrix is needed).

Therefore, transforming a single vertex with surface normal requires $16 + 9 = 25$ multiplications and $12 + 6 = 18$ additions.

Trivial Accept/Reject Classification

In the trivial accept/reject classification stage, primitives (now in world coordinates) are tested to see whether they lie wholly inside or outside the view volume. By identifying primitives that lie outside the view volume early in the rendering pipeline, processing in later stages is minimized. We will clip primitives that cannot be trivially accepted or rejected in the clipping stage.

To trivially accept or reject a primitive, we must test each transformed vertex against the six bounding planes of the view volume. In general, the bounding planes will not be aligned with the coordinate axes. Each test of a vertex against a bounding plane requires multiplications and 3 additions (the dot product of a homogeneous point with a 3D plane equation). A total of 6 • 4 = 24 multiplications and 6 • 3 = 18 additions are required per vertex.

Lighting

Depending on the shading algorithm to be applied (constant, Gouraud, or Phong), illumination model must be evaluated at various locations: once per polygon for Phong shading, once per vertex for Gouraud shading, or once per pixel for Phong shading Ambient, diffuse, and specular illumination models are commonly used in high-performance systems.

In constant shading, a single color is computed for an entire polygon, based on the position of the light source and on the polygon's surface-normal vector and diffuse color. The first step is to compute the dot product of the surface-normal vector and the light vector (3 multiplications and 2 additions for directional light sources). If an attenuation factor based on the distance to the light source is used, we must calculate it and multiply it by the dot product here. Then, for each of the red, green, and blue color components, we multiply the dot product by the light-source intensity and diffuse-reflection coefficient (2 multiplications), multiply the ambient intensity by the ambient-reflection coefficient (1 multiplication), and add the results (1 addition). If we assume a single directional light source, calculating a single RGB triple requires 3 + 3 • (2 + 1) = 12 multiplications and 2+3-1=5 additions. Gouraud-shading a triangle requires three RGB triples—one for each vertex.

Viewing Transformation

In this stage, primitives in world coordinates are transformed to normalized projection (NPC) coordinates. This transformation can be performed by multiplying vertices in world coordinates by a single 4x4 matrix that combines the perspective transformation (if used) and any skewing or nonuniform scaling transformations needed to convert world coordinates to NPC coordinates. This requires 16 multiplications and 12 additions per vertex. Viewing transformation matrices, however, have certain terms that are always zero. If we I take advantage of this, we can reduce the number of computations for this stage by perhaps 25 percent. We will assume that 12 multiplications and 9 additions per vertex are required in the viewing transformation stage.

Note that if a simple lighting model (one that does not require calculating the distance between the light source and primitive vertices) is used; modeling and viewing transformation matrices can be combined into a single matrix. In this case only one transformation stage is required in the display pipeline - a significant savings.

Clipping

In the clipping stage, lit primitives that were not trivially accepted or rejected are clipped to the view volume. Clipping serves two purposes: preventing activity m one screen window from affecting pixels in other windows, and preventing mathematical overflow and underflow from primitives passing behind the eye point or at great distances.

Exact clipping is computationally practical only for simple primitives, such as lines and Polygons. These primitives may be clipped using any of the 3D clipping algorithms. Complicated primitives, such as spheres and parametrically ended patches, are difficult to clip, since clipping can change the geometric nature of the Primitive. Systems designed to display only triangles have a related problem, since a clipped triangle may have more than three vertices.

An alternative to exact clipping is scissoring. Here primitives that cross a clipping boundary are processed as usual until the rasterization stage, where only pixels inside the viewport window are written to the frame buffer. Scissoring is a source of inefficiency, however, since effort is expended on pixels outside the viewing window. Nevertheless, it is the only practical alternative for clipping many types of complex primitives. In the pipeline described here, all clipping is performed in homogeneous coordinates. This is really only necessary for z clipping, since the w value is needed to recognize vertices that lie behind the eye.

Many systems clip to x and y boundaries after the homogeneous divide for efficiency. This simplifies x and y clipping, but still allows primitives that pass behind the eye to be recognized and clipped before w information is lost. The number of computations required for clipping depends on how many primitives cross the clipping boundaries, which may change from one frame to the next. A common assumption is that only a small percentage of primitives (10 percent or fewer) need clipping. If this assumption is violated, system performance may decrease dramatically.

Division by w and Mapping to 3D Viewport

Homogeneous points that have had a perspective transformation applied, in general, have w values not equal to 1. To compute true x, y, and z values, we must divide the x, y, and z components of each homogeneous point by w. This requires 3 divisions per vertex. In many systems, vertex x and y coordinates must be mapped from the clipping coordinate system to the coordinate system of the actual 3D viewport. This is a simple scaling and translation operation in x and y that requires 2 multiplications and 2 additions per vertex.

Rasterization

The rasterization stage converts transformed primitives into pixel values, and generally stores them in a frame buffer. As discussed above, rasterization consists of three subtasks: scan conversion, visible-surface determination, and shading. Rasterization, in principle, requires calculating each primitive's contribution to each pixel, an $O(n \cdot m)$ operation, where n is the number of primitives and m is the number of pixels.

In a software rendering system, rasterization can be performed in either of two orders: primitive by primitive (object order), or pixel by pixel (image order).

Estimation of the Complexity of Geometry Calculations

For the next characteristics:

- 10,000 triangles (none clipped);
- Each triangle covers an average of 100 pixels, one-half being obscured by other triangles;
- Ambient and diffuse illumination models (not Phong);
- Gouraud shading;
- 1280 by 1024 display screen, updated at 10 frames per second.

For each frame, we must process 10,000 • 3 = 30,000 vertices and vertex-normal vectors. In the modeling transformation stage, transforming a vertex (including transforming the normal vector) requires 25 multiplications and 18 additions. The requirements for this stage are thus 30,000 • 25 = 750,000 multiplications and 30,000 • 18 = 540,000 additions.

Trivial accept/reject classification requires testing each vertex of each primitive against the six bounding planes of the viewing volume, a total of 24 multiplications and 18 additions per vertex. The requirements for this stage are thus 30,000 • 24 = 720,000 multiplications and 30,000 • 18 = 540,000 additions, regardless of how many primitives are trivially accepted or rejected.

Lighting requires 12 multiplications and 5 additions per vertex, a total of 30,000 • 12 -360,000 multiplications and 30,000 • 5 = 150,000 additions.

The viewing transformation requires 8 multiplications and 6 additions per vertex, a total of 30,000 • 8 = 240,000 multiplications and 30,000 • 6 = 180,000 additions.

The requirements for clipping are variable; the exact number depends on the number o primitives that cannot be trivially accepted or rejected, which in turn depends on the scene and on the viewing angle. We have assumed the simplest case for our database, that a primitives lie completely within the viewing volume. If a large fraction of the primitive needs clipping, the computational requirements could be substantial (perhaps even more than in the geometric transformation stage).

Division by w requires 3 divisions per vertex, a total of 30,000 • 3 = 90,000 divisions. Mapping to the 3D viewport requires 2 multiplications and 2 additions per vertex, a total 60,000 multiplications and 60,000 additions.

Summing the floating-point requirements for all of the geometry stages gives a total of 120,000 multiplications/divisions and 1,470,000 additions/subtractions per frame. Since anew frame is calculated every 1/10-second, a total of 22.2 million multiplications/divisions and 14.7 million additions/subtractions (36.9 million aggregate floating-point operations) as required per second—a very substantial number.

4.2.2 The Architectural Decisions and Algorithm Approaches for the Real-Time Rendering Systems

A direct way to add concurrency to rasterization calculations is to cast the various steps of a software algorithm into a hardware pipeline. This technique has been used to build a number of inexpensive, moderately high-performance systems. This approach can be used with either of the two main rasterization approaches: object order and image order algorithms.

Pipelined Object-Order Architectures

Object-order rasterization methods include the z-buffer, depth-sort, and binary space partition (BSP) - tree algorithms. The outer loop of these algorithms is an enumeration of primitives in the database, and the inner loop is an enumeration of pixels within each primitive. For polygon rendering, the heart of each of these algorithms is rasterizing a single polygon.

In order to increase the productivity and to widen the functional capability the multichannel systems is employed in object order architectures [3]. These systems consist of common scene manager and some rendering channels. Each channel as well as a system, in general, presents a seven-stage pipeline (Fig. 4.1). The image is put out by means of projection video devices. The images having generated by separate channels, are mixed to a single whole on the screen. In such multichannel systems the scene manager has a particular functional aim to form a few priority lists. Our version of constructing the scene manager for a multichannel system as a special-purpose device is presented in [3]. The following stages were joint to build up the efficient algorithm:

- the definition of potentially visible objects;
- the depth sorting of potentially visible objects with marking the hidden objects;
- sorting out the subobjects of the hidden objects.

Image-Order Architectures

The alternative to object-order rasterization methods is image-order (or scan-line) approach. Scan-line algorithms calculate the image pixel by pixel, rather than primitive by primitive. To avoid considering primitives that do not contribute to the current scan line, most scan-line algorithms require primitives to be transformed into screen space and sorted into buckets according to the first scan line in which they each appear.

Ray tracing, also known as *ray casting*, determines the visibility of surfaces by tracing imaginary rays of light from the viewer's eye to the objects in the scene. A center of projection (the viewer's eye) and a window on an arbitrary view plan are selected. The window may be thought of as being divide into a regular grid, whose elements correspond to pixels at the desired resolution. Then, for each pixel in the window, an *eye ray* is fired from the center of projection through the pixel's center into the scene. The pixel's color is set to that of the object at the closest point of intersection.

Ray tracing was first developed by Appel and by Goldstein and Nagel [2]. Appel used a sparse grid of rays used to determine shading, including whether a point was in shadow. Goldstein and Nagel originally used their algorithm to simulate the trajectories of ballistic projectiles and nuclear particles. The pseudocode for this simple ray tracer is shown below.

Select center of projection and window on view plan;

For (each san pixel line in image){

For (each pixel in scan line) {

Determine ray from center of projection pixel;

For (each object in scene){

If (object is intersected and is closest considered thus far)

Record intersection and object name;

}

Set pixel's color to that at closest object intersection; *}*

}

The best hardware decision for this algorithm is processor-per-pixel architecture. For good screen resolution it needs above million processors or interpolation between results of neighbor-processors [3].

Comparing of Architectures

1. Object-parallel systems typically require specialized processors. This implies heavy reliance on custom VLSI chips, making system design difficult and expensive.
2. Image-parallel systems, on the other hand, place more reliance on frame-buffer memory, which can be built with commercial parts such as VRAMs.
3. The specialized nature of object processors limits the types of primitives that can be displayed and the shading algorithms that can be used.
4. Object-parallel systems have poor overload characteristics. Generally, object-parallel systems perform at full speed as long as there are enough object processors. Special provisions must be made to handle large databases, and performance generally decreases rapidly.
5. Ray-tracing algorithm of image-parallel performing cannot be applied in small personal device.

4.3 Designing of Specialized Processors

To show non-regular structure of specialized processors for rendering this section discusses architectures of specialized processors for some stage of standard graphical pipeline.

4.3.1 Scene Processor

Fig. 4.2 shows a version of the scene processor for the systems which use conventional primitives (lines and polygons) and priority algorithms with several levels of detail. In accordance with the extended algorithm the scene manager includes:

- a matrix formation unit (MFU);
- a visibility detection unit (VDU);
- a detail/loader unit (DLU).

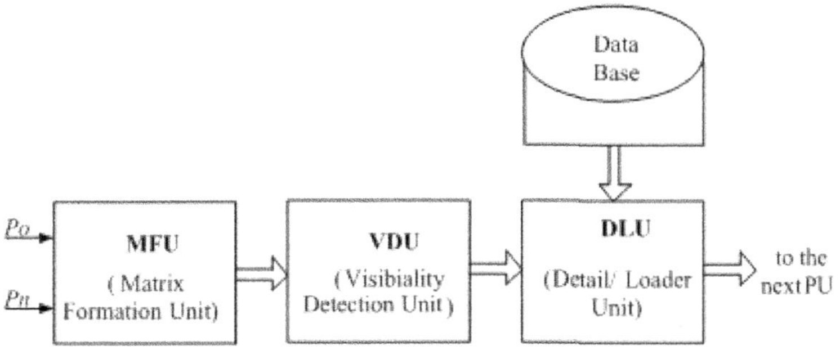

Fig. 4.2 Structure of Scene Processor

The first scene manager action begins with loading the vectors *Po* and *Pn* of the positions and orientations of an observer and an object.

Structure of positional vector is:

$$P = \{x, y, z, \psi, \theta, \gamma\}, \tag{4.1}$$

where $\{x, y, z\}$ - linear coordinate of the center of object/observer coordinate system; $\{\psi, \theta, \gamma\}$ - angle coordinate.

MFU stores angles to object angle registers ROP, ROT, ROG and observer angle registers RNP, RNT, RNG (Fig. 4.3). Using angels as table-pointers MFU takes *sin* and *cos* from the ROM3 and ROM4 and stores them it registers. MUL2 and SUM2 accumulate matrix MA (4.2) and MB (4.3) coefficients.

$$MA=\begin{bmatrix} \cos\psi\cdot\cos\theta & \sin\psi\cdot\sin\gamma-\cos\psi\cdot\cos\gamma\cdot\sin\theta & \cos\psi\cdot\sin\gamma\cdot\sin\theta+\sin\psi\cdot\cos\gamma \\ \sin\theta & \cos\theta\cdot\cos\gamma & -\cos\theta\cdot\sin\gamma \\ -\sin\psi\cdot\cos\theta & \sin\theta\cdot\sin\psi\cdot\cos\gamma+\cos\theta\cdot\sin\gamma & \cos\psi\cdot\cos\gamma-\sin\theta\cdot\sin\psi\cdot\sin\gamma \end{bmatrix} \tag{4.2}$$

$$MB=\begin{bmatrix} \cos\psi\cdot\cos\theta & \sin\theta & -\sin\psi\cdot\cos\theta \\ \sin\psi\cdot\sin\gamma-\cos\psi\cdot\cos\gamma\cdot\sin\theta & \cos\theta\cdot\cos\gamma & \sin\theta\cdot\sin\psi\cdot\cos\gamma+\cos\theta\cdot\sin\gamma \\ \cos\psi\cdot\sin\gamma\cdot\sin\theta+\sin\psi\cdot\cos\gamma & -\sin\gamma\cdot\cos\theta & \cos\psi\cdot\cos\gamma-\sin\theta\cdot\sin\psi\cdot\sin\gamma \end{bmatrix} \tag{4.3}$$

MA is used for translation from the object system to global. For translation from the global system to observer MB-matrix is used. To get MA and MB MFU takes $\{\psi, \theta, \gamma\}$ from *Po* and *Pn*, accordingly.

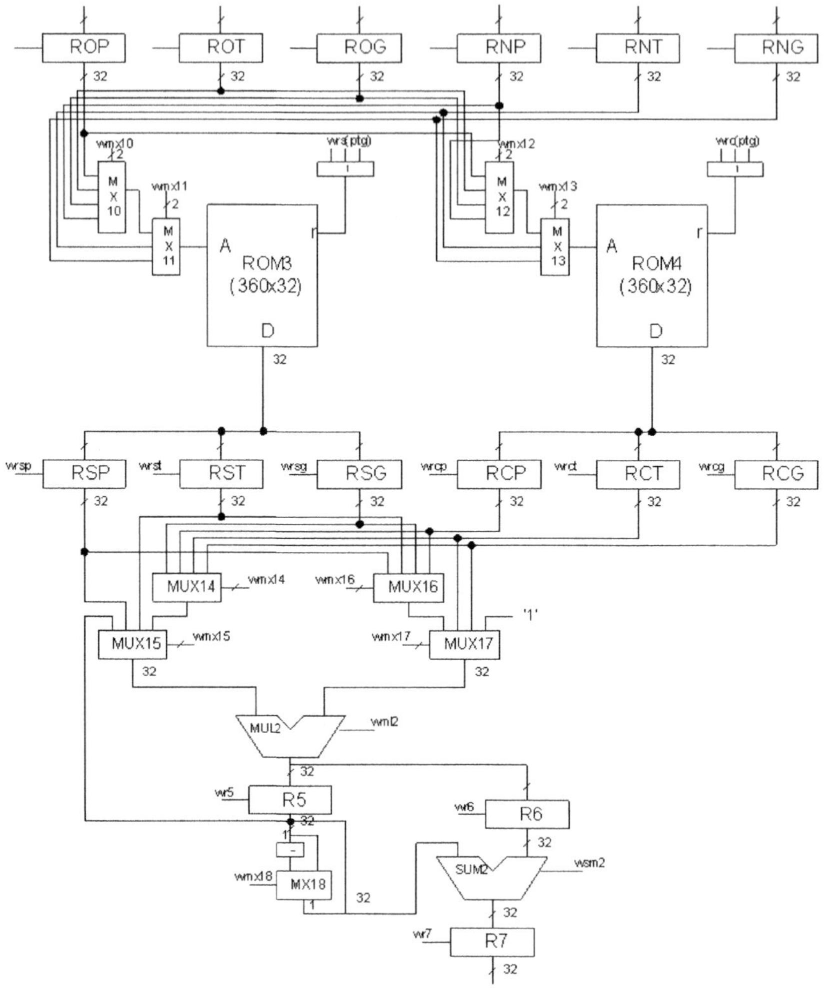

Fig. 4.3 Block diagram of matrix formation part of MFU

To translate center-point of object from the global system to observer expression (4.4) is used (Fig. 4.4).

The VDU sets a "visual" flag for potentially visible objects using its 3D spherical extents. The centers of the objects are transformed into the observer coordinate system for preliminary processing and setting "visual" flags.

$$\begin{bmatrix} x \\ y \\ z \\ 1 \end{bmatrix}^{con} = \left(\begin{bmatrix} x \\ y \\ z \\ 1 \end{bmatrix}^{Po} - \begin{bmatrix} x \\ y \\ z \\ 1 \end{bmatrix}^{Pn} \right) \cdot MB \tag{4.4}$$

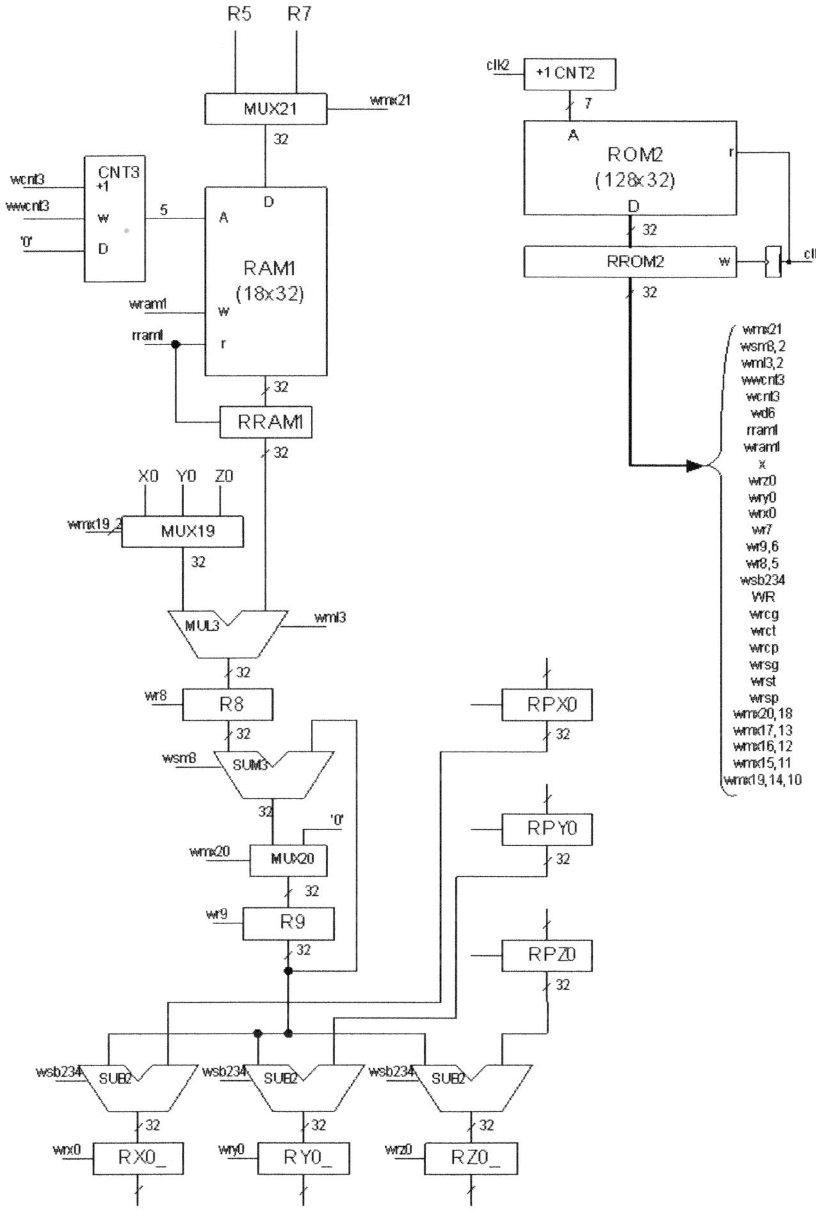

Fig. 4.4 Block diagram of center translation part of MFU

To simplify the process the object's extent is analyzed, as bounded sphere with radius *Ro*. It is obvious; the object is located beyond the scope of visible region, if it is located behind the screen or outside the bounders of the simple viewing pyramid (Fig. 4.5).

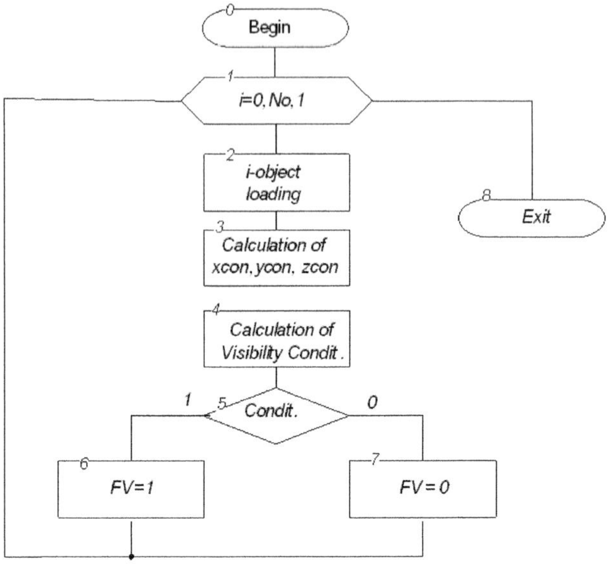

Fig. 4.5 Algorithm of visibility detection

Visibility condition:

$$VisCond = (x_{con} > d_w - R_o) \& (y_{con} < \frac{B_w}{2} + R_o) \& (y_{con} > -\frac{B_w}{2} - R_o) \& \\ (z_{con} < \frac{A_w}{2} + R_o) \& (z_{con} > -\frac{A_w}{2} - R_o), \tag{4.5}$$

where a_w, b_w - size of window; d_w - distance from the observer to the window; A_w, B_w - base of visible pyramid:

$$A_w = \frac{a_w \cdot x_{con}}{d_w}; \quad B_w = \frac{b_w \cdot x_{con}}{d_w}. \tag{4.6}$$

The DLU forms the priority subobjects list for the current level of details (Fig. 4.6).

Algorithm of priority list loading is shown in fig.4.7. Expression for block 2 is:

$$S = (P_o.x - P_n.x) \cdot N_i.x + (P_o.y - P_n.y) \cdot N_i.y + (P_o.z - P_n.z) \cdot N_i.z, \tag{4.7}$$

where $N_i = \{x, y, z\}$ - normal vector for i-node of BSP-tree.

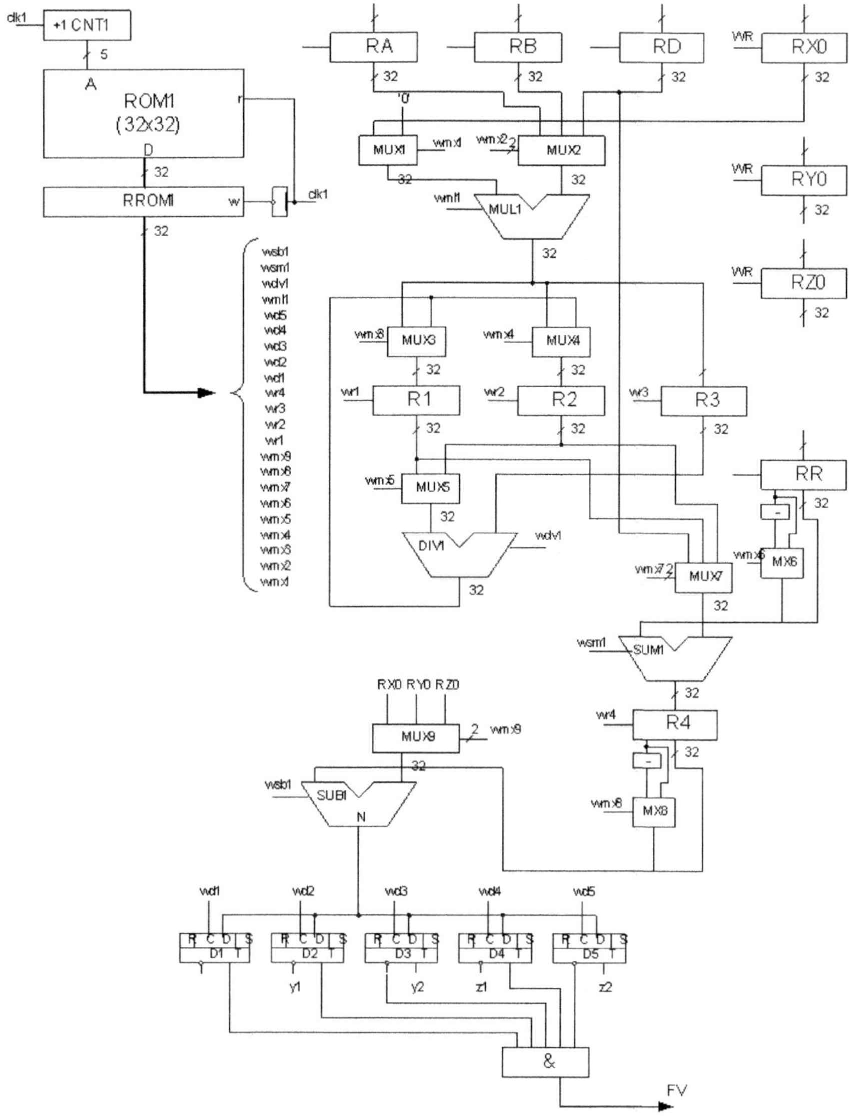

Fig. 4.6 Block diagram of VDU

Block diagram of DLU is shown in Fig.4.8 - 4.9.

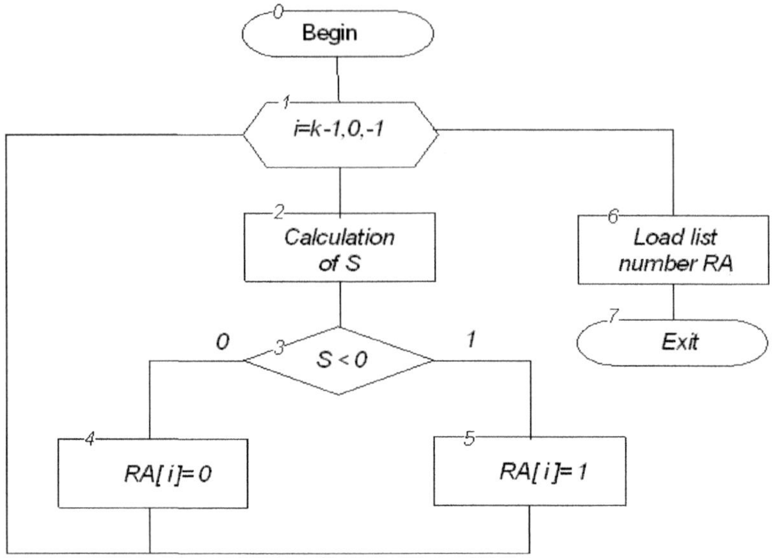

Fig. 4.7 Algorithm of priority-list loading

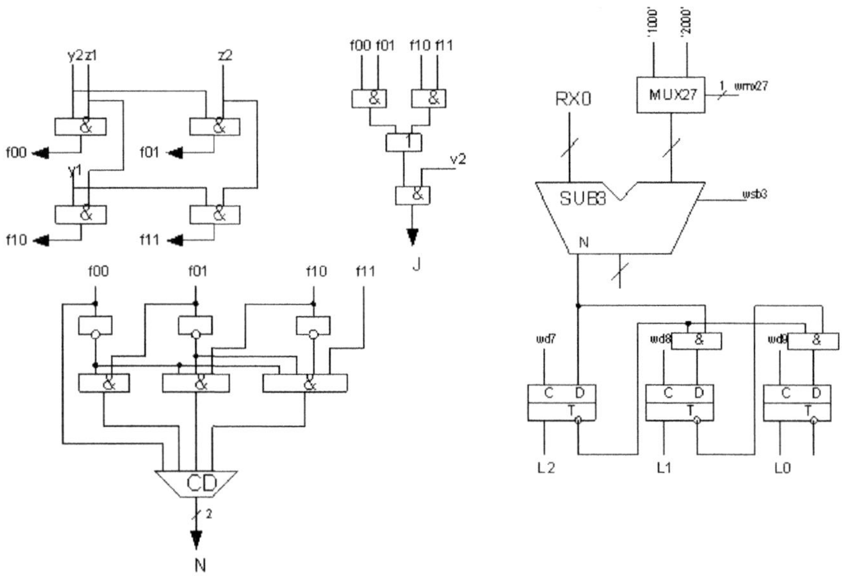

Fig. 4.8 Block diagram of DLU

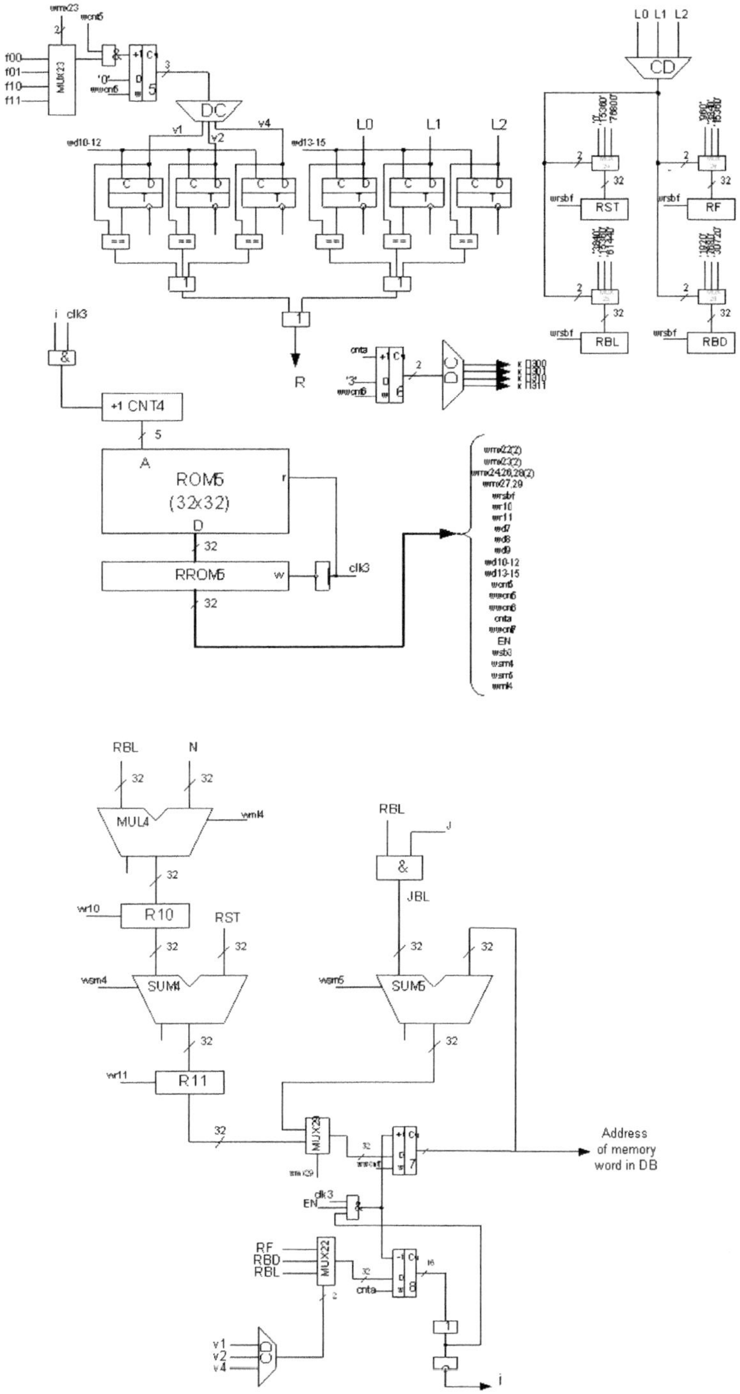

Fig. 4.9 Continues Block diagram of DLU

4.3.2 Clipping Processor

Fig. 4.10 shows an algorithm for 2D clipping of line-segment against the right boundary of screen.

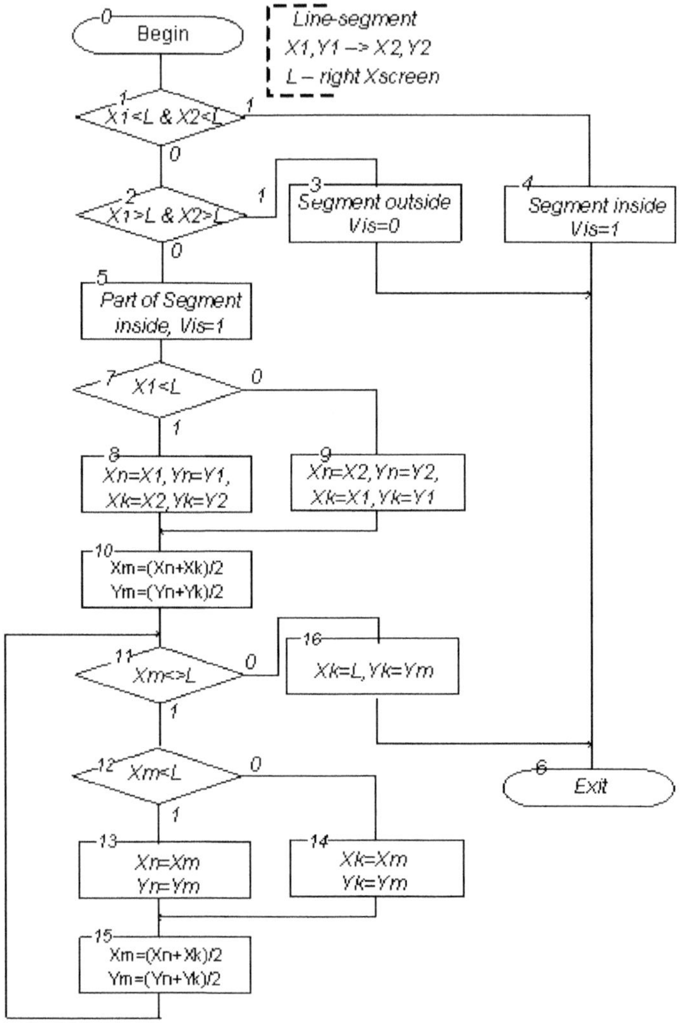

Fig. 4.10 2D clipping algorithm

Lines intersecting a rectangle clip region (or any complex polygon) are always clipped to a single line segment; lines lying on the clip rectangle's border are considered inside and hence are displayed (Vis=1, blocks 1,4). If both endpoints of line are outside the one rectangle's border, it is not displayed (Vis=0, blocks 2,3).

If one endpoint lies inside and one outside, the line intersects the clip rectangle and it is necessary to compute the intersection point using MPSD (Middle-Point Subdivision Algorithms, blocks 5-16).

This algorithm needs only addition and shift operations.

4.4 Application of Runtime Reconfiguration

The main problem of standard pipeline is "information" flow between processors. For complex graphical scene time of data exchange increases proportionally to improvement of quality of the image.

Runtime reconfiguration is a good complement to using specialized processors [5]. First of all, it solves the lack of flexibility of systems using hardware acceleration. The advantage of pure software is that any new application can be loaded at runtime, but the set of hardware-accelerated programs that can be executed in a given system is fixed at design time by the coprocessors included in it. The solution to this drawback is to also modify the hardware at runtime, something that's only possible with reconfiguration.

The best way to exploit these advantages is by using self-reconfigurable systems, where the FPGA can modify its own configuration without using additional external components. This will lead to a true system-on-a-chip solution that doesn't need a microcontroller or any other complex components to implement the reconfiguration.

Run-Time reconfigurable (RTR) systems could by assigned to implement complex algorithms with dynamic behavior using hardware. These systems allow reconfigure whole algorithm or part of them without interrupting computation process and also transparent for hardware environment [6]. RTR systems have several variants of using which reviewed in. One of the variant is implementation of complex computation algorithm using minimal hardware resources. This means algorithm distribution to several parts – computational logical blocks (CLB), and separate configuration/execution of each part. Computational logical blocks configures and later executes sequentially (sequentially managed execution) or using special internal algorithm (algorithmically managed execution). Each computation logical block concerned with other blocks by dataflow and operation unit. Computational logical block – part of algorithm which takes some data as input, process it and put some data as output. So CLB is logical structure with one input and one output. All CLBs could use shared operation unit, which allows reduce quantity of reconfigurable hardware resources and reduce time of each reconfiguration, if partially reconfiguration allowed. CLBs used shared external memory where they stores results of data processing.

System contains set of blocks which ensuring algorithm configuration, execution, data exchange etc. Offered system contains following blocks (Fig. 4.11).

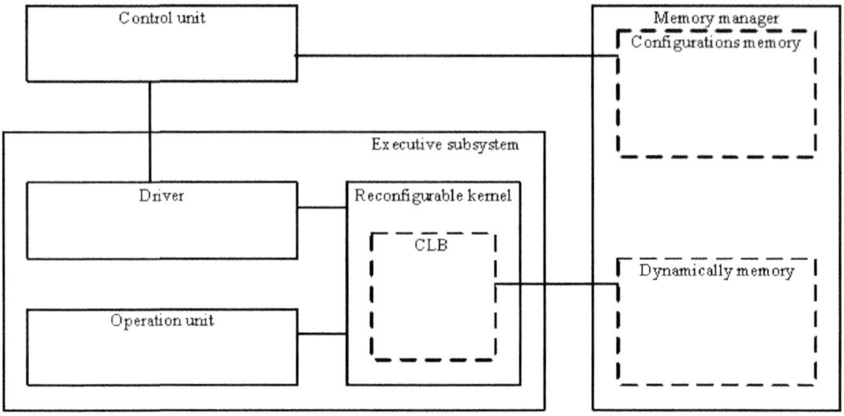

Fig. 4.11 Block diagram of RTR system

Executive subsystem – ensure configuring and reconfiguring CLBs and they execution. Contains: operation unit – non reconfigurable subsystems which implements set of low-level operations (functions) shared by most of CLBs, using of operation unit reduces reconfiguration time by making shared operations with more performance non reconfigurable hardware; reconfigurable kernel – hardware reconfigurable resources which purposed for CLBs configuration and execution; driver – control subsystem for reconfigurable kernel which ensure kernel encapsulation by special hardware interface with hardware environment.

Control unit – ensure execution of managing algorithms (algorithm managing and parallelism support) and interface with external memory where CLB configurations stored (flash memory, for example). Provides user interface for algorithm execution control, data processing etc.

Memory manager – ensure supply of external dynamically memory for executive subsystem and configurations memory for control unit. Implements set of caching or/and swap algorithms etc.

4.5 Application of UML for HDL-Code Creation

Direct transforming of UML state diagram into HDL is the first step in the automated synthesis of an FPGA circuits. UML has emerged as a common foundation for model driven architecture modeling. UML allows to build platform independent descriptions that can be used by designers to make informed decision about their hardware/software tradeoffs. UML is supported by a wide range of tools [7]. The exchange of models between tools is supported by the XML standard, an XML-based description language which captures the details of UML model diagrams in a portable, machine readable format.

HDL is one of a class of computer languages used to provide formal description of electronic circuitry. An HDL standard text-based expression is capable of describing the temporal behavior and/or (spatial) circuit structure of an electronic system. HDL is widely used in hardware design to specify details of chip design for either specialized chips or FPGAs. For custom or standard-cell based integrated circuit, such as a processor or other kind of specialized digital logic chip, HDL specifies a model for the expected behavior of a circuit before that circuit is designed and built. Special logic synthesis tools are then invoked that ultimately provide the geometric information used to produce photolithographic masks necessary for the fabrication of the device.

For programmable logic devices such as FPGAs, HDL code is first delivered to a logic compiler (FPGA synthesis tool), and the output is uploaded into the device. The unique property of this process and of programmable logic in general, is that it is possible to alter the HDL code many times, compile it, and upload into the same device for testing.

The transformation from high-level UML state diagram to HDL is based on a multi-step process, which consists of the following steps.

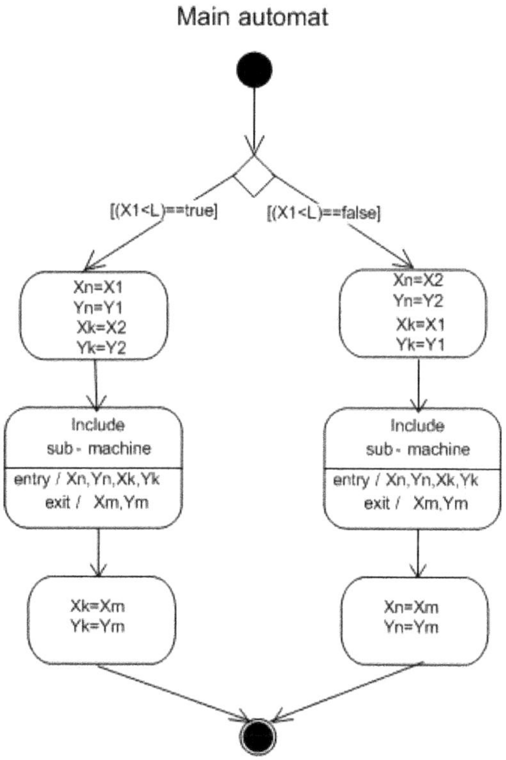

Fig. 4.12 UML state diagrams for 2D clipping algorithm

Step 1. State diagrams are created using UML state diagram notation. These diagrams describe system behavior using states, events and actions and correspond closely with the high-level design approach taken by circuit designers.

Step 2. UML diagrams are exported to XMI, a standard XML-based intermediate form. XMI uses predefined XML elements and attributes to specify the states, events and actions that make up the state diagram. The initial impetus for XMI was to enable UML diagrams to be imported and exported across different UML tools. Our approach uses the XMI as input to the next stage of processing.

Step 3. The XML representation of state machines is parsed by a Java-based XML parsing utility.

Step 4. Data extracted from the XMI by the Java parser is mapped to HDL templates, resulting in HDL suitable for use in FPGA construction.

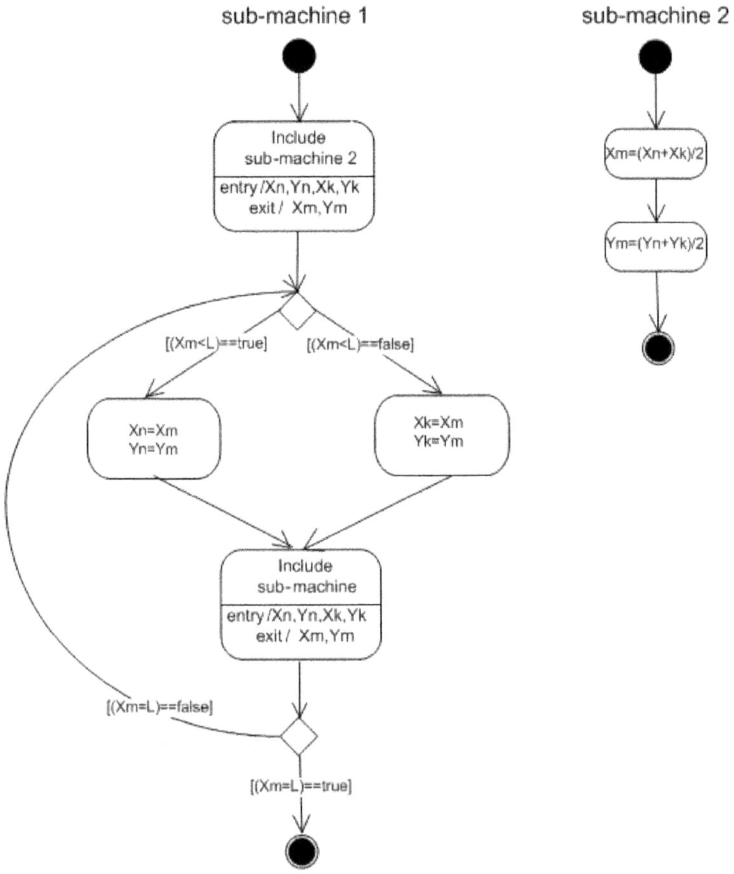

Fig. 4.13 UML state diagrams for MPSD algorithm

4.5.1 Example for 2D Clipping Realization

Fig. 4.12 shows a decomposition of intersection point's computing (blocks 5-16 of Fig. 4.10) and conversion to UML notation.

UML state diagrams for sub-machine 1 and sub-machine 2 are shown in Fig. 4.13. Program code below contains a HDL-code for Sub-machine 2 and 1 simulation in Active-HDL.

```
-- Architecture of middle-point-subdivision block
entity sum_div is
        port(   Xn: in integer;
                Xk: in integer;
        Yn: in integer;
                Yk: in integer;
                Xm: out integer:=0;
                Ym: out integer:=0
                );
end sum_div;
----------------------------------------------------
architecture sum_div1 of sum_div is
begin
        process (Xn,Xk,Yn,Yk)
        begin
                Xm<=(Xn+Xk)/2;
                Ym<=(Yn+Yk)/2;
        end process;
end sum_div1;

------------------------------------------------------------------------------------------
-- Architecture of processor for cross-point calculation
entity cpu is
        port(   Xn: in integer;
                Xk: in integer;
                Yn: in integer;
                Yk: in integer;
                L:  in integer;
                Xc: out integer:=0;
                Yc: out integer:=0;
                Res: out integer:=0 );
end cpu;

architecture cpu1 of cpu is
component sum_div is
        port(Xn,Xk,Yn,Yk: in integer:=0;
             Xm,Ym: out integer:=0);
end component sum_div;
signal Xna,Xka,Yna,Yka,Xca,Yca: integer:=0;
begin
  U1 : sum_div port map(Xna,Xka,Yna,Yka,Xca,Yca);
process
variable Xn1,Xk1,Yn1,Yk1,Xc1,Yc1: integer;
variable Res1: integer:=0;
 begin
 Xn1:=Xn;
 Xk1:=Xk;
```

```
Yn1:=Yn;
Yk1:=Yk;

if Xk /= L and Xn /= L then
m1: loop
--  Xc1:=(Xn1+Xk1)/2;
--  Yc1:=(Yn1+Yk1)/2;
  Xna<=Xn1;
  Xka<=Xk1;
  Yna<=Yn1;
  Yka<=Yk1;
  Xc1:=Xca;
  Yc1:=Yca;
  if Xc1>L then
    Xk1:=Xc1;
    Yk1:=Yc1;
  else
    Xn1:=Xc1;
    Yn1:=Yc1;
  end if;
 wait for 100 ns;
 exit m1 when Xc1 = L;
 end loop m1;
else
 wait for 10 ns;
 if Xk = L             then
   Xc1:=Xk;
   Yc1:=Yk;
 else
           Xc1:=Xn;
           Yc1:=Yn;
 end if;
end if;
Xc<=Xc1;
if Res1=0 then
Yc<=Yc1;
end if;
Res<=1;
Res1:=1;
end process;
end cpu1;
```

4.5.2 Fragment of HDL for Scene Processor Simulation

```
library ieee;
use ieee.std_logic_1164.all;
use ieee.numeric_std.all;
package work is
        type normal is array (0 to 32) of unsigned(0 to 48);
        type matr is array (0 to 8) of unsigned(0 to 48);
end work;
```

```
library ieee;
use ieee.std_logic_1164.all;
use ieee.numeric_std.all;
use WORK.work.all;
entity graf is
       port(k,n : in integer; a,b2,d,Ro : in unsigned(0 to 48);
                 Pnor1,P1 : in normal; B : in matr; clk : in std_logic;
                 FV : out std_logic; Ra : out unsigned(31 downto 0));
end graf;
library ieee;
use ieee.std_logic_1164.all;
use ieee.numeric_std.all;
use WORK.work.all;
architecture graf_ar of graf is
       signal j : integer range 0 to 48;
       signal f : std_logic;
       signal xcon,ycon,zcon : unsigned(0 to 65);
       signal A1,B1 : unsigned (0 to 114);
       signal S : normal;
       signal temp1,d1 : unsigned (0 to 48);
       signal Pnor,P : normal;
begin
       Pnor<=Pnor1;
       P<=P1;
       d1<=d;
       --S=(Po.x-Pn.x)*Pnor.x+(Po.y-Pn.y)* Pnor.y+(Po.z-Pn.z)* Pnor.z;
       --xcon=(Po.x-Pn.x)*matr_b[0]+(Po.y-Pn.y)*matr_b[1]+
       --   (Po.z-Pn.z)*matr_b[2];
       --ycon=(Po.x-Pn.x)*matr_b[3]+(Po.y-Pn.y)*matr_b[4]+
       --   (Po.z-Pn.z)*matr_b[5];
       --zcon=(Po.x-Pn.x)*matr_b[6]+(Po.y-Pn.y)*matr_b[7]+
       --   (Po.z-Pn.z)*matr_b[8];
       --A=a*xcon/d;
       --B=b*xcon/d;
       --xcon>(d-Ro);
       --ycon<(B/2+Ro);
       --ycon>(-B/2-Ro);
       --zcon<(A/2+Ro);
       --zcon>(-A/2-Ro);
       process (clk)
       begin
                 if clk='1' and clk'event then
                           S(32)<=(others=>'0');
                           for i in к-1 downto 0 loop
                           j<=n-i;
                           S(i)(0 to 33)<=((P(i)(0 to 16))-(P(j)(0 to 16)))*
                           (Pnor(i)(0 to 16))+((P(i)(17 to 32))-(P(j)(17 to 32)))*
                           (Pnor(i)(17 to 32))+((P(i)(33 to 48))-(P(j)(33 to 48)))*
                           (Pnor(i)(33 to 48));
                                     if S(i)>S(32) then   Ra(i)<='0';
                                     else Ra(i)<='1';
                                     end if;
```

```
                        end loop;
                        for i in 1 to 8 loop
                                j<=n-i;
                                xcon<=(P(i)(0 to 16)-P(j)(0 to 16))*B(0)+
                                (P(i)(17 to 32)-P(j)(17 to 32))*B(1)+
                                (P(i)(33 to 48)-P(j)(33 to 48))*B(2);
                                ycon<=(P(i)(0 to 16)-P(j)(0 to 16))*B(3)+
                                (P(i)(17 to 32)-P(j)(17 to 32))*B(4)+
                                (P(i)(33 to 48)-P(j)(33 to 48))*B(5);
                                zcon<=(P(i)(0 to 16)-P(j)(0 to 16))*B(6)+
                                (P(i)(17 to 32)-P(j)(17 to 32))*B(7)+
                                (P(i)(33 to 48)-P(j)(33 to 48))*B(8);
                                A1<=(a*xcon)/d1;
                                B1<=(b2*xcon)/d1;
                                f<='0';
                                if (d-Ro)>xcon then f<='1'; end if;
                                if (B1/2+Ro)<ycon then f<='1';        end if;
                                if (((not B1)/2)-Ro)>ycon then f<='1'; end if;
                                if (A1/2+Ro)<zcon then f<='1';        end if;
                                if (((not A1)/2)-Ro)>zcon then f<='1'; end if;
                                if f='1' then        FV<='1';
                                else FV<='0';
                                end if;
                        end loop;
                end if;
        end process;
end graf_ar;
```

4.6 Summary and Future Directions

The ability to move from UML diagrams to HDL hardware descriptors is the first step in an effort to use model-based architecture to further optimize and automate the small systems development. Having made the decision to implement an algorithm using FPGAs, there are numerous decisions concerning the positioning of components that will impact product viability.

By analyzing the whole system in terms of supporting software and hardware, it is possible exploit new opportunities for image generation systems developing.

References

[1] Nokia Internet Tables (2008), `http://europe.nokia.com/` (accessed April 26)

[2] Foley, J.D., et al.: Computer graphics: principal and practice. Addison-Wesley, Reading (1997)

[3] Bashkov, E.A., Malcheva, R.: The Performance Estimation of Scene Manager in Real-Time Scene Generator. Sofia, 123–127 (1995)

[4] Malcheva, R.: The problems of modeling and rendering of the realistic complex scenes. In: Proceedings of ECCPM 2002, Balkema, pp. 537–538 (2002)

[5] Gonzalez, I., Aguayo, E., Lopez-Buedo, S.: Self-reconfigurable embedded systems on Low-Cost-FPGAs. Micro 27(4), 49–57 (2007)
[6] Grytsenko, A.A., Malcheva, R.V., Barkalov, A.A.: Run-time reconfigurable system for distributed algorithm execution. In: LVIV: CSIT 2006, pp. 164–166 (2006)
[7] Mellor, S.J., Balcer, M.J.: Executable UML. A Foundation for Model-Driven Architecture, Indianapolis. Addison-Wesley, Reading (2002)

Part II
Digital Design with Programmable Logic

5 Logic Synthesis Method of Digital Circuits Designed for Implementation with Embedded Memory Blocks of FPGAs

Mariusz Rawski, Paweł Tomaszewicz, Grzegorz Borowik, and Tadeusz Łuba

Warsaw University of Technology, Institute of Telecommunications,
Nowowiejska 15/19, 00-665 Warsaw, Poland

Abstract. The paper presents logic synthesis method targeted at FPGA architectures with specialized embedded memory blocks (EMBs). Existing methods do not ensure effective utilization of the possibilities provided by such modules. The problem of efficient mapping of combinational and sequential parts of design can be solved using decomposition algorithms. The main question of this paper is the application of decomposition based methods for efficient utilization of modern FPGAs. It will be shown that functional decomposition method allows for very flexible synthesis of the designed system onto heterogeneous structures of modern FPGAs composed of logic cells and EMBs. Finally we present results of the experiments, which evidently show, that the application of functional decomposition algorithms in the implementation of typical signal and information processing systems greatly influences the performance of resultant digital circuits.

5.1 Introduction

The technological advancements in Field Programmable Gate Arrays (FPGA) in the past decade have opened new paths for digital system design engineers. An FPGA can be described as an array of programmable logic cells interconnected by programmable connections. Each cell can implement a simple logic function (of a limited number of inputs) defined by a designer's CAD tool. A typical programmable device has a large number (64 to over 300,000) of such cells that can be used to form complex digital circuits. The ability to manipulate the logic at the gate level means that the designer can construct a custom processor to implement the desired function efficiently.

Since FPGA introduction in the 1980's, the manufacturers have been extending their chips' ability to implement digital systems by introducing specialized mechanisms such as low-latency carry-chain-routing lines that speed-up the addition and subtraction operations, dedicated multiplier function blocks or even fully functional MAC blocks called DSP blocks.

M. Adamski et al. (Eds.): Design of Digital Systems and Devices, LNEE 79, pp. 121–144.
springerlink.com

Modern FPGA devices are also equipped with memory-based structures [9, 28]. These specialized embedded memory blocks make it possible to implement data storage modules such as shift registers or RAM blocks. In many cases, though, the designer does not need such elements in his/her design or does not utilize all of such resources. During the mapping stage, these blocks may be considered as logic units unless they are being used for data storage. The memories act as very large logic cells, where the number of inputs is equal the number of address lines. Unfortunately, the existing CAD tools are not well suited to utilize all possibilities that such EMB blocks offer due to the lack of appropriate synthesis methods. Typically, after the logic synthesis stage, technology-dependent mapping methods are used to map design onto available resources [11]. However, such an approach is inefficient due to the fact that the quality of post-synthesis mapping is highly dependent on the quality of technology-independent optimization step.

Recently, efforts have been made to develop methods based on functional decomposition that would allow for efficient utilization of these EMB blocks [6, 24, 25, 26].

Moreover, in modern logic synthesis of PLD, FPGA modules as well as PLA structures, the problem of finite state machine synthesis is significant due to its widespread practical application, but in particular internal states encoding. This encoding influences both the structure of the realization of the FSM (i.e., the connections between the combinational block and the memory block) and the complexity of the combinational block.

Attempts to solve the above problem resulted in many methods for the structural synthesis of FSMs. Their diversity results from different analysis, different assumptions and, subsequently, designing the methods for specific types of target components. Thus, different methods of the synthesis of FSM for PLA structures [9, 12], for ROM memories [1] and PLD modules [4, 10] exist. Unfortunately, the current solutions which concern FPGAs with EMBs are not efficient.

In this paper, new logic synthesis methods that allow for very efficient utilization of embedded memory blocks are presented. Proposed methods are based on functional decomposition [8, 15, 17, 22, 23]. These methods can be used for specific designs, i.e. to implement FIR filters using the concept of distributed arithmetic (DA). In the end, experimental results that prove efficiency of the proposed methods are shown.

5.2 Decomposition of Boolean Functions

In this section, only information that is necessary for an understanding of this paper is reviewed. More detailed description of functional decomposition based on partition calculus can be found in [8, 22].

5.2.1 Functional Decomposition

The set X of input variables of Boolean function is partitioned into two subsets: *free variables* U and *bound variables* V, such that $U \cup V = X$. Assume that the input variables $x_1, \ldots, x_n$ have been relabeled in such a way, that:

$$U = \{x_1, ..., x_r\} \text{ and}$$

$$V = \{x_{n-s+1}, ..., x_n\}. \quad (5.1)$$

Consequently, for an n-tuple x, the first r components are denoted by x^U and the last s components are denoted by x^V.

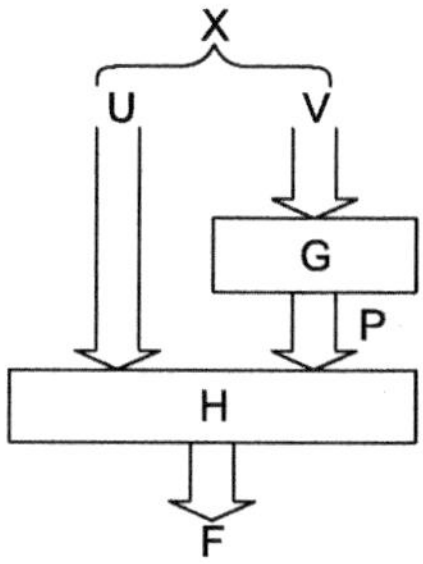

Fig. 5.1 Schematic representation of the serial decomposition

Let F be a Boolean function with n inputs and m outputs and let (U, V) be the pair of sets defined above. Assume that F is specified by a set of the function's cubes. Let G be a function with s inputs and p outputs, and let H be a function with $r + p$ inputs and m outputs. The pair (G, H) represents a serial decomposition of F with respect to (U, V), if for every minterm b relevant to F, $G(b^V)$ is defined, $G(b^V) \in \{0, 1\}^p$, and $F(b) = H(b^U, G(b^V))$. G and H are called blocks of the decomposition (Fig. 5.1).

Theorem 5.1 Let P_V, P_U, and P_F be partitions induced on the function F input cubes by the input sub-sets V and U, and outputs of F, respectively. If there exists a partition P_G on the set of function F input cubes such that $P_V \leq P_G$, and $P_U \cdot P_G \leq P_F$, then F has a serial decomposition with respect to (U, V).

The r-admissibility test allows one to obtain the set U of free variables for which there exists function G (generally with t outputs) such, that $|U| + t < n$, where $n = |X|$.

Let $P_1, ..., P_k$ be partitions on M which is the set of minterms of function F. The set of partitions $\{P_1, ..., P_k\}$ is r-admissible in relation to partition θ if and only if there is a set $\{P_{k+1}, ..., P_r\}$ of two-block partitions such that the product π of partitions $P_1, ..., P_k, P_{k+1}, ..., P_r$ satisfies the inequality $\pi \leq \theta$, and there does not exist any set of $r - k - 1$ two-block partitions which meets this requirement.

The r-admissibility has the following interpretation. If a set of partitions $\{P_1, ..., P_k\}$ is r-admissible, then there might exist a serial decomposition of F (Fig. 5.1) in which component H has r inputs: k primary inputs corresponding to free input variables which induce $\{P_1, ..., P_k\}$ and $r - k$ inputs which are G outputs. Thus, to find a decomposition of F in which component H has r inputs, we must find a set of input variables which induces an r-admissible set of input partitions. The following corollary can be applied to check whether a given set of input partitions is r-admissible.

Corollary 5.1 For $\sigma \leq \tau$, let $\tau|\sigma$ denote the quotient partition and $\eta(\tau|\sigma)$ be the number of elements in the largest block of $\tau|\sigma$. Also, let $e(\tau|\sigma)$ denote the smallest integer equal to or larger than $\log_2\eta(\tau|\sigma)$, i.e. $e(\tau|\sigma) = \lceil \log_2 \eta(\tau|\sigma) \rceil$. Then, $\{P_1, ..., P_k\}$ is r-admissible, if

$$r = k + e(\pi|\pi_F), \tag{5.2}$$

where π is the product of partitions $P_1, ..., P_k$ and $\pi_F = \pi \cdot P_F$.

Corollary 5.1 provides a means to evaluate the admissibility of a set of input variable partitions $P(x_i)$. In searching of a maximum set of variables, which can be connected to circuit H directly, we compute sets of t-admissible partitions $P(x_i)$ only, where t is a given number of inputs of circuit H.

Property 5.1 If a set of partitions $P = \{P_1, ..., P_k\}$ is m-admissible, then each subset of P is m′-admissible, where $m' \leq m$. Thus, the necessary condition for set P to be m-admissible is that each subset P' of P has to have r-admissibility $r(P') \leq m$.

Example 5.1 R-admissibility evaluation.

Table 5.1 Table of example function

	x_1	x_2	x_3	x_4	x_5	x_6	y_1	y_2	y_3
1	1	0	0	1	0	0	0	0	0
2	1	0	1	1	1	0	0	1	0
3	1	1	0	1	1	1	0	1	0
4	1	0	1	0	0	0	0	0	1
5	0	0	1	1	1	0	1	0	0
6	1	1	0	1	0	0	1	0	1
7	1	0	0	1	1	1	1	0	0
8	0	0	1	0	0	1	1	1	1
9	0	0	1	0	1	0	1	1	0

For the function of 6 input variables and 3 output variables described in Table 5.1, we have the following partitions induced by input variables:

$$P_{x_1} = \{\overline{5,8,9}; \overline{1,2,3,4,6,7}\},$$

$$P_{x_2} = \{\overline{1,2,4,5,7,8,9}; \overline{3,6}\},$$

$$P_{x_3} = \{\overline{1,3,6,7}; \overline{2,4,5,8,9}\},$$

$$P_{x_4} = \{\overline{4,8,9}; \overline{1,2,3,5,6,7}\},$$

$$P_{x_5} = \{\overline{1,4,6,8}; \overline{2,3,5,7,9}\},$$

$$P_{x_6} = \{\overline{1,2,4,5,6,9}; \overline{3,7,8}\}$$

and the output partition:

$$P_F = P_{y_1} \cdot P_{y_2} \cdot P_{y_3} = \{\overline{1}; \overline{4}; \overline{2,3}; \overline{5,7}; \overline{6}; \overline{8}; \overline{9}\}.$$

The quotient partitions are as follows:

$$P_{x_1} \mid P_F = \{\overline{(5)(8)(9)}; \overline{(1)(4)(2,3)(6)(7)}\},$$
$$P_{x_2} \mid P_F = \{\overline{(1)(2)(4)(5,7)(8)(9)}; \overline{(3)(6)}\},$$
$$P_{x_3} \mid P_F = \{\overline{(1)(3)(6)(7)}; \overline{(2)(4)(5)(8)(9)}\},$$
$$P_{x_4} \mid P_F = \{\overline{(4)(8)(9)}; \overline{(1)(2,3)(6)(5,7)}\},$$
$$P_{x_5} \mid P_F = \{\overline{(1)(4)(6)(8)}; \overline{(2,3)(5,7)(9)}\},$$
$$P_{x_6} \mid P_F = \{\overline{(1)(2)(4)(5)(6)(9)}; \overline{(3)(7)(8)}\}.$$

Thus, the r-admissibility of partitions induced by input variables is:

$$r(P_{x_1}) = 1 + e(P_{x_1} \mid P_F) = 1 + \lceil \log_2 \eta(P_{x_1} \mid P_F) \rceil = 1 + \lceil \log_2 5 \rceil = 4,$$
$$r(P_{x_2}) = 1 + e(P_{x_2} \mid P_F) = 1 + \lceil \log_2 6 \rceil = 4,$$
$$r(P_{x_3}) = 1 + e(P_{x_3} \mid P_F) = 1 + \lceil \log_2 5 \rceil = 4,$$
$$r(P_{x_4}) = 1 + e(P_{x_4} \mid P_F) = 1 + \lceil \log_2 4 \rceil = 3,$$
$$r(P_{x_5}) = 1 + e(P_{x_5} \mid P_F) = 1 + \lceil \log_2 4 \rceil = 3,$$
$$r(P_{x_6}) = 1 + e(P_{x_6} \mid P_F) = 1 + \lceil \log_2 6 \rceil = 4.$$

5.2.2 Decomposition into EMB Blocks

Functional decomposition relies on breaking down a complex system into a network of smaller co-operating sub-functions in such a way that the original system behavior is preserved. A single step of the functional decomposition replaces function F with two sub-functions (Fig. 5.1). This process is recursively applied to both G and H blocks until a network is constructed where each block can be directly implemented in a single logic cell of target FPGA architecture.

A logic cell can implement any function of a limited number of input variables (typically 4 or 5). Thus, the main effort of logic synthesis methods based on decomposition is to find such a partition of input variables into the free set and the bound set that allows for decomposition with the block G which not exceed the size of the logic cell. Various methods are used, including exhaustive search, since the size of the logic cell is small. Noticeably, the main constraint is the number of inputs to block G and not the number of outputs. This is because the block G with more outputs than available in the logic cell can be implemented with few logic cells operating in parallel.

Since EMB blocks can be configured to work as logic cells of many different sizes, methods intended for logic cells are not efficient. The main reason is that the method has to check decompositions for many different sizes of block G. The second factor is that, in case of EMBs the efficiency of utilization of these blocks depends on carefully selected size of block G. For example, M512 RAM block of Stratix device can be configured among others as an 8-input and 2-output logic cell or 7-input and 4-output logic cell. Let us assume that in the decomposition search, the following solutions are possible: the block G with 8 inputs and 1 output or the block G with 7 inputs and 3 outputs. From EMBs utilization point of view, the second solution is better, since it utilizes 384 bits of total 512 bits available, while the first solution utilizes only 256 bits.

In this paper, we present a method that uses the concept of r-admissibility to efficiently create decompositions that utilize the EMB blocks in a high degree. The method is based on the balanced functional decompositions. Based on redundant variable analysis of each output of a multi-output function, parallel decomposition separates F into two or more functions, each of which has as its inputs and outputs a subset of the original inputs and outputs. Although in this method the crucial point of the whole mapping process is created by the serial decomposition algorithm, the parallel decomposition based on argument reduction process plays a very important role. Thanks to this algorithm the functional decomposition procedure can start directly with a two-level, espresso based specification. Thus, the method itself allows one to develop a uniform autonomous tool for decomposition-based technology mapping for FPGAs.

R-admissibility is used to evaluate serial decomposition possibilities for different sizes of G block, according to possible configuration of EMB blocks. For a function with n input variables there are $\binom{n}{k}$ possible solutions of serial decomposition with k inputs G block. Since an EMB can be configured as a block of many different sizes, the possible solution space is large. Using Property 5.1, the search can be drastically reduced. This will be explained in the following example.

Example 5.2 R-admissibility application to serial decomposition evaluation.

For the function from Example 5.1 we have that the r-admissibility of single input variables $x_1, \ldots, x_6$ is 4, 4, 4, 3, 3 and 4, respectively. This means that only for $U = \{x_4\}$, $V = \{x_1, x_2, x_3, x_5, x_6\}$ and $U = \{x_5\}$, $V = \{x_1, x_2, x_3, x_4, x_6\}$ a decomposition with 2 outputs from block G may exist.

When considering solutions with 4 inputs to block G, according to Property 5.1, only the solution with $U = \{x_4, x_5\}$, $V = \{x_1, x_2, x_3, x_6\}$ should be evaluated. We have:

$$(P_{x_4} \cdot P_{x_5}) \mid P_F = \{\overline{(4)(8)}; \overline{(9)}; \overline{(1)(6)}; \overline{(2,3)(5,7)}\}$$
$$r(P_{x_4} \cdot P_{x_5}) = 2 + e(P_{x_4} \cdot P_{x_5} \mid P_F) = 2 + \lceil \log_2 2 \rceil = 3.$$

This means that for such variable partitioning, a decomposition may exist with 1-output G block.

5.2.3 Parallel Decomposition

Consider a multiple-output function F. Assume that F has to be decomposed into two components, G and H, with disjoint sets Y_G and Y_H of output variables. This problem occurs, for example, when we want to implement a large function using components with a limited number of outputs. Note that such a parallel decomposition can also alleviate the problem of an excessive number of inputs of F. This is because, for typical functions most outputs do not depend on all input variables. Therefore, the set X_G of input variables on which the outputs of Y_G depend, may be smaller than X. Similarly, for the set X_H of input variables on which the outputs of Y_H depend may be smaller than X. As a result, components G and H have not only fewer outputs, but also fewer inputs than F. The exact formulation of the parallel decomposition problem depends on the constraints imposed by the implementation style. One possibility is to find sets Y_G and Y_H, such that, the combined cardinality of X_G and X_H is minimal. Partitioning the set of outputs into only two disjoint subsets is not an important limitation of the method, because the procedure can be applied for components G and H again.

Example 5.3 Consider the multiple-output function given in Table 5.2. The minimal sets of input variables on which each output of F depends are:

y_1: $\{x_1, x_2, x_6\}$
y_2: $\{x_3, x_4\}$
y_3: $\{x_1, x_2, x_4, x_5, x_8\}$, $\{x_1, x_2, x_4, x_6, x_8\}$
y_4: $\{x_1, x_2, x_3, x_4, x_7\}$
y_5: $\{x_1, x_2, x_4\}$
y_6: $\{x_1, x_2, x_6, x_8\}$.

An optimal two-block decomposition, minimizing the card X_G + card X_H (where card X is the cardinality of X), is $Y_G = \{y_1, y_3, y_6\}$and $Y_H = \{y_2, y_4, y_5\}$, with $X_G = \{x_1, x_2, x_4, x_6, x_8\}$ and $X_H = \{x_1, x_2, x_3, x_4, x_7\}$. The truth tables for components G and H are shown in Table 5.3a and 5.3b.

Table 5.2 Function F

	x_1	x_2	x_3	x_4	x_5	x_6	x_7	x_8	y_1	y_2	y_3	y_4	y_5	y_6
1	0	0	0	1	1	1	0	0	0	0	0	0	–	0
2	1	0	1	0	0	0	0	0	0	0	–	1	0	1
3	1	0	1	1	1	0	0	0	0	1	1	0	1	1
4	1	1	1	1	0	1	0	0	0	1	1	1	1	0
5	1	0	1	0	1	0	0	0	0	0	0	–	0	1
6	0	0	1	1	1	0	0	0	1	1	0	1	0	0
7	1	1	1	0	0	0	0	0	1	0	–	0	1	0
8	1	0	1	1	0	1	0	0	1	1	0	0	–	1
9	1	0	1	1	0	1	1	0	–	1	0	1	–	1
10	0	0	0	1	1	1	0	1	0	0	1	0	–	1
11	0	0	0	1	1	0	0	1	–	–	1	0	0	0

Table 5.3a Function G of parallel decomposition

	x_1	x_2	x_4	x_6	x_8	y_1	y_3	y_6
1	0	0	1	1	0	0	0	0
2	1	0	0	0	0	0	0	1
3	1	0	1	0	0	0	1	1
4	1	1	1	1	0	0	1	0
5	0	0	1	0	0	1	0	0
6	1	1	0	0	0	1	1	0
7	1	0	1	1	0	1	0	1
8	0	0	1	1	1	0	1	1
9	0	0	1	0	1	-	1	0

Table 5.3b Function H of parallel decomposition.

	x_1	x_2	x_3	x_4	x_7	y_2	y_4	y_5
1	0	0	0	1	0	0	0	0
2	1	0	1	0	0	0	1	0
3	1	0	1	1	0	1	0	1
4	1	1	1	1	0	1	1	1
5	0	0	1	1	0	1	1	0
6	1	1	1	0	0	0	0	1
7	1	0	1	1	1	1	1	-

The algorithm itself is general in the sense that function to be parallely decomposed can be specified in compact cube notation. Calculation of the minimal sets of input variables for each individual output can be a complex task. Thus, in practical implementation heuristic algorithms are used which support calculations with the help of so called indiscernible variables.

5.2.4 Balanced Decomposition

In the balanced decomposition the serial and parallel decompositions are intertwined in a top-down synthesis process to obtain the required circuit structure. At each step, either parallel or serial decomposition is performed, both controlled by certain input parameters. In the case of serial decomposition the parameters G_{in} and G_{out} denote the number of G inputs and outputs, respectively. In the case of parallel decomposition the parameter G_{out} represents the number of G outputs. Intertwining of serial and parallel decomposition opens up several interesting possibilities in multilevel decomposition. Experimental results show that the right balance between parallel and serial decomposition and the choice of control parameters significantly influence the area and depth of the resultant network.

Example 5.4 The influence of the decomposition strategy on the final result of the FPGA-based mapping process will be explained with function F representing DA logic of a certain wavelet filter described by the following coefficients [1495, −943, −9687, 18270, −9687, −943, 1495].

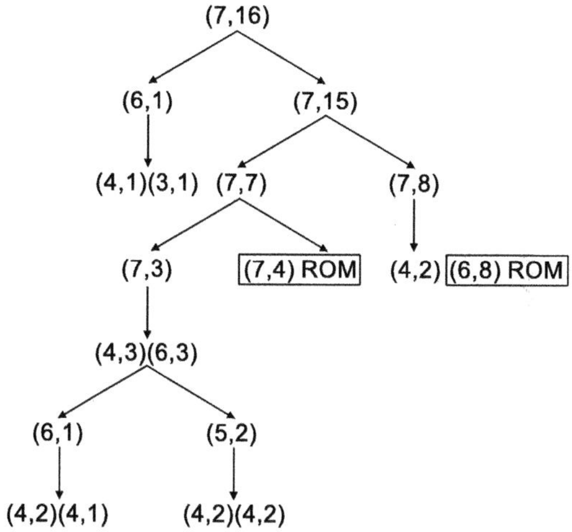

Fig. 5.2 Decomposition process for the ahp (7,16) filter – strategy 1

As F is a 7-input and 16-output function, in the first step of the decomposition, both parallel and serial decomposition can be applied. Let us apply parallel decomposition first (Fig. 5.2). The parallel decomposition with $G_{out} = 1$ generates two components: the first one with 6 inputs and 1 output, and the second with 7 inputs and 15 outputs. This is illustrated by two arrows with the common starting point going to different directions. The smaller component is the subject to two-stage serial decomposition resulting in block G with 4 inputs and 1 output and block H with 3 inputs and 1 output (both G and H blocks are implemented using 2 cells). Notation (4,1)(3,1) at the end of the arrow shows the number of inputs and outputs for functions G(4,1) and H(3,1), respectively. The second component is again decomposed in parallel yielding components (7,7) and (7,8). For the (7,8) component serial decomposition is assumed, now resulting in block G with 4 inputs and 2 outputs (implemented with 2 logic cells). Thus, the next step deals with a 6-input function H, which can be directly implemented in ROM. In the next iterative step, parallel decomposition is applied to split the (7,7) component into (7,3) and (7,4) blocks. It is reasonable to implement the (7,4) block in a ROM. The second block is decomposed serially yielding G(4,3) and H(6,3). As G block can be implemented with 3 logic cells, the next step deals with function H. Parallel decomposition applied to function H generates two components. Each of them is the subject to two-stage serial decomposition. The obtained network can be built of 14 logic cells and 2 M512 ROMs.

If we change the size of the smaller component in the first step of the parallel decomposition, i.e., (7,4) instead of (6,1), then the implementation requires 3

M512 ROMs and 9 logic cells. However, making decision to apply the serial decomposition instead of the parallel decomposition to decompose (7,16), the implementation requires only 3 ROMs. The structure is shown in Fig. 5.3.

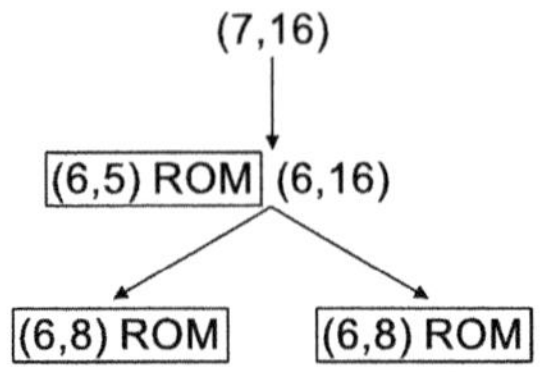

Fig. 5.3 Decomposition process for the ahp (7,16) filter – strategy 2

Balanced decomposition was implemented as software package called DEMAIN [19, 27]. Recently the package was improved to help designers to deal with large truth tables. All described methods of truth tables transformations can be performed easily, and results are shown immediately on the screen for further processing. It is designed for performing manual operations on functions, and therefore is meant to be highly user friendly, as well as cross-platform compatible. After choosing the operation, a dialog pops up which can be used for inputting the parameters of the operation. After the operation is performed, its results are displayed in the project window.

5.3 Sequential Circuits Synthesis

Embedded memory blocks can also be used for implementation of sequential machines in a way that requires fewer logic cells than the traditional flip-flop based implementation. This may be used to implement "non-vital" sequential parts of the design, saving logic cell resources for more important sections [22]. Since the size of embedded memory blocks is limited, such an implementation may require more memory than is available in a device. To reduce the memory usage in ROM-based sequential machine implementations, a structure with next state logic partially implemented in the ROM and partially implemented in logic cells was proposed [22].

In the considered FSM implementation the combinational logic is split into two parts. One part is implemented in embedded memory blocks which are configured as ROM memory, with its content determined at the time of the programming. The second part, called the address modifier, is used to reduce the number of memory address lines (Fig. 5.4). The address modifier is implemented in programmable logic blocks containing LUTs. This proposal is a cross-fertilized approach between recent progress in finite-state machine synthesis and in micro-computer architectures. Similar ideas can be found in [4].

Presented problem is intimately related to the encoding problem of FSM which is of fundamental importance in a sequential synthesis, especially the state-machine synthesis.

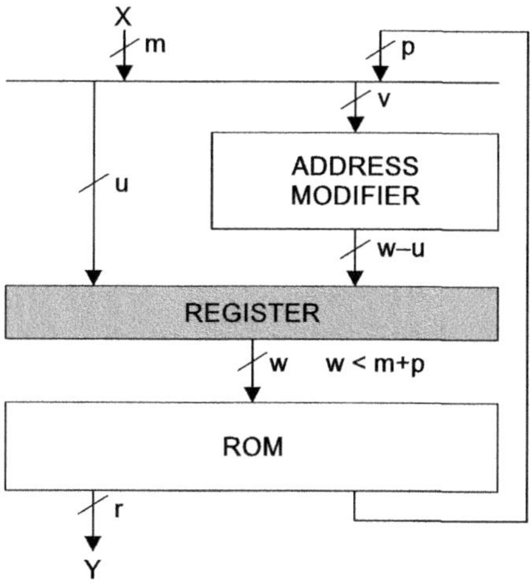

Fig. 5.4 Implementation of an FSM using an address modifier

5.3.1 Basic Information

Let $\boldsymbol{A} = (S, V, \delta, Y, \lambda)$ be an FSM (completely or incompletely specified), where: S – set of internal states, V – set of input symbols, δ – state transition function, Y – set of output symbols, λ – output function, and the values $m \geq \lceil \log_2 |V| \rceil$ and $p \geq \lceil \log_2 |S| \rceil$ denote the number of inputs and state variables, respectively.

Partition description and partition algebra [13] are applied to describe logic dependencies in such an FSM.

Let $\boldsymbol{T}$ be an isomorphic function between the domain D_δ of the transition function and the set $T = 1, \ldots, t$, where $t = |D_\delta|$. Set T represents the ROM cells needed to store the next state pair $\delta(v, s)$ for each pair (v, s). Thus, the characteristic partition P_c of the FSM is defined in the following way:

Each block B_{P_c} of the characteristic partition includes these elements from the set T which correspond to these pairs (v, s) from the domain D_δ which the transition function $\delta(v, s) = s'$ maps onto the same next state s'.

Example 5.5. For the FSM and function $\boldsymbol{T}$ shown in Table 5.4 the characteristic partition is:

$$P_c = \{\overline{1,8,12,14}; \overline{2,7,10,16}; \overline{6,9,13}; \overline{3,5,11,15}; \overline{4}\}.$$

A partition P on the set T is related to a partition π on the states set S if for any inputs v_a, v_b the condition that s_i, s_j belong to one block of the partition π implies that the elements from T corresponding to pairs (v_a, s_i) and (v_b, s_j) belong to one block of the partition P.

A partition P on the set T is related to a partition θ on the input symbols set V if for any state s_a, s_b the condition that v_i, v_j belong to one block of the partition θ

implies that the elements from T corresponding to pairs (v_i, s_a) and (v_j, s_b) belong to one block of the partition P.

Table 5.4 FSM transition table and $\boldsymbol{T}$ mapping

	00	01	11	10	(x_1,x_2)
S	v_1	v_2	v_3	v_4	V
s_1	s_1	s_2	s_4	–	
s_2	–	–	s_5	s_4	
s_3	s_3	s_2	s_1	s_3	
s_4	s_2	–	s_4	s_1	
s_5	s_3	s_1	s_4	s_2	

	V			
S	v_1	v_2	v_3	v_4
s_1	1	2	3	–
s_2	–	–	4	5
s_3	6	7	8	9
s_4	10	–	11	12
s_5	13	14	15	16

In particular, a partition P on the set T is related to the set $\{\pi, \theta\}$ if it is related to both π and θ.

Example 5.6 For FSM from Table 5.4, the partition

$$P_1 = \{\overline{1,2,3,4,5,6,7,8,9};\overline{10,11,12,13,14,15,16}\}$$

is related to the partition $\pi = \{\overline{s_1,s_2,s_3};\overline{s_4,s_5}\}$, while the partition

$$P_2 = \{\overline{1,2,6,7};\overline{10,13,14};\overline{3,4,5,8,9};\overline{11,12,15,16}\}$$

is related to the set $\{\pi, \theta\}$, and $\theta = \{\overline{v_1,v_2};\overline{v_3,v_4}\}$.

5.3.2 Implementation of Finite State Machines in FPGA's

Any FSM, $\boldsymbol{A} = (S, V, \delta, Y, \lambda)$, can be implemented as in Fig. 5.4 using an address modifier.

If $\Pi = \{\pi_1, \ldots, \pi_p\}$ is the set of two-block partitions on S and $\Theta = \{\theta_1, \ldots, \theta_m\}$ is the set of two-block partitions on V, while P_k is a partition on the set T which is related to either π_i or θ_j, then $\boldsymbol{p} = \{P_1, \ldots, P_{m+p}\}$ is the set of all partitions related to partitions $\{\pi_1, \ldots, \pi_p, \theta_1, \ldots, \theta_m\}$. Partitions in Π correspond to the state variables and partitions in Θ correspond to the input variables.

Fact 5.1 To achieve unambiguous encoding of address variables and, at the same time, maintain the consistency relation $\boldsymbol{T}$ with the transition function, two-block partitions $\boldsymbol{P} = \{P_1, \ldots, P_w\}$ have to be found, such that:

$$P_1 \cdot P_2 \cdot \ldots \cdot P_w \leq P_c. \tag{5.3}$$

This is a necessary and sufficient condition for $\{P_1, \ldots, P_w\}$ to determine the address variables. This is because each memory cell is associated with a single block of P_c, i.e., with those elements from T which map the corresponding (v, s) pairs onto the same next state.

Although some of the partitions for the ***P*** set can be selected from the ***p*** set, the selection is made in such a way that the simplest addressing unit (address modifier) is produced. Such a selection is possible thanks to the method [5, 6], based on the notion of r-admissibility [8].

5.3.3 States Encoding

Assume that u partitions $\{\pi_1, \ldots, \pi_l\}$ and $\{\pi_1, \ldots, \pi_{u-l}\}$ were chosen. These partitions correspond to the address lines driven by a single variable, either a state variable q or an external variable x. The result is the state and input symbol partial encoding; e.g.,

$$a_1 = q_1, \ldots, a_l = q_l, a_{l+1} = \theta_1, \ldots, a_u = \theta_{u-l}.$$

The encoding of state variables is possible thanks to the method of construction and coloring of weighted graphs [5].

Corollary 5.2. Inequality (5.3) can be written as:

$$P_{i_1} \cdot P_{i_2} \cdot \ldots \cdot P_{i_u} \cdot P_{i_{u+1}} \cdot \ldots \cdot P_{i_w} \leq P_C, \tag{5.4}$$

where $P_u = P_{i_1} \cdot P_{i_2} \cdot \ldots \cdot P_{i_u}$ is related to the partitions $\{\pi_1, \pi_2, \ldots, \pi_l, \theta_1, \theta_2, \ldots, \theta_{u-l}\}$.

The encoding of the part of the state variables remaining after the partial encoding (input variables, in general) can be obtained from the following rules:

$$\pi_1 \cdot \pi_2 \cdot \ldots \cdot \pi_l \cdot \pi = \pi(\mathbf{0}), \tag{5.5}$$
$$\theta_1 \cdot \theta_2 \cdot \ldots \cdot \theta_{u-l} \cdot \theta = \theta(\mathbf{0}), \tag{5.6}$$

where π and θ represent partitions corresponding to these remaining state variables. $\pi(\mathbf{0})$ as well as $\theta(\mathbf{0})$ are partitions whose blocks are equal to their elements.

Since the design process may be considered as a decomposition of the memory block into two blocks: a combinational address modifier and a smaller memory block, we need to find function G which will determine the second part of the memory address bits.

Inequality (5.4) can be transformed into:

$$P_U \cdot P_G \leq P_c. \tag{5.7}$$

Corollary 5.3 A partition P_G has to be constructed, such that:

$$P_G \geq P_V, \tag{5.8}$$

where $P_G = P_{i_{u+1}} \cdot \ldots \cdot P_{i_w}$ and P_V is related to the partition set $\{\pi, \theta\}$.

Let us assume that input variables are encoded.

Theorem 5.2 Partition P_V can be constructed in the following way:

$$P_V = P_S \cdot P_{V_\theta}, \tag{5.9}$$

where P_S is the partition related to $\pi(\mathbf{0})$ on the set of states S, and P_{V_θ} is the partition related to θ.

Proof. Let assume that $P_V = P_{V_\pi} \cdot P_{V_\theta}$ (Corollary 5.3), where P_{V_π} is related to π, and P_{V_θ} is related to θ. Since P_U and P_V satisfy $P_U \cdot P_V \leq P_c$, we have $P_U \cdot P_{V_\pi} \cdot P_{V_\theta} \leq P_c$. As a result, $P_U \cdot P_S \cdot P_{V_\theta} \leq P_c$.

Let $\langle V, R, E, P\rangle$ be a quadruple, where V is a set of elements, R is an equivalence relation on the set V, E is a set of pairs in relation P on the set V, P is a two-element relation. A triple $M(V|R, E, P)$ is a multi-graph, where $V|R$ is an equivalence class for an equivalence relation on the set V.

Since there exists an isomorphism $V|R \leftrightarrow V'$, we can construct a natural mapping from $M(V|R, E, P)$ multi-graph to $G(V', E', P)$ graph. Such a mapping $\psi: M \to G$ allows for calculation of a chromatic number $\chi(G) = \chi(M)$.

Let us apply these notions to the construction of the P_G partition. Inequality (5.7) allows us to construct a quotient partition $P_U \mid P_c$.

Corollary 5.4 The triple $\langle P_V, E_1, P_1\rangle$ – where P_V is a partition given by equation (5.9), P_1 is a relation which represents incompatibilities in quotient partition $P_U \mid P_c$ on the set T (relation of incompatibility in quotient partition $P_U \mid P_c$ is a relation among all elements in each block of the partition separately) and E_1 is the set of pairs in the relation P_1 – is a multi-graph $M_1(P_V, E_1, P_1)$.

After mapping $\psi_1: M_1 \to G_1$ we calculate a chromatic number $\chi(G_1)$ which is equal to $\chi(M_1)$.

The coloring of the G_1 graph determines the P_G partition.

Example 5.7 Let assume that input variables for the transition table are encoded (Tab. 5.4). Based on Fact 5.1, we obtain set $U = \{q_1, q_2, x_2\}$, where q_1, q_2 are internal variables generating partitions, respectively:

$$\pi_1 = \{\overline{s_1, s_2, s_4}; \overline{s_3, s_5}\}, \pi_2 = \{\overline{s_1, s_3, s_4}; \overline{s_2, s_5}\}.$$

Then

$$P_U \mid P_c = \{\overline{(1,12)(10)}; \overline{(2)(3,11)}; \overline{(5)}; \overline{(4)}; \overline{(6,9)}; \overline{(7)(8)}; \overline{(13)(16)}; \overline{(14)(15)}\}.$$

As

$$P_{V_\theta} = P(x_1) = \{\overline{1,2,6,7,10,13,14}; \overline{3,4,5,8,9,11,12,15,16}\},$$

and

$$P_S = \{\overline{1,2,3}; \overline{4,5}; \overline{6,7,8,9}; \overline{10,11,12}; \overline{13,14,15,16}\},$$

we obtain

$$P_V = \{\overline{1,2}; \overline{3}; \overline{4,5}; \overline{6,7}; \overline{8,9}; \overline{10}; \overline{11,12}; \overline{13,14}; \overline{15,16}\}.$$

According to Corollary 5.4, the M_1 multi-graph and its G_1 image shown in Fig. 5.5 is constructed. $\chi(M_1) = 3$, and thus $\mu = 5$.

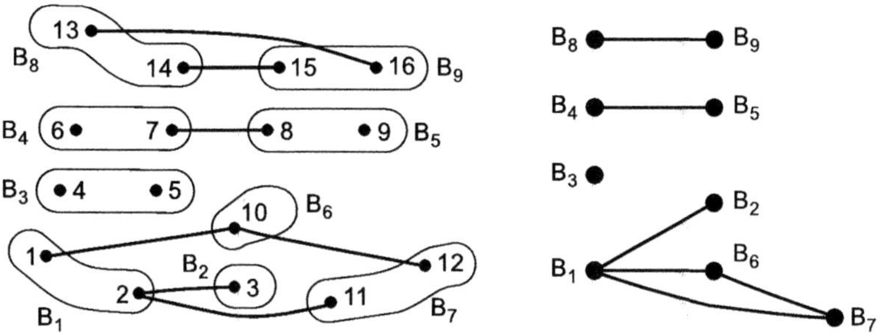

Fig. 5.5 M_1 multi-graph and G_1 graph

The value of

$$\mu = |U| + \lceil \log_2(\chi(M_1)) \rceil \tag{5.10}$$

determines the size of the memory required. In the case of $\mu > w$, a new partition P_V' has to be constructed. Then, P_V has to be multiplied by appropriately chosen two-block partitions related to those which are generated by input variables from the U set. In that case the result is a non-disjoint decomposition.

Example 5.8 Let $w = 4$. Selecting additional external variable to generate partition P_{V_θ}, we obtain:

$$P_{V_\theta} = P(x_1, x_2) = \{\overline{1,6,10,13}; \overline{2,7,14}; \overline{3,4,8,11,15}; \overline{5,9,12,16}\},$$

and then

$$P_V = \{\overline{1}; \overline{2}; \overline{3}; \overline{4}; \overline{5}; \overline{6}; \overline{7}; \overline{8}; \overline{9}; \overline{10}; \overline{11}; \overline{12}; \overline{13}; \overline{14}; \overline{15}; \overline{16}\},$$

and consequently a new multi-graph M_1. As a result $\chi(M_1) = 2$ and $\mu = 4$.

In the next step the remaining state variables are calculated.

Corollary 5.5 The triple $\langle P_S, E_2, P_2 \rangle$ – where P_S is the partition related to $\pi(0)$ on the states set S, P_2 is a relation which represents incompatibilities in quotient partition $P_{V_\theta} | P_G$ and E_2 is the set of pairs in the relation P_2 – is a multi-graph $M_2(P_S, E_2, P_2)$.

Similarly to the case discussed above, coloring of the G_2 image graph for the M_2 multi-graph yields a new partition on the S set.

Finally, this new partition is encoded with a minimal binary code. Value $\lceil \log_2(\chi(M_2)) \rceil$ determines the number of bits needed to encode the remaining state variables. Hence,

$$v = |V_\theta| + \lceil \log_2(\chi(M_2)) \rceil \tag{5.11}$$

determines the number of inputs to the address modifier.

Example 5.9 As a result of coloring the image graph G_1 for the multi-graph M_1 presented in Example 5.8, we obtain partition

$$P_G = \{\overline{1,2,5,7,9,12,14,16}; \overline{3,4,6,8,10,11,13,15}\},$$

and then

$$P_{V_\theta} \mid P_G = \{\overline{(1)(6,10,13)}; \overline{(2,7,14)}; \overline{(3,4,8,11,15)}; \overline{(5,9,12,16)}\}.$$

According to Corollary 5.5, we construct a multi-graph M_2 and its image G_2 (Fig. 5.6).

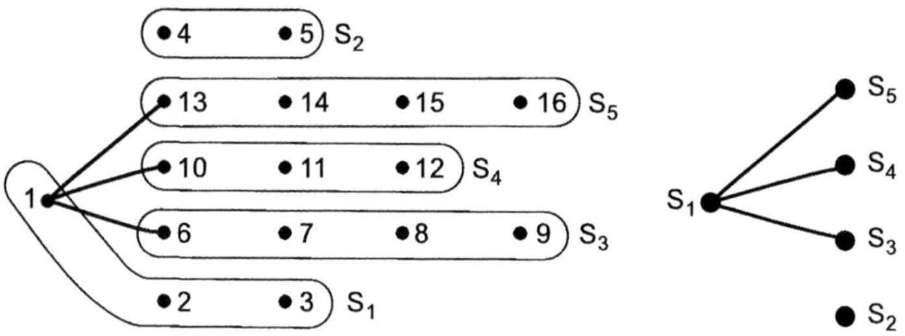

Fig. 5.6 Multi-graph M_2 and graph G_2

By coloring the image graph G_2, we obtain two possible partitions on the set S:

$$\pi^1 = \{\overline{s_1, s_2}; \overline{s_3, s_4, s_5}\}, \pi^2 = \{\overline{s_1}; \overline{s_2, s_3, s_4, s_5}\}.$$

One of those is chosen and encoded with natural binary code.

For example, partition π^1 can be generated by internal variable q_3. We encode partition π^1, so that:

$$\pi^1 = \{\overline{s_1, s_2}(0); \overline{s_3, s_4, s_5}(1)\};$$

thus

$$P_{V_{\pi^1}} = \{\overline{1,2,3,4,5}; \overline{6,7,8,9,10,11,12,13,14,15,16}\}.$$

Consequently

$$P_V = P_{V_\theta} \cdot P_{V_{\pi^1}} = \{\overline{1}; \overline{2}; \overline{3,4}; \overline{5}; \overline{6,10,13}; \overline{7,14}; \overline{8,11,15}; \overline{9,12,16}\}.$$

Finally, we can construct the truth table of address modifier and the memory ROM content.

5.3.4 *Construction of Partition P_G*

The graph G_1 can be colored in many different ways. As a consequence, many partitions P_G could be obtained. Although the number of blocks in all partitions P_G is the same, the number of address modifier inputs (Equation 5.11) could be different. The construction of a partition on the set S is similar to that of the quotient partition $P_{V_\theta} \mid P_G$. It leads to a difference in the number of remaining (after partial encoding) internal variables for different partitions P_G.

Example 5.10 Let us assume that as a result of coloring the image graph G_1 for multi-graph from Example 5.8, we obtain partition,

$$P_G = \{\overline{1,3,5,8,9,11,12,15,16};\overline{2,4,6,7,10,13,14}\};$$

then

$$P_{V_\theta} \mid P_G = \{\overline{(1)(6,10,13)};\overline{(2,7,14)};\overline{(3,8,11,15)(4)};\overline{(5,9,12,16)}\}.$$

By constructing the multi-graph M_2 and coloring its image graph G_2, we obtain the partition on the set S,

$$\pi = \{\overline{s_1};\overline{s_2};\overline{s_3,s_4,s_5}\}.$$

As a result, we need two additional internal variables for encoding this partition. Consequently, it is a worse solution than that used in Example 5.9.

A special construction of P_G partition is proposed.

Noticeably, the number of partition blocks on the states set S is closely related to the incompatibility relation in the quotient partition $P_{V_\theta} \mid P_G$. In consequence of joining P_V blocks to the P_G partition blocks, it has to be done in such a way to obtain as least incompatibilities in $P_{V_\theta} \mid P_G$ as it possibly can.

It is easily seen that $P_{V_\theta} \le P_V$. Let us calculate the quotient partition $P_{V_\theta} \mid P_V$ and remove those elements which are incompatible in the quotient partition $P_U \mid P_c$. One can show that the remaining part of P_{V_θ} could be joined to a new partition (according to $P_U \mid P_c$), that the number of blocks is less or equal $\chi(M_1)$. That new partition is so-called core of P_G. One can observe that graph G_1 has to be colored in accordance with the core of P_G. As a result, we obtain partition P_G whose number of blocks is equal to $\chi(M_1)$.

Example 5.11 Based on Example 5.8

$$P_{V_\theta} \mid P_V = \{\overline{(1)(6)(10)(13)};\overline{(2)(7)(14)};\overline{(3)(4)(8)(11)(15)};\overline{(5)(9)(12)(16)}\}.$$

According to

$$P_U \mid P_c = \{\overline{(1,12)(10)};\overline{(2)(3,11)};\overline{(4)};\overline{(5)};\overline{(6,9)};\overline{(7)(8)};\overline{(13)(16)};\overline{(14)(15)}\},$$

elements (1) and (10) are incompatible, and new P_{V_θ} is:

$$P'_{V_\theta} = \{\overline{2,7,14};\overline{3,4,8,11,15};\overline{5,9,12,16};\overline{6,10,13}\}.$$

Thus, we obtain two solutions:

$$core_1 P_G = \{\overline{2,6,7,10,13,14}; \overline{3,4,5,8,9,11,12,15,16}\},$$
$$core_2 P_G = \{\overline{2,5,7,9,12,14,16}; \overline{3,4,6,8,10,11,13,15}\}.$$

As a result

$$P_{G_1} = \{\overline{1,3,4,5,8,9,11,12,15,16}; \overline{2,6,7,10,13,14}\},$$
$$P_{G_2} = \{\overline{1,2,5,7,9,12,14,16}; \overline{3,4,6,8,10,11,13,15}\}.$$

Solution P_{G_2} was used in Example 5.9. Noticeably,

$$P_{V_\theta} \mid P_{G_1} = \{\overline{(1)(6,10,13)}; \overline{(2,7,14)}; \overline{(3,4,8,11,15)}; \overline{(5,9,12,16)}\}$$

is a similar solution.

5.3.5 Application of the Method

The general idea of the discussed FSM realization in FPGA structures lies in the decomposition into two modules: address modifier and memory microcode. The address modifier can be implemented in logic cells while the memory microcode can be implemented in EMBs configured as ROM memory.

In general, it is possible to treat the address modifier and the memory as separate combinational blocks and implement them independently, with the application of different strategies for decomposition of combinational circuits. Alternating application of serial and parallel decomposition has been shown to be extremely effective strategy to construct a structure utilizing both logic cells and EMBs.

Example 5.12 According to the presented method, in the first stage, a decomposition of the benchmark *tbk* onto two blocks has been made; the address modifier and ROM memory of the capacity of 4096 bits. In result the address modifier has been obtained represented with the truth table of 7-inputs and 5-outputs as well as the memory content of the word length of 8. Subsequently each of them has been decomposed onto the network of embedded memory blocks and logic cells. It was assumed that the EMB block had the built in register and it can also be configured as the typical combinational structure.

Fig. 5.7a exemplifies an implementation in the programmable device that has EMBs of capacity of 2048 bits each. Parallel decomposition onto two blocks of 2048 bits each was applied to realize the memory ROM content. These blocks are accurate for the EMB memories of the configuration of the word length of 4, whereas the address modifier block was implemented in EMB block of the configuration of the word length of 8. Some inputs and outputs of this block were not utilized.

Fig. 5.7b shows another implementation of the benchmark *tbk*. The programmable device used has EMB memories of the 512 bits and 4096 bits built in. Hence, it is possible to realize the memory ROM content in the block of the configuration of the word length equal 8. For the address modifier block implementation the parallel decomposition was applied which results in five one-output functions. After serial decomposition of each onto logic cells, it results: first function – one cell, function second – seven cells, third function – five cells, fourth function – six cells, fifth function – five cells. Finally, linking the second, third, fourth and fifth realizations into one function results in two blocks. The former block was implemented in the EMB memory of the capacity of 512 bits and the word length of 4, and the latter in one logic cell.

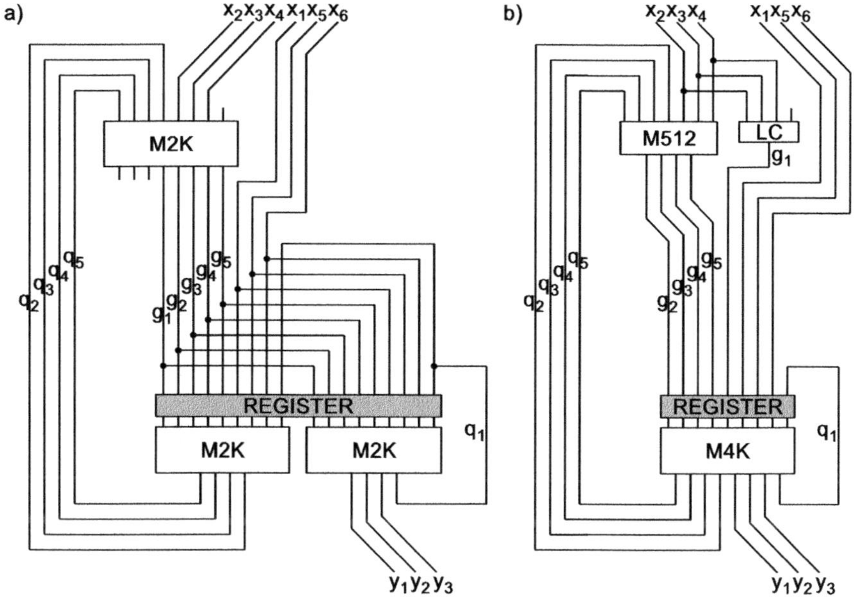

Fig. 5.7 *tbk* benchmark implementation; in programmable device a) with M2K built in memories, b) with M512 and M4K built in memories

5.4 Experimental Results

FPLD devices have a very complex architecture. They combine PLA-like structures with FPGA and memory-based structures. In many cases, designers cannot utilize all the possibilities that such complex architectures provide due to the lack of appropriate synthesis methods. Embedded memory arrays make possible an implementation of memory-like blocks, such as large registers, FIFOs, RAM or ROM modules.

These memory blocks account for a large part of the devices area. For example, Altera [31] EP20K1500E devices provide 51,840 logic cells and 442 Kbit of SRAM. Taking under consideration the conversion factors of logic elements and memory bits to logic gates (12 gates/logic element and 4 gates/memory bit), it turns out that embedded memory arrays account for over 70% of all logic resources. Since not every design consists of such modules as RAM or ROM, in many cases these resources are not utilized. Then embedded memory blocks can be used for implementation of combinational logic in a way that requires less resources than the traditional cell-based implementation. Such blocks may be used to implement "non-vital" sequential parts of the design, saving logic cell resources for more important sections. Since the size of embedded memory blocks is limited, such an implementation may require more memory than is available in a device. To reduce memory usage in ROM-based implementations, a structure with combinational logic partially implemented in the ROM and partially implemented in logic cells was proposed.

In Table 5.5, the experimental results of Daubechies' 9/7-tap bio-orthogonal filter banks are presented. All filters have 16-bit signed samples and have been designed using the DA concept [18, 27]. The presented method was used to increase efficiency of the DA tables implementations.

Table 5.5 presents the results for filter implementations using Stratix EP1S10F484C5 device. In the implementation without decomposing the filters, their functions were modeled in HDL and Quartus2 was used to map the model into the target structure. In the implementation using decomposition (denoted *dec*), a software tool implementing the described method was used to initially decompose DA tables and then the Quartus2 system was applied to map the filters into FPGA.

Table 5.5 Implementation results of 9/7 filters.

Filter	LC	EMB	Bits
alp	236	7×M512, 1×M4K	8192
alp dec	248	1×M4K	4096
ahp	204	4×M512	2048
ahp dec	210	2×M512	1024
slp	204	4×M512	2048
slp dec	211	2×M512	1024
shp	236	7×M512, 1×M4K	8192
shp dec	246	1×M4K	4096

The implementation of filters is characterized by the number of logic cells (LC) and the number of memory modules (EMB). Memory bits are also presented to give the memory usage. In all cases, decomposition reduces the size of memory and the number of memory modules. For example, an implementation of ahp filter requires 204 LCs and 4 M512 embedded memories if performed by the Quartus2 software. The presented method allows DA logic of this filter to be implemented with 2 M512 memories and 11 LCs and the whole filter with 210 LCs and 2 M512 memories.

The next proposed method for sequential machines was applied to implement several FSM's from standard benchmark set [30] in Flex10K and Stratix devices using Quartus2 (v6.0 SP1) system. In Table 5.6 and 5.7, a comparison of different FSM implementation techniques is presented.

Table 5.6 Implementation results comparing in the Flex10K structure (EPF10K10LC84-3)

Benchmark	Quartus2 Encoding (bits)	Quartus2 LUT [LC]	AM/ROM [LC/bit]	AM/ROM (new method) [LC/EMB[1)]] (Encoding bits)
cse	Auto (16)	129	2 / 5632	21 / 1 (8),
	Minimal (4)	92		9 / 2 (7)
mark1	Auto (15)	38	2 / 5120	16 / 1 (5),
	Minimal (4)	40		9 / 2 (6)
s1	Auto (20)	174	96 / 5632	107 / 1 (7),
	Minimal (5)	164		69 / 2 (7)
s386	Auto (13)	54	9 / 5632	35 / 1 (8)
	Minimal (4)	55		
tbk	Auto (32)	895[2)]	333 / 4096	266 / 1 (5),
	Minimal (5)	1077[2)]		25 / 2 (5)

[1)] device has 3 EMB blocks with 2048 bits each.
[2)] implementation not possible – not enough CLB resources.

Table 5.7 Implementation results comparing in the Stratix structure (EP1S10F484C5)

Benchmark	Quartus2 Encoding (bits)	Quartus2 LUT [LC]	AM/ROM [LC/bit]	AM/ROM (new method) [LC/EMB[1)]] (Encoding bits)
cse	Auto (16)	112	2 / 5632	77 / 1 M512s (4),
	Minimal (4)	90		8 / 1 M4Ks (7)
mark1	Auto (15)	32	2 / 5120	8 / 1 M4Ks (6)
	Minimal (4)	38		
s1	Auto (20)	168	96 / 5632	129 / 1 M512s (5),
	Minimal (5)	152		71 / 1 M4Ks (7)
tbk	Auto (32)	902	333 / 4096	261 / 1 M512s (5),
	Minimal (5)	959		21 / 1 M4Ks (5)

[1)] device has 920448 bits of memory in 512 bit blocks (M512s) as well as in 4096 bit blocks (M4Ks).

The 'LUT' column (Tab. 5.6, 5.7) shows the number of logic cells required to implement a given FSM in the "traditional" way using logic cells only. In this case two different state encoding methods available in Quartus2 were applied. In column 'AM/ROM', the results of implementation of a given FSM using the concept of address modifier are presented. In this approach, the address modifier can be implemented using logic cells, and ROM can be implemented with EMB blocks. The number of logic cells and the number of memory bits are presented in this column. It can be easily observed that decomposition improves the quality of implementation in ROM resources, as well as the quality of implementation in logic cells only. An improvement of this method by using decomposition into the mixed structure built of LCs and EMBs with application of state encoding method presented in this article is shown in the last column (Tab. 5.6, 5.7). The number of state encoding bits is presented in brackets.

For some examples more than one solution is possible, e.g., benchmark *cse* can be implemented in memory of size 32,768 bits (not available in small programmable devices) or with the use of 90 logic cells (Stratix), 92 logic cells (Flex10K). However, application of the new method allows for an implementation with 21 logic cells and 1 EMB or with 9 logic cells and 2 EMBs when using a Flex10K device or 77 logic cells and 1 M512 or 8 logic cells and 1 M4K when using a Stratix device.

Noticeably, the new approach allows for much more efficient utilization of available resources. It is also possible to trade off the number of logic cells used with the number of embedded memory blocks.

5.5 Conclusions

The modern programmable structures deliver the possibilities to implement digital circuits in dedicated embedded blocks. This makes designing of such circuits an easy task. However the flexibility of programmable structures enables more advanced implementation methods to be used. In particular, best results can be obtained by utilizing the parallelisms in implemented algorithms and by applying advanced synthesis based on decomposition methods. In case, the designed circuit contains complex combinational blocks, the influence of the design methodology and decomposition synthesis methods on the efficiency of practical digital circuit implementation is extremely significant. This is typical for many practical designs i.e. when implementing digital filters using the DA concept or ROM-based FSMs with address modifier.

The most efficient approach to logic synthesis of combinational and sequential circuits relies on the effectiveness of the functional decomposition synthesis methods. Although these methods were already used in decomposition algorithms, they were never applied with a technology specific mapper targeted at FPGA structure together. This paper shows that it is possible to apply the functional decomposition method for the synthesis of FPGA-based circuits directed towards area and delay optimization.

References

[1] Adamski, M., Barkalov, A.: Architectural and Sequential Synthesis of Digital Devices. University of Zielona Góra Press, Zielona Góra (2006)
[2] Ashar, P., Devadas, S., Newton, A.R.: Sequential Logic Synthesis. Kluwer Academic Publishers, Norwell (1992)

[3] Astola, J.T., Stankovi , R.S.: Fundamentals of Switching Theory and Logic Design. Springer, Heidelberg (2006)
[4] Barkalov, A., W grzyn, M.: Design of control units with programmable logic. University of Zielona Góra Press, Zielona Góra (2006)
[5] Borowik, G.: Finite state machines synthesis for FPGA structures with embedded memory blocks. PhD thesis, Faculty of Electronics and Information Technology, Warsaw University of Technology, Warsaw (2007) (in Polish)
[6] Borowik, G., Falkowski, B., Łuba, T.: Cost-efficient synthesis for sequential circuits implemented using embedded memory blocks of FPGAs. In: Design and Diagnostics of Electronic Circuits and Systems, pp. 99–104. IEEE, Kraków (2007)
[7] Borowik, G., Rawski, M.: Effective implementation of sequential circuits using FPGAs with embedded memory blocks. In: Programmable Devices and Systems, pp. 175–180. IFAC, Kraków (2004)
[8] Brzozowski, J.A., Łuba, T.: Decomposition of Boolean Functions Specified by Cubes. Journal of Multiple Valued Logic and Soft Computing 9(4), 377–418 (2003)
[9] Cong, J., Yan, K.: Synthesis for FPGAs with embedded memory blocks. In: Proceedings of the 2000 ACM/SIGDA Eighth International Symposium on Field Programmable Gate Arrays, pp. 75–82 (2000)
[10] Czerwinski, R., Kania, D.: State assignment for PAL-based CPLDs. In: Proceedings of the 8th Euromicro Conference on Digital System Design, pp. 127–134 (2005)
[11] De Micheli, G.: Synthesis and Optimization of Digital Circuits. McGraw-Hill Higher Education (1994)
[12] De Micheli, G., Brayton, R.K., Sangiovanni-Vincentelli, A.L.: Optimal state assignment for finite state machines. IEEE Trans. on CAD of Integrated Circuits and Systems 4(3), 269–285 (1985)
[13] Hartmanis, J., Stearns, R.E.: Algebraic structure theory of sequential machines. Prentice-Hall international series in applied mathematics. Prentice-Hall, Inc., Upper Saddle River (1966)
[14] Krishnamoorthy, S., Tessier, R.: Technology mapping algorithms for hybrid FPGAs containing lookup tables and PLAs. IEEE Trans. on CAD of Integrated Circuits and Systems 22(5), 545–559 (2003)
[15] Łuba, T.: Multi-level logic synthesis based on decomposition. Microprocessors and Microsystems 18(8), 429–437 (1994)
[16] Łuba, T., Górski, K., Wronski, L.B.: ROM-based finite state machines with PLA address modifiers. In: EURO-DAC, pp. 272–277. IEEE Computer Society Press, Los Alamitos (1992)
[17] Łuba, T., Selvaraj, H.: A general approach to boolean function decomposition and its applications in FPGA-based synthesis. VLSI Design 3(3-4), 289–300 (1995)
[18] Meyer-Baese, U.: Digital Signal Processing with Field Programmable Gate Arrays. In: Signals and Communication Technology, 2nd edn., Springer-Verlag New York, Inc., Secaucus (2004)
[19] Nowicka, M., Łuba, T., Rawski, M.: FPGA-based decomposition of Boolean functions. algorithms and implementation. Advanced Computer Systems, 502–509 (1999)
[20] Rawski, M., Falkowski, B.J., Łuba, T.: Digital signal processing designing for FPGA architectures. Facta Universitas 20(3), 437–459 (2007)

[21] Rawski, M., Łuba, T., Jachna, Z., Tomaszewicz, P.: The influence of functional decomposition on modern digital design process. In: Adamski, M.A., Karatkevich, A., Węgrzyn, M. (eds.) Design of Embedded Control Systems, pp. 193–203. Springer, US (2005)
[22] Rawski, M., Selvaraj, H., Łuba, T.: An application of functional decomposition in rom-based fsm implementation in fpga devices. Journal of Systems Architecture 51(6-7), 424–434 (2005)
[23] Rawski, M., Selvaraj, H., Łuba, T., Szotkowski, P.: Multilevel synthesis of finite state machines based on symbolic functional decomposition. International Journal of Computational Intelligence and Applications 6(2), 257–271 (2006)
[24] Rawski, M., Tomaszewicz, P., Selvaraj, H., Łuba, T.: Efficient implementation of digital filters with use of advanced synthesis methods targeted FPGA architectures. In: DSD, pp. 460–466. IEEE Computer Society, Washington (2005)
[25] Sasao, T., Iguchi, Y., Suzuki, T.: On LUT cascade realizations of FIR filters. In: DSD 2005, pp. 467–475. IEEE Computer Society, Washington (2005)
[26] Scholl, C.: Functional Decomposition with Application to FPGA Synthesis. Kluwer Academic Publishers, Norwell (2001)
[27] Tomaszewicz, P., Nowicka, M., Falkowski, B.J., Łuba, T.: Logic synthesis importance in FPGA-based designing of image signal processing systems. In: MIXDES, pp. 141–146 (2007)
[28] Villa, T., Sangiovanni-Vincentelli, A.L.: NOVA: state assignment of finite state machines for optimal two-level logic implementation. IEEE Trans. on CAD of Integrated Circuits and Systems 9(9), 905–924 (1990)
[29] Wilton, S.J.E.: SMAP: Heterogeneous technology mapping for area reduction in FPGAs with embedded memory arrays. In: FPGA, pp. 171–178 (1998)
[30] Yang, S.: Logic synthesis and optimization benchmarks user guide version 3.0. Tech. rep., Microelectronics Center of North Carolina (1991)
[31] Altera Corporation (2010), `http://www.altera.com`

6 Efficient Technology Mapping Method for PAL-Based Devices

Dariusz Kania

Silesian University of Technology, Institute of Electronics,
ul. Akademicka 16, 44-100 Gliwice, Poland
e-mail: dkania@polsl.pl

Abstract. The core of a contemporary CPLD device is a PAL-based logic block which consists of a programmable AND matrix and a fixed OR matrix. A new technology mapping method for PAL-based devices based on the analysis of graph of outputs is described. The presented approach uses original method for illustrating a minimized form of a multi-output Boolean function. Graph node represents groups of multiple-output implicants with common output part. The essence of the method is the process of searching for appropriate multi-output implicants that can be shared by several functions. A new method for the description of cascaded feedback connections is presented. The experimental results show that the proposed algorithm leads to significant reduction of chip area used by resulting circuits.

6.1 Introduction

The most of commercially available CPLDs (Complex Programmable Logic Devices) consist of PAL-based logic blocks (Fig. 6.1). This type of circuits will be referenced as PAL-based CPLDs in contrast to less popular group of a PLA-based CPLDs [22].

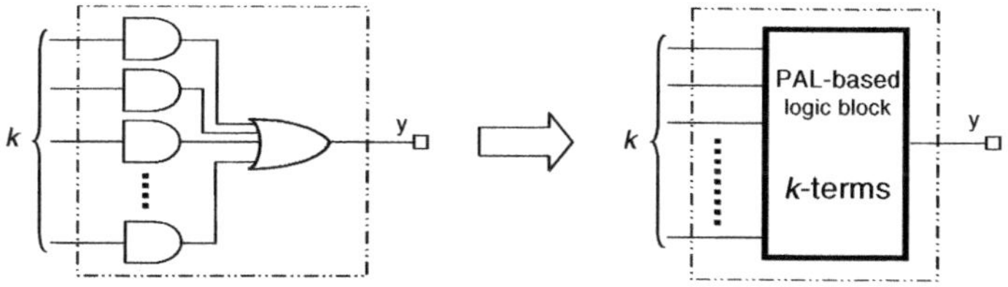

Fig. 6.1 The structure and block diagram of PAL-based logic block with k-terms

The classical method of logic synthesis, dedicated for PAL-based CPLDs, consists of two steps. First a two-level minimisation is applied separately to every

M. Adamski et al. (Eds.): Design of Digital Systems and Devices, LNEE 79, pp. 145–163.
springerlink.com

single-output function. Next implementation of the minimised functions in PAL-based blocks containing a predefined number of product terms is performed. The two-level minimization algorithms based on Quine-McCluskey approach, like e.g. Espresso do not support any technology mapping features. Technology mapping is done afterwards independently. If the number of implicants p, representing a function after minimisation, is greater than the number of product terms k, available in a logic block (Fig. 6.1), a greater number of logic blocks have to be used to implement the function. The classical product term expansion method consists in introducing cascaded feedback connections, increasing propagation delays between inputs and outputs [3]. Some other concepts of product term expansion are also presented in literature, e. g. in [3, 14].

Some minimization, decomposition and partitioning methods dedicated for PLDs, especially for FPGA and PLA devices, together with algorithms and results can be found in [1, 4, 7, 9, 10, 13, 14, 20, 24, 25]. Sometimes, algorithms developed for LUT-based FPGAs are directly adapted to other PLD architectures [1, 4]. In methods based on single-level decomposition of Boolean function minimization and partitioning problems are considered simultaneously [9, 11]. The partitioning is based on finding the minimum number of input variables needed to produce a group of output function. The primary objective of these methods is minimization of PLA area. This idea is conceptually similar to that of decomposition of PLA matrices [5, 6, 18]. In both approaches a single two-level Boolean function (PLA) is decomposed into two stages of cascaded PLA's such that their total area is smaller than that of the original PLA [5, 6, 7, 8, 20, 24, 25].

Sometimes, decomposition strategy leads to a multi-level implementation. Decomposition of Boolean function for different universal logic blocks based on PLA architecture is also known [10, 12, 13]. This decomposition consists of partitioning the set of outputs into two or more disjoint subsets. Such decomposition separates a multiple output Boolean function into two or more component, so that each component function can be implemented with separate building block. This strategy is dedicated for PLA building blocks. All experimental results were presented for hypothetical PLA blocks that do not exist in commercially available CPLDs.

Sometimes decomposition methods are dedicated for FSMs [6, 12, 20, 21, 26]. A characteristic feature of algorithms of this kind is a process of appropriate coding of inputs and outputs, which significantly influences minimisation of product term numbers in blocks obtained as the result of decomposition [6, 20, 21, 26]. Problems of appropriate input and output coding are widely discussed in connection with issues concerning coding of internal states in FSMs (Finite State Machine) [2, 19, 23]. The problems are among other things related to symbolic internal state coding, the theory of dichotomy, multi-valued function minimisation, analysis of output dominance, etc. [2, 19].

The main limitation of PAL-based logic blocks is not the number of inputs but the number of multi-input product terms available in one block. Mapping a large number of Boolean functions to minimal number of technology-oriented logic blocks is a difficult task. Various technology-mapping tools are available for look-up-based FPGAs or PLA-based CPLDs [10]. Most of these tools allow the number of inputs per look-up table or number of inputs, outputs and terms per PLA block to be specified by the user, but none of them map to PAL-based logic blocks directly. They are not suitable for commercially available CPLDs, such as the for example Altera and Cypress MAX family.

The essence of synthesis dedicated for PAL-based structures comprises two major tasks: minimising the number of PAL-based logic blocks used and adjusting the designed circuit to fit the structures of PAL-based blocks best. The proposed technology mapping concerns simultaneously above issues. The objective of this chapter is to present technology mapping method for PAL-based CPLDs. The method consists in searching for the common multi-output implicants [19] that is carried out after having completed the classical two-level minimization of the multi-output function by means of the Espresso algorithm.

6.2 Theoretical Backgrounds

Let f be a multi-output logic function $f:B^n \to B^m$, where $B=\{0,1\}$. The classical implementation of the function $f:B^n \to B^m$ within the PAL-based structures is related to implementation of the minimized functions $f_o:B^n \to B^1$ ($o=1,2,\ldots,m$) by means of the PAL-based logic blocks. Let the discriminant Δ_{fo} be the decimal number equal to the number of those implicants, for which single-output function $f_o:B^n \to B^1$ constitutes true $\{1\}$ values.

Let δfo denote the number of logic blocks necessary for implementation of the oth function. In the case when $\Delta fo > k$, implementation of the fo function by means of the PAL-based logic blocks consisting of k terms needs the realization of cascaded feedback connections.

The number of $\delta_{fo}=\lceil(\Delta_{fo}-k)/(k-1)\rceil+1$ PAL-based logic blocks consisting of k-terms will be used, where symbol $\lceil x \rceil$ denotes the lowest integer number not less than x. For implementation of m-functions (every function has been minimized separately), implementation of δ^1_f PAL-based logic blocks is necessary, where $\delta^1_f = \sum_{o=1}^{m}\left(\left\lceil \frac{\Delta_{f_o}-k}{k-1} \right\rceil +1\right)$.

The minimized form of multi-output functions $f:B^n \to B^m$ can be described by a set of multi-output implicants, including an input part consisting of components {0,1,-} and an output part consisting of {0,1} components [19]. Let y be an m-component output vector that is associated with the output part of the multi-output implicant. The number of the same y vectors that constitute the subset of multi-output implicants defining the $f:B^n \to B^m$ function will be called the discriminant Δ_y. Let $\mu(\Delta_y)$ (range of Δ_y discriminant) be a decimal number equal to the number of {1} components included in the y vector. Let's assume, that $G<Y,\vec{U}>$ is the primary directed graph (Fig. 6.2), where Y is the set of all the graph nodes Δ_y, while $\vec{U}$ is a set of graph edges connecting such nodes of the graph Δ_{ys}, Δ_{yr} that the code distance between the y_s, y_r vectors is 1, and $\mu(\Delta_{ys})+1=\mu(\Delta_{yr})$.

f:$B^5 \rightarrow B^4$ (*f*.pla)

```
.i 5
.o 4
.ilb a b c d e
.ob f4 f3 f2 f1
.p 10
00-01 1111
01000 1111
01011 1111
00010 1110
-111- 1110
1100- 1010
10011 1010
-0101 1000
11-00 0100
0010- 0010
.e
```

$\mu(\Delta_y)=4 \; \Delta_{1111}=3$

$\mu(\Delta_y)=3 \; \Delta_{1110}=2; \Delta_{1101}=0; \Delta_{1011}=0; \Delta_{0111}=0$

$\mu(\Delta_y)=2 \; \Delta_{1100}=0; \Delta_{1010}=2; \Delta_{1001}=0; \Delta_{0110}=0; \Delta_{0101}=0; \Delta_{0011}=0$

$\mu(\Delta_y)=1 \; \Delta_{1000}=1; \Delta_{0100}=1; \Delta_{0010}=1; \Delta_{0001}=0$

$$\mathbf{Y}=\{\Delta_{1111},\Delta_{1110},\Delta_{1101},\Delta_{1011},\Delta_{0111},\Delta_{1100},\Delta_{1010},\Delta_{1001},\Delta_{0110},\Delta_{0101},\Delta_{0011},\Delta_{1000},\Delta_{0100},\Delta_{0010},\Delta_{0001}\}$$

$$\vec{\mathbf{U}}=\{(\Delta_{1000},\Delta_{1100}); (\Delta_{1000},\Delta_{1010}); (\Delta_{1000},\Delta_{1001}); (\Delta_{0100},\Delta_{1100}); (\Delta_{0100},\Delta_{0110}); (\Delta_{0100},\Delta_{0101}); (\Delta_{0010},\Delta_{1010}); (\Delta_{0010},\Delta_{0110}); (\Delta_{0010},\Delta_{0011}); (\Delta_{0001},\Delta_{1001}); (\Delta_{0001},\Delta_{0101}); (\Delta_{0001},\Delta_{0011}); (\Delta_{1100},\Delta_{1110}); (\Delta_{1010},\Delta_{1110}); (\Delta_{0110},\Delta_{1110}); (\Delta_{1100},\Delta_{1101}); (\Delta_{0101},\Delta_{1101}); (\Delta_{1001},\Delta_{1101}); (\Delta_{1010},\Delta_{1011}); (\Delta_{1001},\Delta_{1011}); (\Delta_{0011},\Delta_{1011}); (\Delta_{0110},\Delta_{0111}); (\Delta_{0101},\Delta_{0111}); (\Delta_{0011},\Delta_{0111}); (\Delta_{1110},\Delta_{1111}); (\Delta_{1101},\Delta_{1111}); (\Delta_{1011},\Delta_{1111}); (\Delta_{0111},\Delta_{1111})\}$$

$$G < Y, \vec{U} >$$

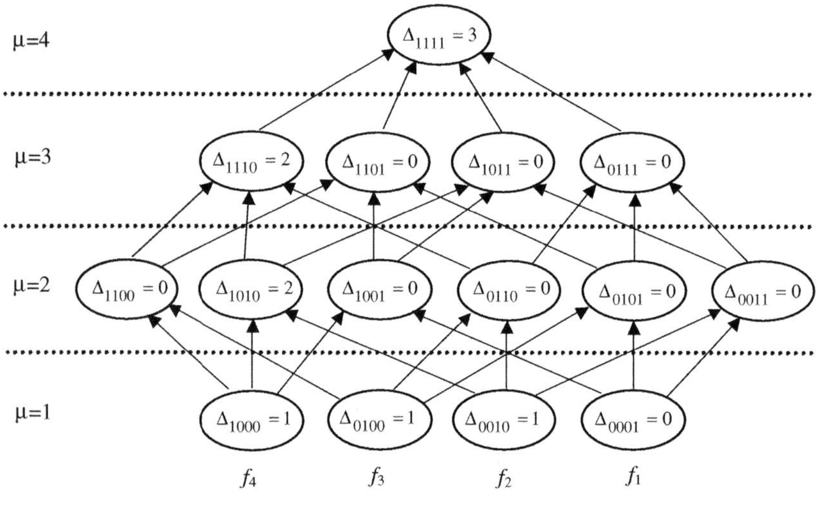

Fig. 6.2 Representation of the minimized function *f*:$B^5 \rightarrow B^4$ by means of the primary graph of outputs

By means of elimination from the primary graph such nodes, where $\Delta_y = 0$, we obtain the reduced graph presented in the Fig. 6.3a. For simplification, the nodes of the graph contain decimal value of discriminants only (Fig. 6.3b). In subsequent part of this paper such reduced graph of outputs will be called graph of outputs.

Every node of the first range that is related to the implicants of the oth output of the m-output function can be associated with the decimal value of Δ_o^m equal to the sum of discriminants included in nodes covered by all the paths starting from this node and ending in nodes of the upper ranges (Fig. 6.3).

Based on values of discriminants Δ_o^m, the number of PAL-based logic blocks, which are necessary for implementation of the multi-output function, can be calculated (for every function separately, after having minimized the multi-output function). This number is equal to $\delta_f = \sum_{o=1}^{m} \left(\left\lceil \frac{\Delta_o^m - k}{k-1} \right\rceil + 1 \right)$ and, for most cases, is greater than δ^1_f. For implementation of four-output function presented in Fig.6.3 the number of $\delta_f = \sum_{o=1}^{4} \left(\left\lceil \frac{\Delta_o^4 - k}{k-1} \right\rceil + 1 \right) = 4+3+4+1 = 12$ PAL-based logic blocks with three terms (k=3) is necessary.

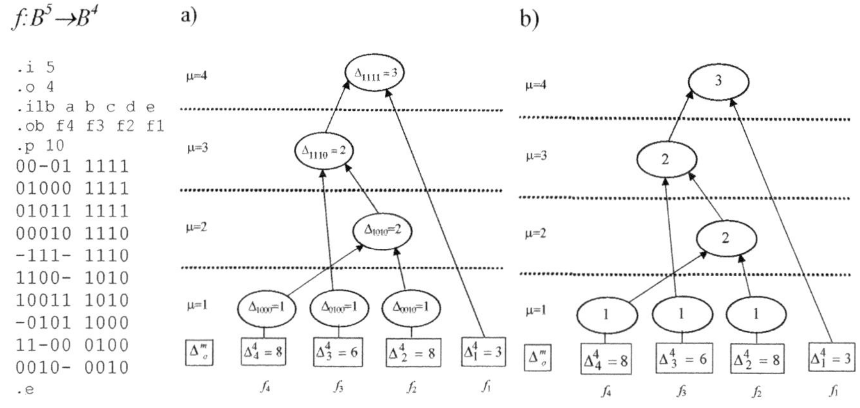

Fig. 6.3 The reduced graph of outputs of the minimized function $f{:}B^5 \rightarrow B^4$ with associated values of discriminants Δ_o^m

Apparently, on the basis of analysis of the graph, solutions that use fewer logic blocks can be found. Nodes of the graph correspond to the number of multi-output implicants associated with one of the multi-output vectors. For example, when a node of the μth range belongs to the graph and for that node $\Delta_y = k$, implementation of k implicants constituting common resources of the μ functions, is possible within one block (Fig. 6.4b). After selection of the node transformation of the graph is made leading to reduction of the Δ_o^m coefficients (Fig. 6.4a,c).

As a result, the selection of a certain node introduces cascaded feedback connection, which is shown on the reduced graph by the fourth-range nodes marked on the graph with the dashed line (Fig.6.4c).

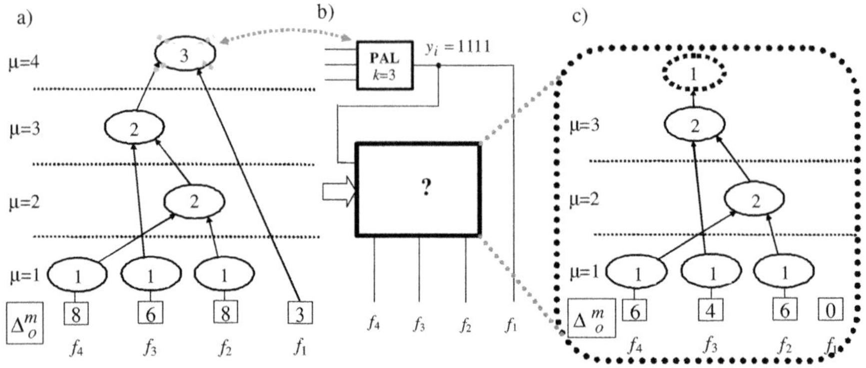

Fig. 6.4 a) Example of reduced graph of outputs; b) Implementation of the implicants defined by the fourth-range node; c) The graph after reduction

6.3 Technology Mapping Algorithm

Let ${}^{i}\Delta_y$ be the discriminants that correspond to the node, which is chosen during i^{th} step of the algorithm of implementation of the multi-output function. Implementation of the group of implicants which correspond to the selected node ${}^{i}\Delta_y$, during the ith step of the technology mapping algorithm, may lead to minimization of the number of used PAL-based blocks consisting of k terms, if the requirement ${}^{i}\delta_f - {}^{i+1}\delta_f > \lceil ({}^{i}\Delta_y - k)/(k-1) \rceil + 1$ is met.

Since selection of the node ${}^{i}\Delta_y$ affects $\mu({}^{i}\Delta_y)$ discriminants ${}^{i}\Delta_o^m$, the condition for minimization of the PAL-based logic blocks (after having the discriminants re-ordered in such a way, that the selected node affects the consecutive ${}^{i}\Delta_j^{\mu({}^{i}\Delta_y)}$ discriminants), can be shown in the following form:

$$\sum_{j=1}^{\mu({}^{i}\Delta_y)} \left(\left\lceil \frac{{}^{i}\Delta_j^{\mu({}^{i}\Delta_y)} - k}{k-1} \right\rceil + 1 \right) - \sum_{j=1}^{\mu({}^{i}\Delta_y)} \left(\left\lceil \frac{{}^{i+1}\Delta_j^{\mu({}^{i}\Delta_y)} - k}{k-1} \right\rceil + 1 \right) > \left\lceil \frac{{}^{i}\Delta_y - k}{k-1} \right\rceil + 1 \quad (6.1)$$

Let $r_j^{\mu({}^{i}\Delta_y)}$ be numbers calculated from the congruence: ${}^{i}\Delta_j^{\mu({}^{i}\Delta_y)} - 1 \equiv r_j^{\mu({}^{i}\Delta_y)} (\mathrm{mod}(k-1))$ where $j=1,2,\ldots,\mu({}^{i}\Delta_y)$.

Theorem: (*about selection of the node of the graph of outputs*)

If there exists a node of the graph (i.e. discriminant) ${}^i\Delta_y$, for which

1. the range $\mu({}^i\Delta_y)\geq 2$ and ${}^i\Delta_y \geq k$ or
2. the range $\mu({}^i\Delta_y)\geq 2$ and within the set of remainders $\mathbf{R} = \{r_j^{\mu({}^i\Delta_y)};\ j \in <1, \mu({}^i\Delta_y)>\}$ there exist at least two such remainders, that $0 < r_a^{\mu({}^i\Delta_y)} < {}^i\Delta_y < k$ and at the same time $0 < r_b^{\mu({}^i\Delta_y)} < {}^i\Delta_y < k$ or
3. the range $\mu({}^i\Delta_y) = 2$ and ${}^i\Delta_y = k$ or ${}^i\Delta_y \geq 2k-1$ or
4. the range $\mu({}^i\Delta_y) = 2$ and $k < {}^i\Delta_y < 2k-1$ and within the set of remainders $\mathbf{R} = \{r_j^{\mu({}^i\Delta_y)};\ j \in <1, \mu({}^i\Delta_y)>\}$ there exists at least one such remainders, that $0 < r_a^{\mu({}^i\Delta_y)} < {}^i\Delta_y - (k-1)$

then implementation of the implicants associated with that particular discriminant leads to a reduced number of the PAL-based blocks consisting of k terms necessary for implementation of given multi-output function.

Proof of above theorem is presented in appendix.

The essence of the theorem presented above consists in determination, which are the provisions that should be imposed onto the discriminant in order to provide that realization of the multi-output implicants associated with it, would lead to minimization of the number of PAL-based logic block in use. The theorem serves as a background to draw up a technology-mapping algorithm for multi-level implementation of multi-output logic functions by means of PAL-based logic blocks.

As number of logic blocks $\delta_f = \sum_{o=1}^{m}\left(\left\lceil \frac{\Delta_o^m - k}{k-1} \right\rceil + 1\right)$ is generally greater than the value of $\delta_f^1 = \sum_{o=1}^{m}\left(\left\lceil \frac{\Delta_{f_o} - k}{k-1} \right\rceil + 1\right)$, the main concept of the algorithm proposed consists in analysis of the graph nodes and searching for the nodes that are associated with possible large groups of common implicants. Let us assume that the graph of outputs describes multi-output implicants that are accomplished by means of ${}^i\delta_f$ PAL-based logic blocks containing k terms each. Selection, at the ith step of iteration, the ${}^i\Delta_y$ node of the graph implies utilization of ${}^i\gamma = \lceil({}^i\Delta_y\text{-}k)/(k\text{-}1)\rceil + 1$ PAL-based logic blocks. This leads to reduction of the graph of outputs, and the graph, after reduction, describes the implicants that are covered by ${}^{i+1}\delta_f$ PAL-based logic blocks. The higher is the value of the expression ${}^i\delta_f - ({}^{i+1}\delta_f + {}^i\gamma)$ the better is the selection of the node in question.

The rules for selection of the node can be deduced directly from the theorem about selection of the node of the graph of outputs and can be listed in the following way:

1. Firstly, at the very beginning, one has to choose the ${}^{i}\Delta_{y}$ node, for which $\mu({}^{i}\Delta_{y}) = \max$ (obviously, the node must meet provisions of the theorem).
2. From the nodes of the same range, further selection must be carried out depending on values of discriminants:
   2.a. if there exist nodes, for which ${}^{i}\Delta_{y} \geq k$ - the node, for which the discriminant ${}^{i}\Delta_{y} = \max$
   2.b. if values of all the discriminants are lower than k - the node, for which within the set of remainders $\mathbf{R} = \{r_j^{\mu({}^{i}\Delta_y)};\ j \in <1, \mu({}^{i}\Delta_y)>\}$ there exists the maximum number of remainders $r_x^{\mu({}^{i}\Delta_y)}$ that meet the condition $0 < r_x^{\mu({}^{i}\Delta_y)} < {}^{i}\Delta_y < k$

The theorem and conclusions about choosing nodes of the graph of outputs serve as the basis for the technology-mapping algorithm of multi-level synthesis implemented in PALDec system.

Example

Let us consider the multi-output function $f{:}B^5 \rightarrow B^4$ which, after minimization (Espresso) can be depicted in the file *f.pla* (Fig. 6.5a). The reduced graph of outputs, associated to it, is shown in Fig. 6.5.b. Direct realization of implicants with PAL-based logic blocks that contain k=3 terms each (by means of the classical method, after minimization the multi-output function) needs $^{0}\delta_f = \sum_{o=1}^{4}\left(\left\lceil \frac{{}^{0}\Delta_o^4 - k}{k-1} \right\rceil + 1\right) = 4+3+4+1 = 12$ PAL-based logic blocks. During the first step of the proposed algorithm, the implicants associated with the node Δ_{1111}=3 are realized. This step involves only one PAL-based logic block that contains 3 terms (${}^{1}\gamma = 1$) and leads to significant lowering of the number of blocks that are necessary for direct realization of implicants depicted by the reduced graph ${}^{1}\delta_f + {}^{1}\gamma = \sum_{o=1}^{4}\left(\left\lceil \frac{{}^{1}\Delta_o^4 - k}{k-1} \right\rceil + 1\right) + {}^{1}\gamma = 3+2+3+0+1 = 9 \leq {}^{0}\delta_f$. This graph is obtained after removing the node Δ_{1111}=3 and introducing an additional node, which is connected with the node Δ_{1110} and represents the cascaded feedback connection (Fig. 6.5c).

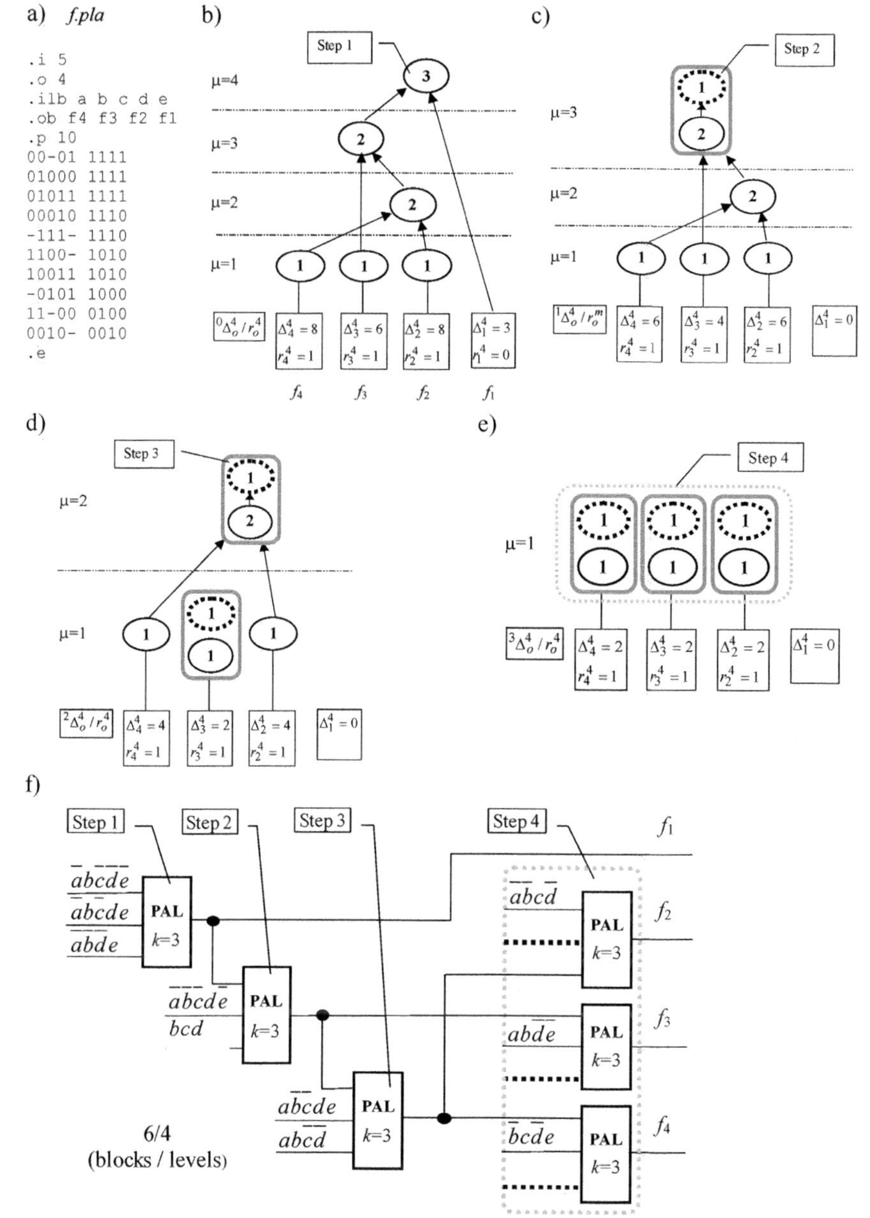

Fig. 6.5 Realization of function $f{:}B^5 \to B^4$ a) description of the function by the *f.pla* file; b) reduced graph of outputs, c,d,e) graphs of outputs that correspond to successive steps of technology-mapping process, f) final structure of the circuit

The implicants that correspond to the nodes Δ_{1110}, Δ_{1010} (Fig.6.5c,d) are realized during the subsequent steps of circuit synthesis. Realization of those implicants leads to further reduction of the required PAL-based logic blocks number

$$^2\delta_f + {}^2\gamma = \sum_{o=1}^{4}\left(\left\lceil \frac{{}^2\Delta_o^m - k}{k-1} \right\rceil + 1\right) + {}^2\gamma = 2+1+2+0+1 = 6 \le {}^1\delta_f \,;$$

$$^3\delta_f + {}^3\gamma = \sum_{o=1}^{4}\left(\left\lceil \frac{{}^3\Delta_o^m - k}{k-1} \right\rceil + 1\right) + {}^3\gamma = 1+1+1+0+1 = 4 \le {}^2\delta_f \,.$$

The nodes that correspond to the 1st range (Fig.6.5e) are implemented at the last stage of the synthesis process. The final circuit representation that uses $\delta_f = {}^3\delta_f + \sum_{i=1}^{3} {}^i\gamma < \delta_f^1$ PAL-based logic blocks is shown in Fig. 6.5f.

6.4 Experimental Results

Some experimental results that were obtained when using the software implementation of the proposed technology mapping method are presented in Table 6.1. The columns contain numbers of PAL-based logic blocks that have *k* terms, which are necessary for implementation of respective benchmarks (columns marked B) as well as numbers of logic levels (columns marked L). The results from new method (PALDec) are compared with classical technology mapping approach (Classical).

Within the set of 88 experiments for which results of technology mapping had been compared for the both methods, the proposed approach (PALDec) led to 75 solutions (85%) that used less number of logic blocks than the classical technique (Classical). Significant discrepancies can be observed if the logic blocks with low number of terms have been exploited.

Having compared the overall number of logic blocks that are used by all the benchmarks under experiments, one can observe that the PALDec method is more efficient if the PAL-based logic blocks with less number of terms are used (k = 3, 4, 5). The above observation seems to be intuitively obvious, as smaller groups of multi–output implicants that have the same output part, more frequently are obtained as a result of minimization. The common coverage of such implicants brings more benefits if the structures that contain small PAL-based logic blocks are in use.

Disadvantageously, all the circuits that were examined by means of the classical technology mapping method proved to contain less or equal number of logic levels as compared to the results of the PALDec method.

Table 6.1 Results of benchmark synthesis for the PAL-based logic blocks with *k*-terms (*see in the text*)

Classical	k= 3		K= 4		k= 5		K= 6		k= 7		k= 8		k= 9		k= 10	
	B	L	B	L	B	L	B	L	B	L	B	L	B	L	B	L
Alu4	315	5	211	4	159	4	128	3	107	3	91	3	82	3	72	3
Clip	73	4	49	3	38	3	30	3	26	2	22	2	21	2	20	2
Duke2	99	3	72	3	61	2	49	2	44	2	41	2	39	2	36	2
misex3	612	5	410	4	309	4	249	3	208	3	178	3	157	3	143	3
Rd73	70	4	47	3	36	3	29	3	24	3	20	2	19	2	16	2
Rd84	142	5	95	4	72	4	58	3	49	3	42	3	28	2	21	2
sao2	36	3	24	3	19	2	15	2	14	2	11	2	11	2	10	2
seq	691	5	469	4	355	3	287	3	244	3	212	3	188	3	172	3
spla	425	6	293	5	227	4	188	4	163	3	144	3	128	3	120	3
table3	264	4	181	4	135	3	110	3	94	3	80	3	71	2	65	2
table5	272	4	183	4	141	3	112	3	95	3	82	3	74	2	65	2
Σ	2999		2034		1552		1255		1068		923		818		740	
Σ		48		41		35		32		30		29		26		26

PALDec	k= 3		k= 4		k= 5		k= 6		k= 7		k= 8		K= 9		k= 10	
	B	L	B	L	B	L	B	L	B	L	B	L	B	L	B	L
alu4	290	8	193	6	148	6	119	5	98	5	85	4	77	4	67	4
clip	66	5	45	4	35	4	29	4	26	3	23	2	21	2	20	2
duke2	100	4	76	4	64	3	55	3	49	2	46	2	43	2	40	2
misex3	450	8	316	7	257	7	219	6	189	6	170	6	153	4	136	4
rd73	63	6	44	5	32	4	26	4	23	3	19	3	18	3	16	3
rd84	130	8	87	6	67	5	54	4	45	4	40	4	27	3	20	3
sao2	29	5	20	4	16	4	13	4	11	3	10	3	9	3	9	3
Seq	295	8	208	7	165	7	142	6	129	5	114	6	109	6	105	6
Spla	285	6	190	4	158	4	137	4	124	4	115	4	105	4	97	3
table3	154	8	110	7	95	6	80	6	73	6	70	5	63	4	63	5
table5	143	7	104	6	87	6	73	6	69	5	64	5	61	5	56	5
Σ	2005		1393		1124		947		836		756		686		629	
Σ		73		60		56		52		46		44		40		40

Classical – Classical approach, PALDec – proposed technology mapping synthesis based on analysis of nodes of the graph of outputs

The results of experiments are presented in a synthetic way on Fig. 6.6. The values represented on the axis of ordinates in Fig. 6.6a were calculated from the rational formula shown on the graph. $\Sigma blocks_{(\mathrm{classical})}$ and $\Sigma blocks_{(\mathrm{PALDec})}$ denote the relevant total sums of block counts obtained using the corresponding technology mapping methods and presented in Table 6.1. The values represented on Fig. 6.6b were calculated in a similar manner (the formula on Fig. 6.6b, where $\Sigma levels_{(\mathrm{classical})}$ and $\Sigma levels_{(\mathrm{PALDec})}$ denote the relevant total counts of logic levels).

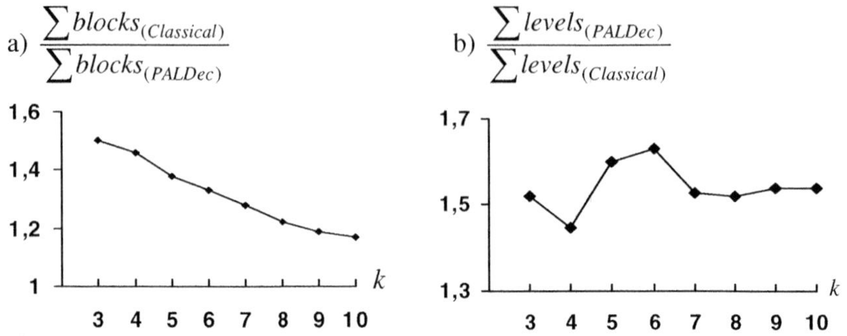

Fig. 6.6 A comparison of the PALDec algorithm with the classical method with respect to the number of logic blocks a); levels b)

Analysis of the benchmarks allows us to state, that in most cases reduction of logic block counts by using the new algorithm is obtained at the expense of certain expansion of logic levels. The proposed method is especially efficient, if $k = 3$ or 4. In this case a significant reduction of block counts, while preserving a comparable expansion number of logic levels, was observed. The rate of total synthesis time for the circuits under experiments, calculated for the proposed method and for the classical approach is about 1.25.

In Table 6.2, some technology mapping results are presented for large benchmarks. The results from the presented method are compared to the results from commercially available software (Quartus). All experiments were executed for EPM3512 produced by Altera. EPM3512 is a classical CPLD device with macrocells, shareable and parallel expanders, programmable interconnect array and I/O blocks. Combinatorial logic is implemented in the logic array, which provides 5 product terms per macrocells. The product-term select matrix allocates these product terms to the OR and XOR gates. The Quartus development system automatically optimises product terms allocation, using XOR gates, shareable and parallel expanders.

Individual columns of Tab.6.2 contain the following elements: i - number of inputs, o - number of outputs, p - number of products, NM - number of macrocells, which are necessary for implementation of the appropriate benchmark (% of all macrocells), t_{pd} (worst-case) - worst-case input to output delay. The columns marked "Quartus" show results of logic synthesis implemented by the Quartus system. For Quartus the typical options of area optimisation technique were selected. All logic resources were used: shareable expanders, parallel expanders and XOR gates. The columns marked "PALDec+Quartus" show results of two steps logic synthesis. In the first step, system PALDec searches for groups of shared multi-output implicants and the PAL-oriented technology mapping is executed for PAL-based logic block with five terms. The result of technology mapping is VHDL structural description [17]. In the second step, placement and mapping to EPM3512 by Quartus is executed. In all the cases the PALDec+Quartus strategy was able to find the best solution. All the results were obtained in very short time (total time of PALDec technology mapping is shorter than 2 seconds for all cases using a computer with Intel Celeron 1GHz).

Table 6.2 Experimental results for large benchmarks implemented in EPM3512 (*see in the text*) i - number of inputs, o - number of outputs, p - number of products, NM - number of macrocells, t_{pd} - delay time (worst-case)

				Quartus		PALDec + Quartus	
EPM3512	i	o	p	NM	t_{pd} worst-case	NM	t_{pd} worst-case
Seq	41	35	1459	257 (50%)	29,7ns	201 (39%)	28,8ns
Spla	16	46	2307	105 (21%)	20,5ns	98 (19%)	19,6ns
Table3	14	14	175	115 (22%)	22,3ns	104 (20%)	19,8ns
Table5	17	15	158	112 (22%)	26,7ns	73 (14%)	19,6ns

The obtained results also have been compared to other academic and firmware tools [14, 15, 16, 17].

6.5 Conclusions

The presented method is an alternative to the classical approach based on two-level minimization of individual single-output functions. The essence of the proposed method is to search for implicants that can be shared by several functions. Subsequent steps of the technology mapping process are adapted to logical resources of PAL-based CPLDs. Adjusting elements of technology mapping to logical resources characteristic for a PAL-based logic block allows for significant improvement of synthesis effectiveness in relation to the classical approach. If compared with another multi-level technology mapping methods, the proposed algorithm seems to be especially advantageous, because it leads to minimization of the number of logic blocks in the synthesised structures.

Simplicity of the algorithms that are based on graph analyzing methods and are useful for technology mapping of multi-output logic functions within the PAL-based structures results in their applicability as an alternative for the other methods.

Investigation carried out by author show that, apart from exploiting device specific features (Expanders, XOR gates), great majority of commercial tools use the classical algorithm as the main synthesis method. Comparison of the proposed technology mapping approach against the classical method may thus be considered as a comparison against a generalised and idealised commercial tool in its most basic form. Results of experiments prove that PALDec compares favourably with Altera tools.

References

[1] Anderson, J.H., Brown, S.D.: Technology mapping for large complex PLDs. In: Proc. of Des. Autom. Conf., pp. 698–703 (1998)

[2] Ashar, P., Devadas, S., Newton, A.R.: Sequential logic synthesis. Kluwer Academic Publisher, London (1992)

[3] Bolton, M.: Digital systems design with programmable logic. Addison-Wesley Publishing Company, New York (1990)

[4] Chen, S.L., Hwang, T.T., Liu, C.L.: A technology mapping algorithm for CPLD architectures. In: IEEE Int. Conf. on Field Program Technol., Hong Kong, pp. 204–210 (2002)
[5] Ciesielski, M.J., Yang, S.: PLADE: A two-stage PLA decomposition. IEEE Trans. on Comput. Aided Des. 11, 943–954 (1992)
[6] Devadas, S., Newton, A.R.: Exact algorithms for output encoding, state assignment, and four-level Boolean minimization. IEEE Tran. on Comput. Aided Des. 10, 13–27 (1991)
[7] Devadas, S., Wang, A.R., Newton, A.R., et al.: Boolean decomposition in multi-level logic optimization. In: IEEE Int. Conf. on Comput. Aided Des., pp. 290–293 (1988)
[8] Devadas, S., Wang, A.R., Newton, A.R., et al.: Boolean decomposition of Programmable Logic Arrays. In: IEEE Cust. Integr. Circuits Conf., pp. 2.5.1–2.5.5 (1988)
[9] Fišer, P., Kubátová, H.: Flexible two-level Boolean minimizer BOOM II and its applications. In: Proc. 9th Euromicro Conf. on Digit. System Des., pp. 369–376 (2006)
[10] Hasan, Z., Hurison, D., Ciesielski, M.: A fast partitioning method for PLA-based FPGAs. IEEE Des. and Test of Comput., 34–39 (1992)
[11] Hlavicka, J., Fišer, P.: BOOM - a heuristic Boolean minimizer. In: Proc. Int. Conf. on Comput. Aided Des., pp. 439–442 (2001)
[12] Józwiak, L., Volf, F.: An efficient method for decomposition of multiple-output Boolean functions and assigned sequential machines. In: European Conf. on Des Autom., pp. 114–122 (1992)
[13] Józwiak, L., Volf, F.: Efficient decomposition of assigned sequential machines and Boolean functions for PLD implementation. In: IEEE Int. Conf. on Electron Technol. Dir., pp. 258–266 (1995)
[14] Kania, D.: Two-level logic synthesis on PALs. Electron Lett. 35, 879–880 (1999)
[15] Kania, D.: A technology mapping algorithm for PAL-based devices using multi-output function graphs. In: Proc. of 26th Euromicro Conf., pp. 146–153 (2000)
[16] Kania, D.: An efficient approach to synthesis of multi-output Boolean functions on PAL-based devices. IEE Comput. and Digit. Tech. 150, 143–149 (2003)
[17] Kania, D.: The logic synthesis for the PAL-based Complex Programmable Logic Devices. Silesian University of Technology, Gliwice (2004)
[18] Malik, A., Harrison, D., Brayton, R.K.: Three-level decomposition with application to PLDs. In: Proc. IEEE Int. Conf. on Comput. Des., pp. 628–633 (1991)
[19] Micheli, G.: Synthesis and optimization of digital circuits. McGraw-Hill, New York (1994)
[20] Sasao, T.: Application of multiple-valued logic to a serial decomposition of PLAs. In: Proc. 19th Int. Symp. on Mult. Valued Log, pp. 264–271 (1989)
[21] Saldanha, A., Katz, R.H.: PLA optimization using output encoding. In: IEEE Int. Conf. on Comput. Aided Des., pp. 478–481 (1988)
[22] Sharma, K.: Programmable logic handbook, PLDs, CPLDs & FPGAs. McGraw-Hill, New York (1998)
[23] Shi, C.J., Brzozowski, J.A.: An efficient algorithm for constrained encoding and its applications. IEEE Tran. on Comput. Aided Des. 12, 1813–1826 (1993)
[24] Wang, L., Almaini, A.E.A.: Optimisation of Reed-Muller PLA implementations circuits. In: IEE Proc. Devices and Syst., vol. 149, pp. 119–128 (2002)
[25] Yang, C., Ciesielski, M.: PLA decomposition with generalized decoders. In: IEEE Int. Conf. on Comput. Aided Des., pp. 312–315 (1989)
[26] Yang, S., Ciesielski, M.: Optimum and suboptimum algorithms for input encoding and its relationship to logic minimization. IEEE Trans. on Comput. Aided Des. 1, 4–12 (1991)

Appendix

THEOREM ABOUT SELECTION OF A NODE OF THE GRAPH OF OUTPUTS

If there exists a node of the graph (i.e. discriminant) ${}^{i}\Delta_y$, for which

1. the range $\mu({}^{i}\Delta_y)\geq 2$ and ${}^{i}\Delta_y\geq k$

   or

2. the range $\mu({}^{i}\Delta_y)\geq 2$ and within the set of remainders $\mathbf{R}=\{r_j^{\mu({}^{i}\Delta_y)};\ j\in<1,\mu({}^{i}\Delta_y)>\}$ there exist at least two such remainders, that $0<r_a^{\mu({}^{i}\Delta_y)}<{}^{i}\Delta_y<k$ and at the same time $0<r_b^{\mu({}^{i}\Delta_y)}<{}^{i}\Delta_y<k$

   or

3. the range $\mu({}^{i}\Delta_y)=2$ and ${}^{i}\Delta_y=k$ or ${}^{i}\Delta_y\geq 2k-1$

   or

4. the range $\mu({}^{i}\Delta_y)=2$ and $k<{}^{i}\Delta_y<2k-1$ and within the set of remainders $\mathbf{R}=\{r_j^{\mu({}^{i}\Delta_y)};\ j\in<1,\mu({}^{i}\Delta_y)>\}$ there exists at least one such remainders, that $0<r_a^{\mu({}^{i}\Delta_y)}<{}^{i}\Delta_y-(k-1)$

then implementation of the implicants associated with that particular discriminant leads to a reduced number of the PAL-based blocks consisting of k terms necessary for implementation of given multi-output function.

Proof:

The value ${}^{i}\Delta_y$ assigned to the selected node affects the number of $\mu({}^{i}\Delta_y)$ of discriminants denoted as ${}^{i}\Delta_j^{\mu({}^{i}\Delta_y)}$, where $j\in<1;\mu({}^{i}\Delta_y)>$. If the selected node ${}^{i}\Delta_y$ is connected with $\mu({}^{i}\Delta_y)$ lower range's nodes then the graph modification process leads to the value increasing of not more than $\mu({}^{i}\Delta_y)$ discriminants. In this case, the discriminats after graph modification can be calculated as follows:

$$\overset{\wedge}{1\leq j\leq \mu({}^{i}\Delta_y)} \qquad {}^{i+1}\Delta_j^{\mu({}^{i}\Delta_y)}=\Delta_j^{\mu({}^{i}\Delta_y)}-{}^{i}\Delta_y+1\,.$$

It is the worst case of PAL-based logic blocks number reduction. The condition of minimization (1) can be expressed by the following equation

$$\sum_{j=1}^{\mu({}^{i}\Delta_y)}\left(\left\lceil\frac{{}^{i}\Delta_j^{\mu({}^{i}\Delta_y)}-k}{k-1}\right\rceil+1\right)-\sum_{j=1}^{\mu({}^{i}\Delta_y)}\left(\left\lceil\frac{{}^{i}\Delta_j^{\mu({}^{i}\Delta_y)}-{}^{i}\Delta_y+1-k}{k-1}\right\rceil+1\right)>\left\lceil\frac{{}^{i}\Delta_y-k}{k-1}\right\rceil+1 \tag{a1}$$

1°. ${}^{i}\Delta_y \geq k$

Let ${}^{i}\Delta_y \in < p(k-1)+1, (p+1)(k-1) >$, where $p \in N$. Since for every $p \in N$

$$\sum_{j=1}^{\mu({}^{i}\Delta_y)}\left(\left\lceil \frac{{}^{i}\Delta_j^{\mu({}^{i}\Delta_y)} - k}{k-1} \right\rceil - \left\lceil \frac{{}^{i}\Delta_j^{\mu({}^{i}\Delta_y)} - k - {}^{i}\Delta_y + 1}{k-1} \right\rceil\right) \geq$$

$$\geq \sum_{j=1}^{\mu({}^{i}\Delta_y)}\left(\left\lceil \frac{{}^{i}\Delta_j^{\mu({}^{i}\Delta_y)} - k}{k-1} \right\rceil - \left\lceil \frac{{}^{i}\Delta_j^{\mu({}^{i}\Delta_y)} - k - [p(k-1)+1]+1}{k-1} \right\rceil\right) =$$

$$= \sum_{j=1}^{\mu({}^{i}\Delta_y)}\left(\left\lceil \frac{{}^{i}\Delta_j^{\mu({}^{i}\Delta_y)} - k}{k-1} \right\rceil - \left\lceil \frac{{}^{i}\Delta_j^{\mu({}^{i}\Delta_y)} - k - p(k-1)}{k-1} \right\rceil\right) = \mu({}^{i}\Delta_y)p$$

and

$$\left\lceil \frac{{}^{i}\Delta_y - k}{k-1} \right\rceil + 1 = \left\lceil \frac{{}^{i}\Delta_y}{k-1} - \frac{k}{k-1} \right\rceil + 1 \leq \left\lceil \frac{(p+1)(k-1)+1}{k-1} - \frac{k}{k-1} \right\rceil + 1 =$$

$$= \left\lceil p+1-\frac{k-1}{k-1} \right\rceil + 1 = p+1$$

then the condition of minimization (a1) is reduced to the inequality $\mu({}^{i}\Delta_y)p > p+1$, which is true for every value $p \in \mathrm{N}$ (on the assumption that $\mu({}^{i}\Delta_y) > 2$).

2°. Since $1 < {}^{i}\Delta_y < k$, the condition in minimization (a1) can be transformed into

$$\sum_{j=1}^{\mu({}^{i}\Delta_y)}\left(\left\lceil \frac{{}^{i}\Delta_j^{\mu({}^{i}\Delta_y)} - k}{k-1} \right\rceil + 1\right) - \sum_{j=1}^{\mu({}^{i}\Delta_y)}\left(\left\lceil \frac{{}^{i}\Delta_j^{\mu({}^{i}\Delta_y)} - {}^{i}\Delta_y + 1 - k}{k-1} \right\rceil + 1\right) > 1$$

and written in the form:

$$\sum_{j=1}^{\mu({}^{i}\Delta_{o_y})}\left(\left\lceil \frac{{}^{i}\Delta_j^{\mu({}^{i}\Delta_{o_y})} - k}{k-1} \right\rceil - \left\lceil \frac{{}^{i}\Delta_j^{\mu({}^{i}\Delta_{o_y})} - {}^{i}\Delta_{o_y} + 1 - k}{k-1} \right\rceil\right) > 1 \qquad \text{(a2)}$$

Let $r_j^{\mu({}^{i}\Delta_y)}$ be numbers calculated from the congruence:

$$ {}^{i}\Delta_j^{\mu({}^{i}\Delta_y)} - 1 \equiv r_j^{\mu({}^{i}\Delta_y)} (\mathrm{mod}(k-1)) \tag{a3} $$

where j=1,2,..., $\mu({}^{i}\Delta_y)$.

Taking into account (a3)

$$ {}^{i}\Delta_j^{\mu({}^{i}\Delta_y)} - k \equiv r_j^{\mu({}^{i}\Delta_y)} (\mathrm{mod}(\,k-1)) $$

there is such a value $p \in N$ that

$$ \frac{{}^{i}\Delta_j^{\mu({}^{i}\Delta_y)} - k}{k-1} = p + \frac{r_j^{\mu({}^{i}\Delta_y)}}{k-1}\,, \tag{a4} $$

then

$$ \left\lceil \frac{{}^{i}\Delta_j^{\mu({}^{i}\Delta_y)} - k}{k-1} \right\rceil = p + 1 \tag{a5} $$

Let

$$ {}^{i}\Delta_y - 1 \equiv r_l (\mathrm{mod}(k-1))\,. \tag{a6} $$

When for the selected remainder $r_a^{\mu({}^{i}\Delta_y)}$ the inequalities $0 < r_a^{\mu({}^{i}\Delta_y)} < {}^{i}\Delta_y < k$ are true, then

$$ r_l \geq r_a^{\mu({}^{i}\Delta_y)} \tag{a7} $$

Considering the inequalities $0 < r_a^{\mu({}^{i}\Delta_y)} < {}^{i}\Delta_y < k$ as well as (a4) and (a6),

$$ \frac{{}^{i}\Delta_a^{\mu({}^{i}\Delta_y)} - k}{k-1} - \frac{{}^{i}\Delta_y - 1}{k-1} = p + \frac{r_a^{\mu({}^{i}\Delta_y)}}{k-1} - \frac{r_l}{k-1}\,. $$

From the foregoing, having substituted (a7)

$$ \left\lceil \frac{{}^{i}\Delta_a^{\mu({}^{i}\Delta_y)} - k}{k-1} - \frac{{}^{i}\Delta_y - 1}{k-1} \right\rceil = p \tag{a8} $$

Due to (d5) and (d8) the following are obtained:

$$ \left\lceil \frac{{}^{i}\Delta_a^{\mu({}^{i}\Delta_y)} - k}{k-1} \right\rceil - \left\lceil \frac{{}^{i}\Delta_a^{\mu({}^{i}\Delta_y)} - k}{k-1} - \frac{{}^{i}\Delta_y - 1}{k-1} \right\rceil = 1 \tag{a9} $$

Similarly

$$\left\lceil \frac{{}^{i}\Delta_b^{\mu({}^{i}\Delta_y)} - k}{k-1} \right\rceil - \left\lceil \frac{{}^{i}\Delta_b^{\mu({}^{i}\Delta_y)} - k}{k-1} - \frac{{}^{i}\Delta_y - 1}{k-1} \right\rceil = 1 \tag{a10}$$

Equations (a9), (a10) are the reason for the minimization condition (a2) to be fulfilled because

$$\sum_{j=1}^{\mu({}^{i}\Delta_y)} \left(\left\lceil \frac{{}^{i}\Delta_j^{\mu({}^{i}\Delta_y)} - k}{k-1} \right\rceil - \left\lceil \frac{{}^{i}\Delta_j^{\mu({}^{i}\Delta_y)} - {}^{i}\Delta_y + 1 - k}{k-1} \right\rceil \right) \geq 2 > 1 \,.$$

3.°For the second-range nodes, for which ${}^{i}\Delta_y = k$, the minimization condition (a1) is fulfilled because

$$\sum_{j=1}^{2} \left\lceil \frac{{}^{i}\Delta_j^{\mu({}^{i}\Delta_y)} - k}{k-1} \right\rceil - \sum_{j=1}^{2} \left\lceil \frac{{}^{i}\Delta_j^{\mu({}^{i}\Delta_y)} - k - (k-1)}{k-1} \right\rceil = 2 > 1 = \left\lceil \frac{k-k}{k-1} \right\rceil + 1 = \left\lceil \frac{{}^{i}\Delta_y - k}{k-1} \right\rceil + 1$$

For the nodes, for which ${}^{i}\Delta_y \geq 2k-1$, the minimization condition (a1) is fulfilled also

$$\sum_{j=1}^{2} \left\lceil \frac{{}^{i}\Delta_j^{\mu({}^{i}\Delta_y)} - k}{k-1} \right\rceil - \sum_{j=1}^{2} \left\lceil \frac{{}^{i}\Delta_j^{\mu({}^{i}\Delta_y)} - k - (2k-1) + 1)}{k-1} \right\rceil = 4 > 2 = \left\lceil \frac{2k-1-k}{k-1} \right\rceil + 1 = \left\lceil \frac{{}^{i}\Delta_y - k}{k-1} \right\rceil + 1$$

4. For the second-range nodes, for which $k < {}^{i}\Delta_y < 2k-1$, the minimization condition (a1) can be transformed into:

$$\sum_{j=1}^{2} \left\lceil \frac{{}^{i}\Delta_j^{\mu({}^{i}\Delta_y)} - k}{k-1} \right\rceil - \sum_{j=1}^{2} \left\lceil \frac{{}^{i}\Delta_j^{\mu({}^{i}\Delta_y)} - k - {}^{i}\Delta_y + 1}{k-1} \right\rceil > 2 \tag{a11}$$

Based on 2 (equations a4, a5)

$$\frac{{}^{i}\Delta_j^{\mu({}^{i}\Delta_y)} - k}{k-1} = p + \frac{r_j^{\mu({}^{i}\Delta_y)}}{k-1} \tag{a12}$$

and
$$\left\lceil \frac{{}^{i}\Delta_{j}^{\mu({}^{i}\Delta_{y})} - k}{k-1} \right\rceil = p + 1 \tag{a13}$$

Let r_l be number calculated from the congruence ${}^{i}\Delta_y - 1 \equiv r_l(\mathrm{mod}(k-1))$, then

$$ {}^{i}\Delta_y - (k-1) - 1 \equiv r_l(\mathrm{mod}(k-1)) \tag{a14}$$

When the inequalities $0 < r_a^{\mu({}^{i}\Delta_y)} < {}^{i}\Delta_y - (k-1) < k$ are true, then

$$ r_l \geq r_a^{\mu({}^{i}\Delta_y)} \tag{a15}$$

Considering the inequalities $0 < r_a^{\mu({}^{i}\Delta_y)} < {}^{i}\Delta_y - (k-1) < k$ as well as (a12) and (a14),

$$\frac{{}^{i}\Delta_a^{\mu({}^{i}\Delta_y)} - k}{k-1} - \frac{{}^{i}\Delta_y - 1}{k-1} = p + \frac{r_a^{\mu({}^{i}\Delta_y)}}{k-1} - 1 - \frac{r_l}{k-1}.$$

From the foregoing, having substituted (a15)

$$\left\lceil \frac{{}^{i}\Delta_a^{\mu({}^{i}\Delta_y)} - k}{k-1} - \frac{{}^{i}\Delta_y - 1}{k-1} \right\rceil = p - 1 \tag{a16}$$

Due to (a13) and (a16) the following is obtained:

$$\left\lceil \frac{{}^{i}\Delta_a^{\mu({}^{i}\Delta_y)} - k}{k-1} \right\rceil - \left\lceil \frac{{}^{i}\Delta_a^{\mu({}^{i}\Delta_y)} - k}{k-1} - \frac{{}^{i}\Delta_y - 1}{k-1} \right\rceil = p + 1 - (p-1) = 2 \tag{a17}$$

Equation (a17) and inequalities $k < {}^{i}\Delta_y < 2k - 1$ are the reason for the minimization condition (a11) to be fulfilled because

$$\sum_{j=1}^{2} \left\lceil \frac{{}^{i}\Delta_j^{\mu({}^{i}\Delta_y)} - k}{k-1} \right\rceil - \sum_{j=1}^{2} \left\lceil \frac{{}^{i}\Delta_j^{\mu({}^{i}\Delta_y)} - k - {}^{i}\Delta_y + 1}{k-1} \right\rceil \geq 2 + 1 > 2$$

7 Reliable FPGA-Based Systems Out of Unreliable Automata: Multi-version Design Using Genetic Algorithms

Nataliya Yakymets and Vyacheslav Kharchenko

National Aerospace University "KhAI", Department of Computer Systems and Networks, Centre of Dependable Systems, Services and Technologies (DESSERT-Centre)

Abstract. This chapter introduces the principles of multi-version digital system design and describes the concept of developing a reliable and robust system out of unreliable parts. We started with the state of the art in the area of multi-version design and explore the motivations for using different approaches to development of digital projects. A few techniques to manage design diversity for FPGA-based systems are proposed. These techniques are based on the use of genetic algorithms (GAs), and partially correct and partially definite automata obtained with GAs. Finally, we suggested GA-based method of multi-version fault-tolerant systems synthesis and discuss case-study for on-board device implementation.

7.1 Introduction

Field experience with design and exploration of digital systems shows that their automation complexity is growing simultaneously with reducing the chip area. Obviously, it makes the properties of fault tolerance, survivability, safety, availability more and more critical.

During the last two decades, different aspects of dependability, principles and techniques for dependable digital systems development were explored in detail. The key paper is [1], where A. Avizienis and J.-C. Laprie worked out a concept of "dependable computing". This work initiated development of the approaches, directed on overcoming of dualism in the tools of evaluation and providing of the required reliability on lines "hardware-software", "development processes - products", "physical faults - design faults".

In the same year J. Dobson and B. Randell published work [2], where the concept of "secure fault tolerance" and principle of its realization for different computing systems were proposed. Thus, the removal of dualism on "reliability - security" was initiated.

After eighteen years, A. Avizienis, J.-C. Laprie, B. Randell and C. Landwehr summarized the results of dependable computing evolution in [3]. Authors defined

M. Adamski et al. (Eds.): Design of Digital Systems and Devices, LNEE 79, pp. 165–192.
springerlink.com

sufficiently complete and balanced system of concepts and taxonomies. By that time definitions of dependability had been given in a number of standards and reports [4].

Meanwhile, the most challenging issues concerning system dependability arise in case of so-called safety-, mission- or business-critical systems such as nuclear power plants safety systems, airborn control systems, customer accounting system in a bank, etc [5, 6, 7]. Failures occurring in these systems may pose lives in danger, environment damage or high economic losses.

According to statistics the main causes of failure in such systems [8, 9]: can be listed as follows:

- hardware quality;
- faults that occurred during the design flow;
- non-compatibility between actual operating conditions and initial technical requirements;
- environmental influence.

Considering the factors listed above one can say that the main reliability characteristics of the system under consideration are mostly innate and rooted deep into the design process itself. Thus, system fault tolerance can be provided by choosing the most appropriate hardware and architecture as well as by using approaches that tend to eliminate or at least cut down the number of errors introduced as a part of design process.

There is a definite tendency to develop digital control systems as complex System-on-a-Chip (SoC) designs implemented on Application Specific Integration Circuits (ASIC), Complex Programmable Logic Devices (CPLD) or Field Programmable Gate Arrays (FPGA) [10, 11]. The concept of SoC refers to integrating all components of a system into a single circuit. To reduce the number of faults resulted from errors that occur during the SoC design process, different approaches and techniques should be used. One of the most important approaches to creating dependable digital systems relies upon the idea of their multi-version or diversity implementation, where versions must be as different as possible. Application of multi-version approach and its attributes are discussed in [1, 3, 12].

To obtain alternate versions and diversify the design, several efficient techniques use essential characteristics of design flow for SoC-oriented architectures provided by the standard Computer Aided Design (CAD) tools [13, 14]. Nevertheless, despite the effectiveness of these techniques most of those manage the life cycle diversity within a single design concept, so it is not possible to make a system robust to the design faults that are common to each version since widely used CAD tools usually have only standard algorithms and conventional logic implemented. Code inaccessibility and unavailability make it impossible to predict a behavior of CAD tool and prevent faults.

To avoid those problems, the application of non-classical design is suggested in order to develop non-conventional digital systems in [15, 16, 17]. It follows rather natural ways of thinking instead of the apparatus of integral and differential calculus. Nowadays, evolvable hardware applications are maturing and seeing shift into the real-world. In 1948 A. Turing suggested using artificial neural network based on very simple elements [18]. Today evolutionary neural networks combine evolutionary principles and those of artificial neural networks [19, 20]. Another research direction is evolutionary algorithms which refer back to the work of Ch. Darwin [21]. Later, the idea of Genetic Algorithms (GAs) was extensively represented by J. Holland in [22]. A number of works discuss the utilization of GAs in digital [17, 23] and analog [24] system design, robotics [25] and even in the industrial design [26]. In [27], it has been pointed out that one of the most effective ways to develop a multi-version fault tolerant system is to combine classical design based on using CAD tools with non-classical ones in order to significantly increase a possibility of receiving the least correlated versions. Nevertheless, the actual problem is how to elaborate the strategy which allows obtaining the highest diversity level in a multi-version system.

A good alternative to the classical design could be GAs that are heuristic by the nature can provide simple and non-trivial solutions as opposed to the classical design. On the other hand, complete correctness of solutions can not always be guaranteed for the same reasons. It is worth to mention that in case of complex systems and critical time needed for design only a selected set of input/output data can undergo testing to define the fitness of versions. This leads to partially definiteness of the evolved versions. Therefore a particular attention must be paid to the issues of developing reliable systems out of such unreliable (partially correct and partially definite) parts.

The remainder of the chapter is organized as follows. Section 7.2 elaborates the strategy to achieve the least correlation level within a multi-version system. Section 7.3 outlines and categorizes digital automata obtained with GA. Section 7.4 assesses reliability of digital systems out of unreliable automata. Section 7.5 continues with the case study describing the development of such systems. Finally, Section 7.6 discusses a practical application that makes use of most of the approaches considered.

7.2 External and Internal Design Diversity

The application of multi-version approach to the system design assumes obtaining version redundancy by varying the set of resources used in design process to receive the most alternate versions of the same project. For example, several CAD packages or several developer groups can be involved into the design flow.

In order to get an n-version project, it is necessary to isolate n subsets of resources that allow implementing the same functionality. The versions obtained from the different subsets will be less correlated than those that are received by varying resources within the only one subset. For example, the lesser correlation can be achieved with several CAD tools rather than with a single CAD package. Thus, the design diversity can be observed from the several levels of system design (Fig. 7.1). So-called *internal* design diversity assumes obtaining alternate versions from only one isolated subset of resources used in design. The *external* one means application of several sets of resources.

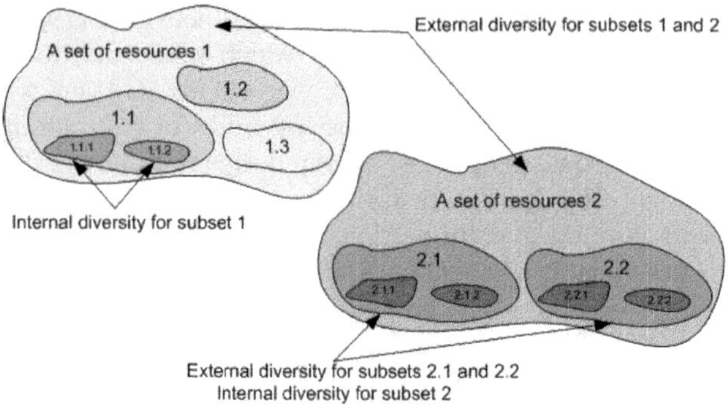

Fig. 7.1 External and internal design diversity

In fact, the major challenges in multi-version system design are:

- isolation of the subsets with the maximum cardinality;
- selection of the diversity metrics, which allow comparing system versions;
- risk analysis while estimating the compatibility of versions received with the different subsets of resources.

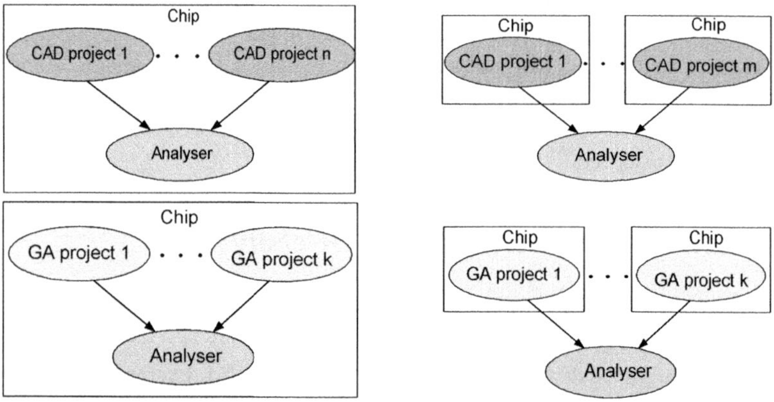

Fig. 7.2 Internal diversity that uses different design approaches

At this point, it seems that one of the effective ways to obtain the most diversified project is to exploit the external design diversity based on the different approaches to system design and use several non-classical approaches such as GAs or neural networks along with the classical one. GA can be viewed as a good-enough alternative to the classical system design, mainly because of the possibility to get simple and non-conventional solutions and going towards the reliability of digital systems developed with GA.

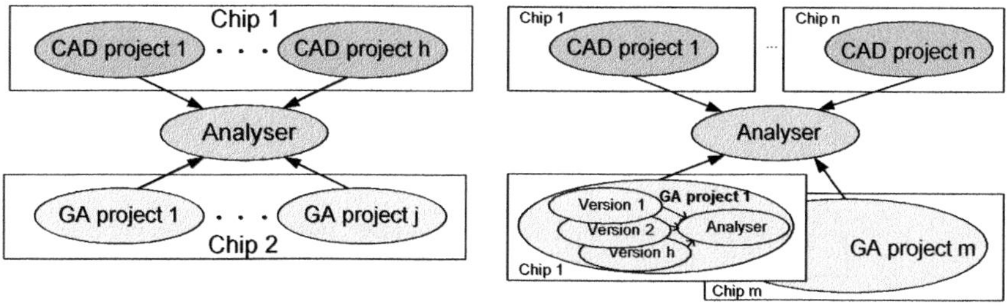

Fig. 7.3 External diversity that uses different design approaches

If standard CAD tools are used, the different versions can be obtained at the following phases: hardware selection, project entrance, compilation, testing and verification. In case of the GA application, the system diversity can be achieved at the phases of GA presetting, selection, crossover, mutation and inversion of individuals. Combination of CAD-based and GA-based approaches allows considering cases where both internal and external design diversities are possible (Fig. 7.2 – 7.3). Moreover, version implementation for a single chip or for a number of chips gives an additional opportunity to utilize spatial diversity in system design.

Because of their simplicity, versions evolved with GA can be even used as an additional feature to control the functionality of a multi-version system developed with the standard tools (Fig. 7.4).

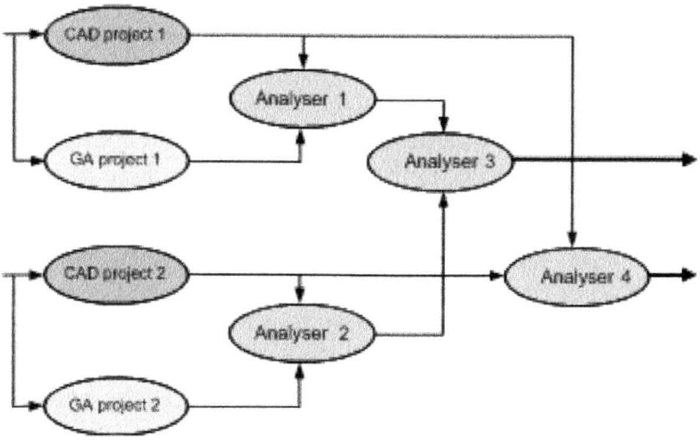

Fig. 7.4 Using versions obtained with GA as the control ones in a multi-version system

To assess a diversity of versions in such multi-version systems, each version that is a digital system with SoC architecture can be represented as a spatial configuration, which is constituted by N logic cells on a chip. A set of logic cells corresponds to set $S = \{s_i\}_{i=1}^{N}$, where s_i are the coordinates of the i-th cell. A degree of distinction for logic cell topology in a chip can be used as a diversity metrics (Fig. 7.5):

$$M_2 = \frac{|S_1 \cap S_2|}{\frac{|S_1| + |S_2|}{2}}, \tag{7.1}$$

where $|S_1|$, $|S_2|$, $|S_1 \cap S_2|$ are the capacities of correspondent sets.

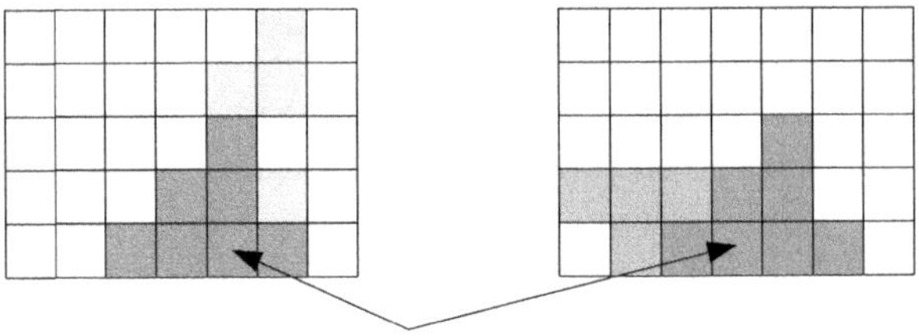

Fig. 7.5 Topology of logic cells in a chip

7.3 Partially Definite and Partially Correct Automata

In this section, we classify versions (automata) that can be evolved with GAs, and investigate the ways to develop fully correct and definite system.

There are four basic types of automata that can be obtained using GA. The first and the simplest type is a *Fully Correct Automaton* (FCA) that is an automaton where each input state x_i in the set of input data X corresponds to output state y_i in the set of output data Y_c for an arbitrary time moment $t_i \in T$, i.e. $\forall\ t_i \in T, x_i \in X: x_i \to y_i,\ y_i \in Y_c, Y_c = Y$, where t_i is an arbitrary time moment; x_i is a current input state; X is a set of input data; y_i is a current output state that corresponds to x_i; Y_c is a set of correct output data; Y is a complete set of output data.

Because of its heuristics, sometimes GA gives only an approximate solutions or *Partially Correct Automata* (PCA) that are automata, where at least one time moment $t_i \in T$ for which there is at least one input state x_i in the set of input data X that does not correspond to correct output state y_i in the set of correct output data Y_c, i.e. $\exists\ t_i \in T, x_i \in X: x_i \to y_i, y_i \notin Y_c \subset Y$.

Since it is known for every pair of input and output state whether an automaton is correct or not, FCA can be composed of the several PCAs in such a way that the complete set of correct output states of automata covers the complete set of input data X for an arbitrary time moment $t_i \in T$ (Fig. 7.6), i.e. $\forall\ t_i \in T, x_i \in X: x_i \to y_i$, $y_i \in Y_c^{(FCA)}; Y_c^{(FCA)} = Y^{(FCA)}$, $Y_c^{(FCA)} = \bigcup_{i=1}^{n} Y_c^{(PCA_i)}$.

Hence, FCA can keep its correctness even though one or more of its PCAs are not correct. A set of PCAs constitutes the complete functional basis if its elements are able to form FCA. A set of PCAs constitutes the minimal functional basis if the incorrectness of at least one PCA included into FCA results in the incorrectness of this FCA.

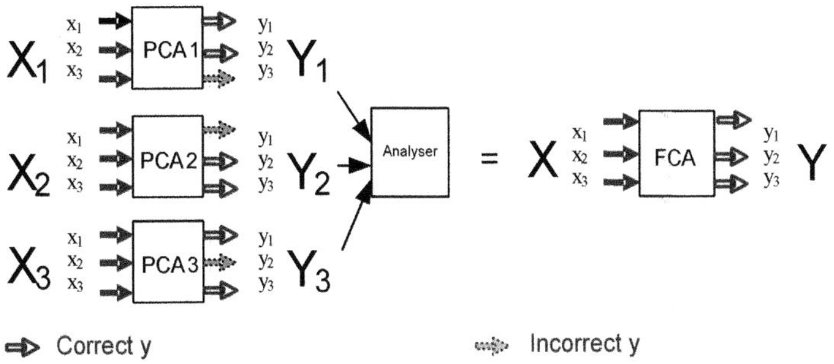

Fig. 7.6 Fully correct automaton composed of several partially correct automata

If we develop a complex digital system with GA and time required to obtain FCA is critical, only a certain number of inputs and outputs can be tested to calculate the fitness of versions [28]. In this case the evolved versions will be partially definite. Thus, *Partially Definite Automata* (PDA) are automata, where at least one time moment $t_i \in T$ for which there is at least one input state x_i in the set of input data X that corresponds to such an output state y_i where information about its correctness (whether y_i is in Y_c or not) is not available, i.e. $\exists\ t_i \in T, x_i \in X: x_i \rightarrow y_i, y_i \notin Y_d, Y_d \subset Y$, where Y_d is a set of definite output data.

Meanwhile, by testing all inputs and outputs to assess fitness of every version, we obtain *Fully Definite Automata* (FDA) that are automata, where each input state x_i in the set of input data X corresponds to such an output state y_i where information about its correctness (whether y_i is in Y_c or not) is available for an arbitrary time moment $t_i \in T$, i.e. $\forall\ t_i \in T, x_i \in X: x_i \rightarrow y_i,\ y_i \in Y_d, Y_d = Y$.

As it is known which input and output states have not be tested, FDA can be composed of several PDAs in such a way that the complete set of definite output states of automata covers the complete set of input data X for an arbitrary time moment $t_i \in T$ (Fig. 7.7), i.e. $\forall\ t_i \in T,\ x_i \in X: x_i \rightarrow y_i,\ y_i \in Y_d^{(FDA)};\ Y_d^{(FDA)} = Y^{(FDA)};$

$$Y_d^{(FDA)} = \bigcup_{i=1}^{n} Y_d^{(PDA_i)}.$$

FDA can keep its definiteness even though one or more of its PDAs are not definite. A set of PDAs constitutes the complete functional basis if its elements are able to form the required correct FDA. A set of PDAs constitutes the minimal functional basis if the indefiniteness of at least one PDA included into FDA results in the indefiniteness of this FDA.

From the definitions given above it is clear that digital automata can be fully definite fully correct, fully definite partially correct, partially definite fully correct or partially definite partially correct.

Fully Definite Fully Correct Automaton (FDFCA) is an automaton where each input state x_i in the set of input data X corresponds to such an output state y_i, where $\forall\ t_i \in T, x_i \in X: x_i \rightarrow y_i,\ y_i \in Y_d$ and $y_i \in Y_c$; $Y_d = Y_c = Y$ for an arbitrary time moment $t_i \in T$.

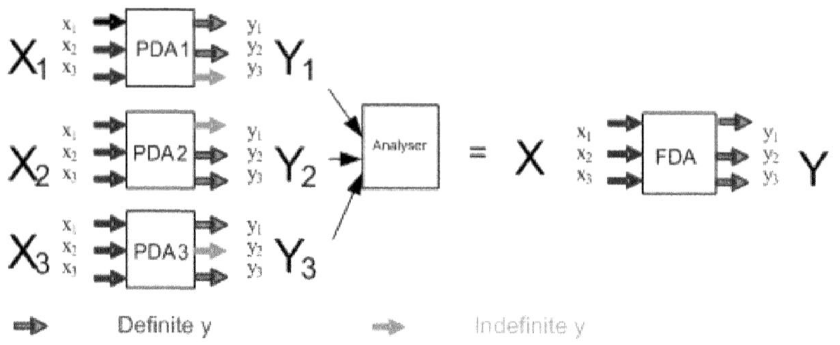

Fig. 7.7 Fully definite automaton composed of several partially definite automata

Fully Definite Partially Correct Automaton (FDPCA) is an automaton which for at least one time moment $t_i \in T$ has at least one input state x_i in the set of input data X that corresponds to such an output state y_i, where $\exists\ t_i \in T, x_i \in X: x_i \rightarrow y_i$, $y_i \in Y_d$ and $y_i \notin Y_c$; $Y_d = Y, Y_c \subset Y$.

Partially Definite Fully Correct Automaton (PDFCA) is an automaton where each input state x_i in the set of input data X corresponds to such an output state y_i, where $\forall\ t_i \in T, x_i \in X_d: x_i \rightarrow y_i,\ y_i \in Y_d$ and $y_i \in Y_c$; $Y_d = Y_c, Y_d \subset Y, Y_c \subset Y; X_d \subset X$ for an arbitrary time moment $t_i \in T$, where X_d is a set of input data where automaton is definite.

Partially Definite Partially Correct Automaton (PDPCA) is an automaton which for at least one time moment $t_i \in T$ has at least one input state x_i in the set of input data X corresponds to such an output state y_i, where $\exists\ t_i \in T, x_i \in X_d: x_i \rightarrow y_i$, $y_i \in Y_d$ and $y_i \notin Y_c$; $Y_c \subset Y_d, Y_d \subset Y, X_d \subset X$.

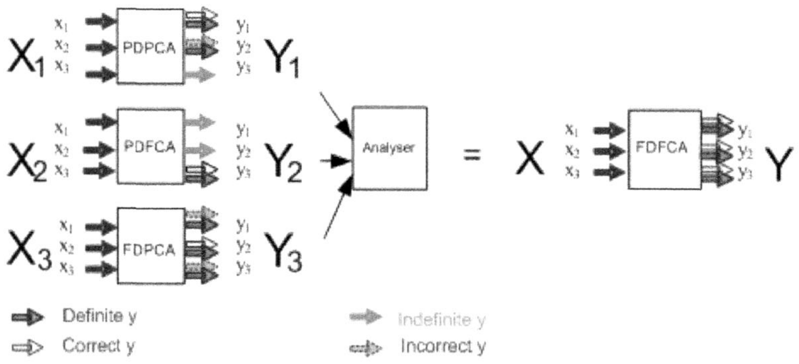

Fig. 7.8 Fully definite fully correct automaton composed of several partially definite and partially correct automata

In fact, the fully definite and correct system or FDFCA can be composed of several PDPCAs (and/or FDPCAs, PDFCAs) in such a way that the complete set of correct and definite output states of automata covers the complete set of input data X for an arbitrary time moment $t_i \in T$ (Fig. 7.8), i.e. $\forall\ t_i \in T,\ x_i \in X\colon x_i \rightarrow y_i$, $y_i \in Y_d^{(FDA)}$ and $y_i \in Y_c^{(FDA)}$; $Y_d^{(FDA)} = Y_c^{(FDA)} = Y^{(FDA)}$; $Y_d^{(FDA)} = \bigcup_{i=1}^{n} Y_d^{(PDA_i)}$; $Y_c^{(FDA)} = \bigcup_{i=1}^{n} Y_c^{(PDA_i)}$.

7.4 Reliability of Digital Systems Out of Unreliable Automata

The architecture of digital systems evolved with GA assumes possible redundancy that occurs because of overlapping correct and definite output data if a system consists of several PCA or/and PDA. This property allows estimation of such systems reliability using the apparatus of structural reliability theory. The analysis of reliability includes the following steps:

- divide a set of input and output data of automata into groups as shown in Fig. 7.9;

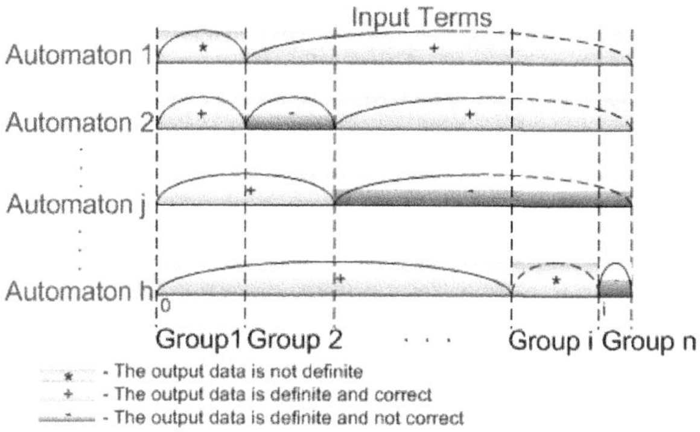

Fig. 7.9 Allocating the groups within the set of input and output terms if a system consists of the several PDA and/or PCA

- estimate the influence of automata failures on the entire system;
- select a primitive object to be used in the reliability analysis: the primitive objects can be automata, groups of definite and correct output data, automata and groups of definite and correct output data;
- develop a reliability block diagram (RBD);
- form the equations to estimate system reliability with regard to its RBD.

If a digital system consists of one fully definite and fully correct version, its reliability is the same as the reliability of a non-redundant non-recoverable system. If a digital system is composed of several PDA or/and PCA it can be considered as a non-recoverable system with passive redundancy. To estimate the reliability of such systems, the additional analysis of their functionality must be performed to prove the choice of the primitive object and develop RBD more precisely.

If the primitive objects of RBD are *automata*, we assume that the failure of every automaton results in a complete loss of its functionality. Such an assumption is reasonable for a few automata as the process of RBD development is quite complicated and non-conventional for each particular system. So, RBD is formed according to already known information about the automata behavior. The example in Fig. 7.10 shows that the system fails if Automaton 1 fails or Automaton 3 fails or Automaton 2 and Automaton 4 fail. The probability of no-failure for such a system is calculated with the following formula: $P(t) = p_1(t)\times p_3(t)\times(1-(1-p_4(t))\times(1-p_2(t)))\times p_K(t)$, where p_i is a probability of no-failure for automaton i; $p_K(t)$ is a probability of no-failure of the subsystem that implements switching between automata.

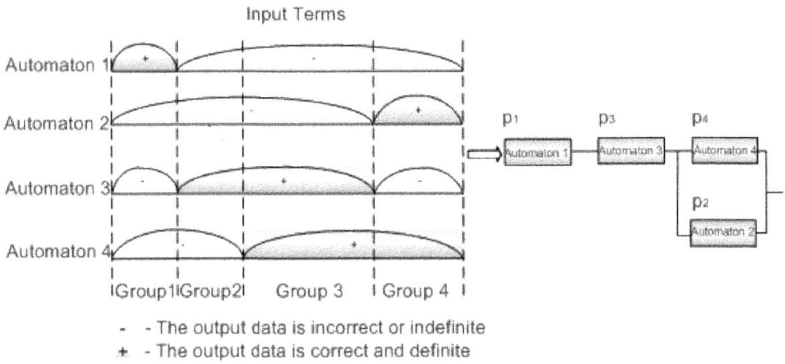

Fig. 7.10 RBD for a system out of four automata if primitive objects are automata

If the primitive objects of RBD are *groups of definite and correct output data*, the assumption is that the failure of an automaton does not result in the complete loss of its functionality and it can produce correct output data for some input terms. In contrast to the previous assumption, this one is reasonable in case a significant number of automata constitute a whole system. Nevertheless, there is one serious disadvantage: during the synthesizing with GA, a digital system is considered as a "black box" with certain inputs and outputs without any information about its internal structure. Therefore, information about the nature and consequences of the automaton's failure is not available. The probability of no-failure for such a system is calculated with the following formula:

$$P(t)=\prod_{i=1}^{n}\left[1-\prod_{j=0}^{m}\left[1-p_{ij}(t)\right]\right]\times p_K(t), \tag{7.2}$$

where $p_{ij}(t)$ is a probability of no-failure for automaton j within the group of correct and definite output data i; n is a number of groups of correct and definite output data; m is a number of automata that have correct and definite output data within the group of correct and definite output data i.

For the example in Fig. 7.11, $P(t) = p_{11}(t) \times p_{23}(t) \times (1-(1- p_{33}(t)) \times (1- p_{34}(t))) \times (1-(1- p_{42}(t)) \times (1- p_{44}(t))) \times p_K(t)$.

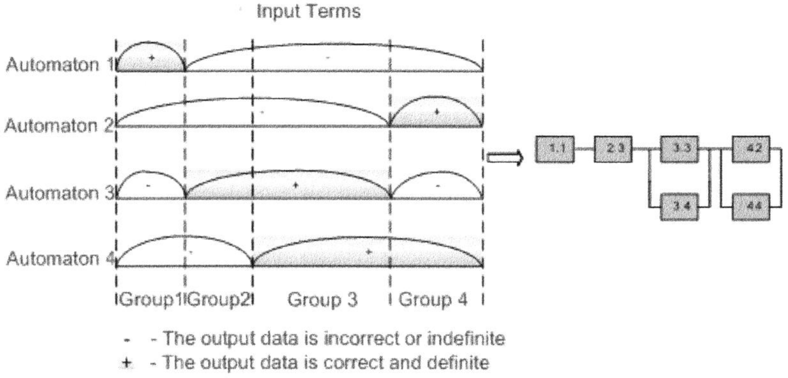

Fig. 7.11 RBD for a system out of four automata if primitive objects are groups of output data

By assuming the primitive objects of RBD to be *automata and groups of definite and correct output data*, we mean that the failure of an automaton might result in either the overall or partial loss of its functionality with the certain probability. The probability of no-failure for such a system is calculated using the following formula:

$$P(t) = P'(t) \times P''(t) \times p_K(t) = P'(t) \times \left(\prod_{i=1}^{n} \left(1 - \prod_{j=0}^{m} \left(1 - q_{ij}(t) \right) \right) \right) \times p_K(t), \quad (7.3)$$

where $P'(t)$ is a probability of no-failure for those parts of automata which incorrect functioning results in an complete functionality loss of the appropriate automaton; $P''(t)$ is a probability of no-failure for those parts of automata which incorrect work results in a functionality loss of the appropriate automaton in the appropriate groups; $q_{ij}(t)$ is the probability of no-failure for automaton j within group i in case of no-failure operation of its part which fault results in the complete functionality loss of automaton j.

The example in Fig. 7.12 shows how the probability of no-failure is calculated for such systems: $P(t) = p'_1(t) \times p'_3(t) \times (1-(1-p'_4(t)) \times (1-p'_2(t))) \times q_{11}(t) \times q_{23}(t) \times (1 - (1 - q_{33}(t)) \times (1 - q_{34}(t))) \times (1 - (1 - q_{42}(t)) \times (1 - q_{44}(t))) \times p_K(t)$.

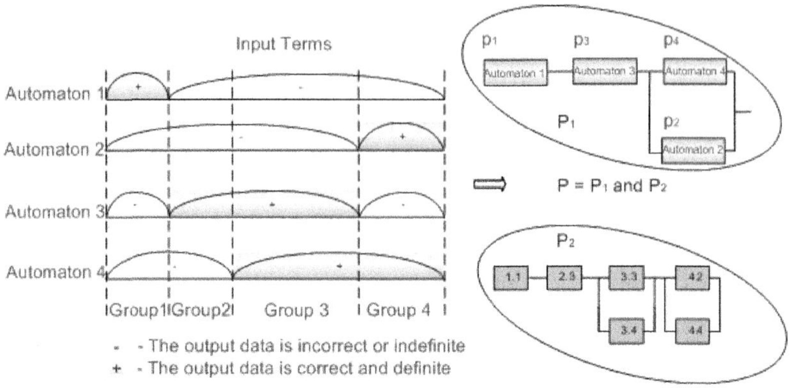

Fig. 7.12 RBD for a system out of four automata if primitive objects are both automata and groups of output data

7.5 Designing Digital Systems Out of Unreliable Automata

7.5.1 General GA-Based Approach

One of the most critical engineering decisions to be taken as a part of the design process with GAs is about properly identifying the level of abstraction for design representation. The process of applying GA at the very low level of system implementation on a FPGA, practically combines two stages of the life cycle – designing and direct implementation. One of the main disadvantages here is a difficulty in reproducing evolved system with other FPGAs [17]. Therefore the main attention was paid to the gate level (Fig. 7.13) that is the one of the low levels of abstraction.

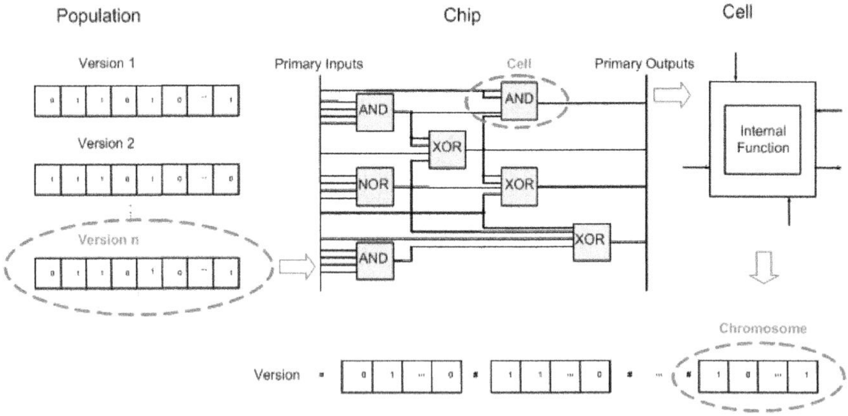

Fig. 7.13 GA applications at a gate level

At the gate level each version is represented as a graph that shows interconnections of cells in FPGA. We assume that each cell can realize one of the following internal functions: AND, OR, XOR, and it has one output and four inputs connected either to one of a primary inputs (input variables from the truth table given for a system) or to an output of any cell. The genotype of an initial population (Table 7.1) is formed by random numbers generator and represents a set of binary strings coding various variants of mapping. Each string consists of several chromosomes.

Table 7.1 Definitions Connected with Genetic Algorithms

GA term	Definition in GA terms	Definition in system design terms
Chromosome	Vector of ones and zeros	Vector of ones and zeros coding internal function and cell interconnections
Individual	Set of chromosomes = possible solution	Version of digital system = graph that shows internal functions of cells and their interconnections
Population	Set of individuals (set of possible solutions)	Set of versions of digital system with the same functionality but different implementations
Fitness function	Function that reflects a degree of how individual corresponds to the required solution.	Fitness is calculated according to the overall number of input data that correspond to the required (as in a truth table) output data

The number of chromosomes is equal to the overall number of cells in the considered PLD area. The following information is coded in a chromosome:

- the probabilities to connect each cell's input to one of the primary inputs;
- the numbers of primary inputs, which can be potentially connected to the cell's inputs;
- the numbers of cells, which outputs are connected to the current cell;
- the internal function of a cell; the probability that a cell is connected to one of the primary outputs (output variables from the truth table given for a system).

The fitness f has been calculated for each individual by comparing its output data with the existing specifications (truth table). It equals a ratio between the number of correct terms of the individual and the overall number of terms from the truth table.

$$f_i = \frac{X_r \times 100\%}{2^n}, \tag{7.4}$$

where i is a number of individual in a population; X_r is a total number of input terms that corresponds to the correct output data; n is a number of inputs.

Digital system design with the use of GA includes tree phases [29]. The initial data for the *first phase* is a truth table of the required system as well as the type of automata. According to the initial data GA evolves a model that represents a digital system at the gate level providing information about the interconnections between logic cells in a chip and their internal functions (Fig. 7.14). During the *second phase* (Fig. 7.15), the obtained model can be improved in order to reach a sufficient reliability level. In the *third phase*, a special subsystem is developed to implement switching between automata and achieve the correct functionality of the whole system. The initial data for this phase is a number of automata utilized in the model and information about their functionality. Finally, a system that includes several PCA or/and PDA and switching subsystem, is implemented on a chip with any standard CAD tool.

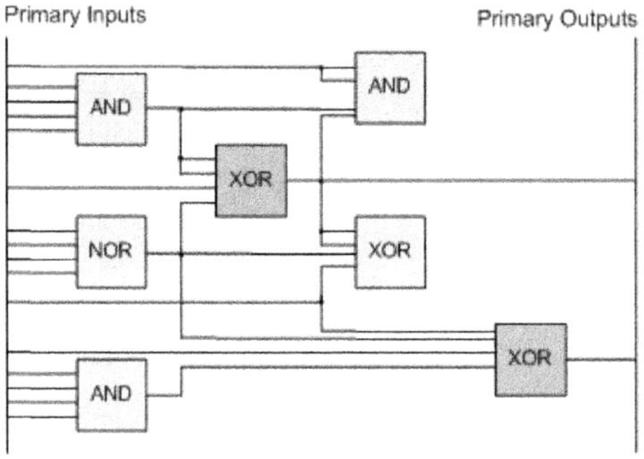

Fig. 7.14 Graphical representation of individual evolved with GA

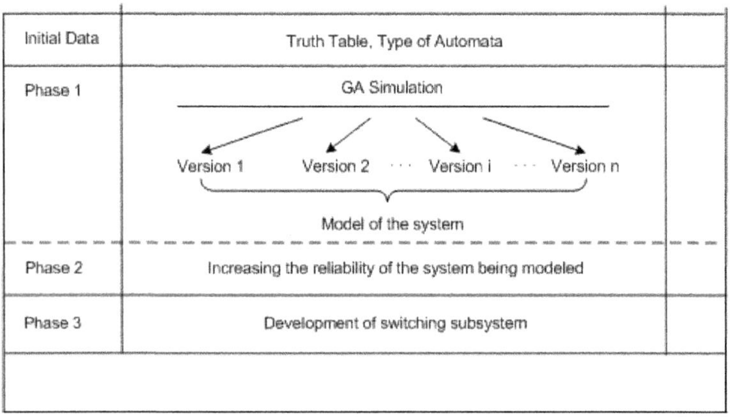

Fig. 7.15 Process of digital system design with GA

7.5.2 Phase 1: Obtaining a System Model

First of all, a level of the automata correctness and definiteness should be proved depending on the system complexity, cost and time constraints. The compactness is the main requirement for the simple systems. Therefore they can be designed with FDFCAs due to their simplicity. In case the compactness and development time are of primary importance, PDFCAs or FDPCAs should be used as well. Implementing system with PDFCAs we reduce development time at the expense of increasing system complexity, whereas using FDPCAs we obtain a simple system but synthesizing becomes more time-consuming. Complex systems, where time constraints are critical, can be developed with PDPCAs.

The process of developing a digital system with FDFCAs includes several runs of GA until the version with 100% fitness has been found. While calculating fitness of every version all input and output states should be tested.

If a digital system is implemented with FDPCAs, GA executes until it has evolved several partially correct solutions. As information about the correct or incorrect output state of each input term of every automaton is available, FDPCAs are gathered in a redundant scheme in such a way that all output states of the scheme have to be correct. System reconfiguration is implemented depending on known information about the correct or incorrect output states of each automaton or if one or more FDPCAs fail. It guaranties a high level of system reliability due to flexible controlling FDPCAs.

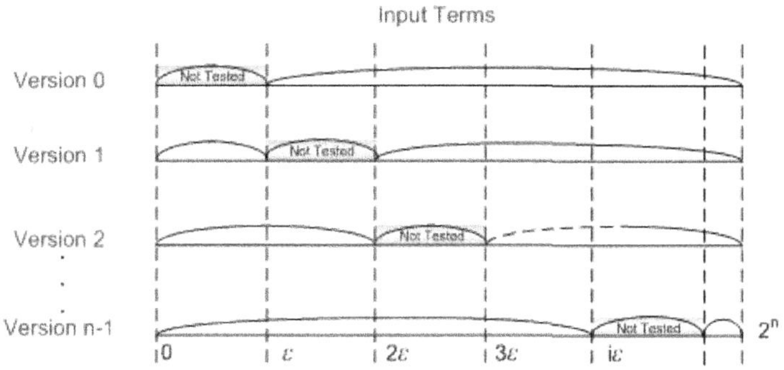

Fig. 7.16 "Sliding Testing" technique

If a digital system is developed with PDFCAs the design flow means running GA until the fully correct versions are obtained. To reduce time required to evolve versions if time constraints are critical, we suggest a so-called "Sliding Testing" technique [28]. It means testing only a part of input and output data to estimate versions' fitness. In other words, the "Sliding Testing" is a tradeoff between development rigour and development time. The application of such a technique requires determining a width of the untestable interval ε (Fig. 7.16). Thus, the "Sliding Testing" technique assumes several steps:

- determine a width of the untestable interval ε;
- set a position of an untestable interval as $[i\times\varepsilon, (i+1)\times\varepsilon]$, i – a number of version in the current population;
- estimate the fitness of each version in the population. It equals a total number of terms which correspond to the correct output data of automaton. Terms restricted by the untestable interval should not be tested.

GA works until several fully correct versions have been obtained. The versions can be included in the redundant schemes in such a way that all output states of the scheme have to be definite.

The optimal value of ε is calculated as $\varepsilon = \left[\frac{2^n}{m} \right]$, where n is an input data capacity; 2^n is an amount of input terms; m is an amount of versions in the current population.

If $\varepsilon < \frac{2^n}{m}$ the dependability of such a PDA is growing at the expense of reducing the amount of untestable input terms (a degree of automaton's definiteness is increasing). If $\varepsilon > \frac{2^n}{m}$ a level of automaton's dependability is becoming lower because of reducing its definiteness.

System reconfiguration is implemented depending on information about the width and position of the untestable intervals of PDFCAs or if one or more PDFCAs fail.

Development of digital systems with PDPCAs involves two design techniques: designing system with FDPCAs and designing system with PDFCAs. This technique, just as the previous one, requires the "Sliding Testing" to be used during the design flow. The difference is that GA proceeds until several partially correct versions, sufficient to design a fully definite and correct system, are obtained. While forming a whole system, the output states, which correspond to the indefinite terms, should be marked as the incorrect ones that are known beforehand. System reconfiguration is implemented depending on information about the width and position of the untestable intervals of PDFCAs and their correct output states or if one or more PDPCAs fail.

7.5.3 Phase 2: Increasing the Reliability of Digital Systems Out of Unreliable Automata

In order to improve fault tolerance characteristics of digital systems the classical solution is to gather several identical or different system versions in duplex or majority architecture. Such an approach can be applied to increase the fault tolerance of digital systems based on PDA or/and PCA. Estimating the reliability of such a multi-version system which is the several fully correct automata consisting of several PDA or/and PCA, the standard formulas from the reliability theory must be used. However, several other methods can be used together with the classical one to increase the fault tolerance of digital systems implemented with PDA or/and PCA [30]. The application of these methods is possible because of the essential architectural features of such systems if the groups of definite and correct output data are considered as primitive objects.

One of the simplest methods that allows development of fault tolerant digital systems with the certain degree of redundancy is the *"Constant" method.* It consists of two steps:

- selection of the minimum number of primitive objects n that have to be present in all groups of output data to provide the required level of system reliability;
- adding new automata in such a way that the number of redundant elements in every group m_i satisfies the following condition: $n \leq m_i$.

While estimating the reliability of such systems *P(t)*, the equations given in section 7.4 must be used. Despite of it simplicity, applications of the method are limited only to specific set of case. Mainly because of the fact that a significant number of automata might be required to satisfy the condition $n \leq m_i$.

For example, there is a need to develop a system that has the minimum number of logic cells K_s but the certain level of reliability $P_s^{init}(t)$ or a system with the maximum level of reliability but the fixed number of logic cells K^{init} then *"Optimal Reservation" method* can be used. It is based on the algorithm of shortest descent [31] which is widely applied for the system optimization according to the criteria "reliability – cost". The proposed method uses the functionality of automata obtained with GA (available information about the correctness and definiteness of each automaton) and allows the architecture of such systems to be optimized according to the criteria "reliability-complexity" as well.

The process of system optimization includes two phases.

Phase 1:

- the set of PDA or/and PCA that constitutes the minimum functional basis of a system is considered as an initial architecture;
- for the initial architecture, calculate the probabilities of no-failure of automata for each group of output data p_i as well as calculate the quantity of logic cells that are essential to the functionality of automata for group i, k_i. Since GA synthesis an automaton is considered as a "black box", we assume the equal probabilities p_i which can be calculated using the value of *P(t)* obtained with the simulation or statistics. The same assumption is used for k_i:

$$k_i = \frac{\sum_{j=1}^{h} l_j}{n \times h}, \tag{7.5}$$

where l_j is a number of logic cells allocated for automaton j within the initial architecture; n is a number of groups of correct and definite output data within the initial architecture; h is an overall number of automata in the initial architecture;

- condition $\forall i = \overline{1,n} : P_i \geq P_s^{init}(t)$ is guaranteed by adding new automata into the initial system architecture (P_i is the probability of no-failure of a system within group i that is calculated using p_i values);
- if a new automaton has been added, the probabilities p_i must be recalculated and the condition $\forall i = \overline{1,n} : P_i \geq P_s^{init}(t)$ must be checked again as the new automaton might have definite and correct output data within the other groups too;

- estimate the probability of no-failure of a system P_s that is calculated according to the equations given in section 7.4;
- estimate the number of logic cells allocated for a system using the following formula:

$$K_s = \sum_{i=1}^{h} l_i \,, \tag{7.6}$$

Phase 2:

- at each step of phase 2, the selection of the new automaton is made in such a way that it must have correct and definite output data within that group, for which it results in maximum reliability increase per unit of occupied area in accordance with index j:

$$j_{i,j} = \frac{p_i(g_{i,j+1}) - p_i(g_{i,j})}{k_i \times p_i(g_{i,j+1})} \,, \tag{7.7}$$

where $g_{i,j}$ is a current group of output data;

- the probabilities p_i are recalculated for each group as well as P_s and K_s for the whole system.

7.5.4 Phase 3: Development of Switching Subsystem

This phase includes the following steps:

- work out the ideal model of system behavior that is based on information about the functionality of each automaton. In this model, every input signal x_i corresponds to the set of ideal (known beforehand) output signals of automata, Y_i^{ideal};
- compare the real input signals of automata with one another and form a set of real output signals, Y_i^{real};
- compare the set of real output signals Y_i^{real} with the set of ideal ones Y_i^{ideal} and choose such a couple $(y_j^{real}, y_j^{ideal})$ that $y_j^{real} = y_j^{ideal}$, where j is a number of automaton which is acceptable for switching.

The described above algorithm can be implemented using Table 7.2.

Table 7.2 Switching logic for two automata

	Information about current behavior of automata		
Known information about definiteness and correctness of automata	A - Automaton	A1 = A2	A1 ≠ A2
	A1 and A2 have definite and correct output	"OK" Switch to A1 or A2	"Fail"
	A1 has definite and correct output	"Risk" Switch to A1	"OK" Switch to A1
	A2 has definite and correct output	"Risk" Switch to A2	"OK" Switch to A2

The top row includes all possible combinations formed by n automata if they lose their correct functionality that can be revealed during the system exploration. The left column includes all possible combinations formed by n automata according to the information from the ideal model. The rest of cells are filled in as follows. If a current combination of the top row fits to the current combination of the left column then the cell has value "Fail". In other case, the cell is filled with value "OK".

7.5.5 Implementation

Implementation of a solution on a chip implies several alternatives. The choice lies between ASIC, PLA, CPLD or FPGA.

The choice of an ASIC involves a lot of restrictions arising from the fact that it is "Application Specific" and quite expensive. Then between PLA, CPLD and FPGA it is the last one that offers the highest flexibility that is in primary importance when we deal with GAs.

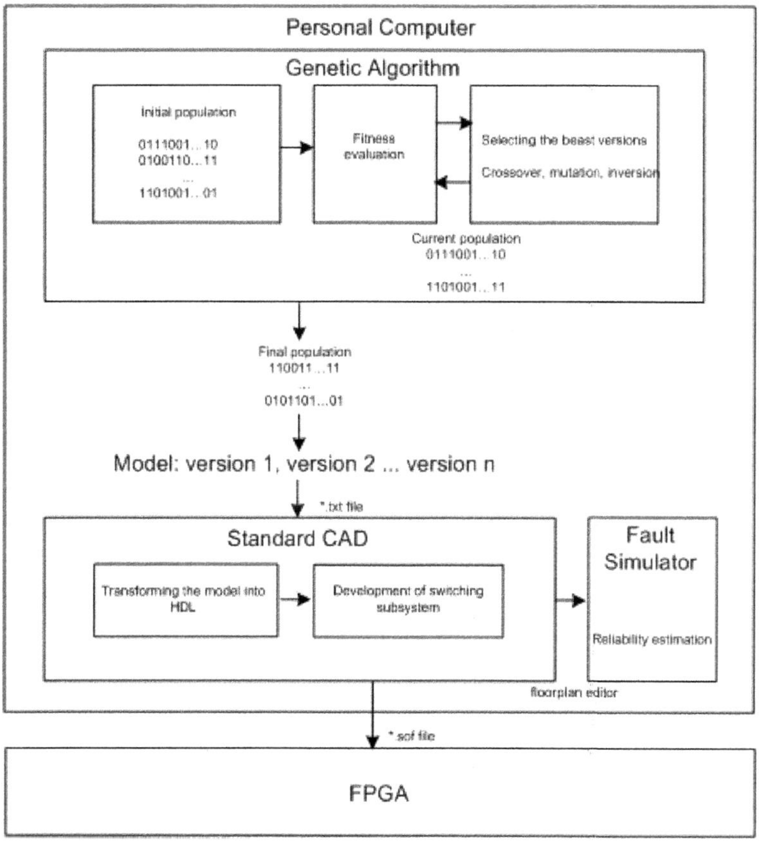

Fig. 7.17 Software and hardware architecture for system design with GA

To translate the model into the acceptable format in order to exploit the standard CAD tool, each automaton needs a description in any of Hardware Description Languages (HDL) as follows. The codes of the internal functions of each cell are to be extracted from binary string that describes the behavior of automata as well as codes that show cell interconnections. Then the appropriate variable for each cell is defined to implement its internal function and set the connections between these variables, inputs and outputs of automaton. As a result, a hierarchical project is obtained by applying the same approach to every automaton of the model: on the top level there is a switching subsystem while automata are represented as "include files" (Fig. 7.17).

7.6 Experimental Application

Digital heating controller developed by means of the standard CAD tool, and currently used in AN-70 plane was chosen to verify our approach. It was implemented using C++, AHDL languages and Quartus II tool on a Pentium IV with 1500 MHz clock and 1 GB RAM.

A simple GA with a population size of 50, GA cycles of 1000, crossover probability of 0.75 and mutation probability of 0.25 was applied. Modeling area was determined as 4×4 array of logic cells of FPGA.

Input data for the heating controller: 1-st bit determines a sign; 2-7 bits determine a value of the temperature (°C). Output data for the heating controller: '01' – the temperature is lower than +15°C, '10' – the temperature from +15°C up to +35°C, '11' – the temperature is higher than +35°C.

During the *first phase* of design flow, a model of heating controller that involves 9 logic cells and includes two PCA evolved in 405 and 789 populations, was obtained. Information about their functioning is represented in Fig. 7.18.

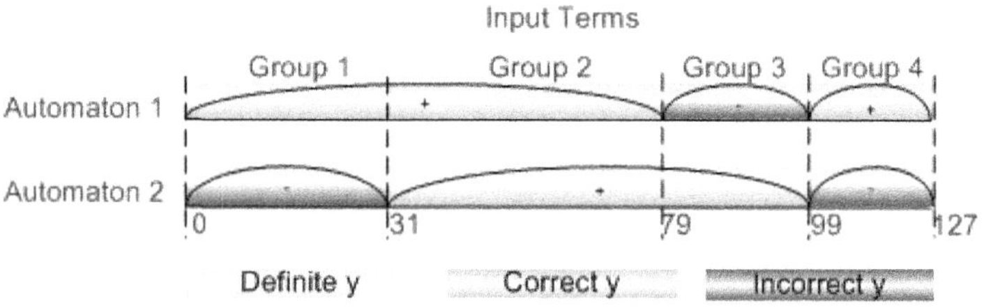

Fig. 7.18 The scheme of term overlapping in the heating controller model implemented with partially correct automata

To assess the reliability of the developed controller, the probability of no-failure was chosen. It was calculated for FPGA failure rate $\lambda=10^{-7}$ 1/hour, $p_K = 1$, time intervals t= $\{10^1, 10^2, 10^3, 10^4, 10^5\}$ hours and overall number of logic cells N=256. Time till the FPGA fault has been given by the exponential distribution.

If the primitive objects are automata, the probability of no-failure is calculated with the following formula: $P(t) = p_1(t) \times p_2(t) \times p_K(t)$. We have assumed that

$p_1(t)=p_2(t)$, so $P(t) = p(t)^2 \times p_K(t)$. The probability of no-failure for a single automaton is equal to $\lambda_a = (\lambda \times N_a)/N$, where N_a is a number of logic cells that constitute single automaton. $N_a = N_s/h$, where N_s is a number of cells allocated for the whole system and h is a number of automata. Therefore, the probability of no-failure for the whole system is the following: $P(t) = e^{-2\times\frac{10^{-7}\times\left[9/2\right]}{256}\times t}$.

If the primitive objects are groups of definite and correct output data, the probability of no-failure is calculated as $P(t) = p_{11}(t)\times(1-(1- p_{12}(t) \times(1- p_{22}(t)) \times p_{23}(t) \times p_{14}(t) \times p_K(t)$. We have assumed that all probabilities p(t) are equal, so $P(t)=p(t)^3\times(1-(1- p(t)^2)\times p_K(t)$. The probability of no-failure for a single group of correct and definite data of automaton equals $p(t)=e^{-\lambda_g t}$, where λ_g is a failure rate of automaton for a group g. λ_g is calculated as $\lambda_g = (\lambda \times N_g)/N$, where $N_g = N_s/(h\times n)$ and n is a number of groups. Therefore, the probability of no-failure for the whole system is the following: $P(t)= e^{-3\times\frac{10^{-7}\times\left[\frac{9}{2\times4}\right]}{256}\times t} \times(1-(1-e^{-\frac{10^{-7}\times\left[\frac{9}{2\times4}\right]}{256}\times t})^2)$.

To compare the developed heating controller with the prototype, we assume prototype to be a single FDFCA. Thus, $P(t)= e^{-\lambda_p t}$, where λ_p is a failure rate of the prototype. λ_p is equal $\lambda_g = (\lambda \times N_s)/N$, so the probability of no-failure for the prototype is $P(t) = e^{-\frac{10^{-7}\times 74}{256}t}$.

The values of P(t) for the obtained controller and its prototype are given in Table 7.3. The gain in reducing the probability of no-failure for both versions of controller is shown in Table 7.4.

Table 7.3 The probability of no-failure for heating controller and its prototype

-	Primitive Object	Time, hour				
-	-	10	10^2	10^3	10^4	10^5
New System	Automata	0.999999965	0.999999648	0.999996484	0.999964844	0.999648499
	Groups of output data	0.999999986816	0.999999868164	0.999998681641	0.99998681647	0.999868170829
Prototype	-	0.999999710938	0.999997109379	0.999971094168	0.99971097927	0.997113548834

Table 7.4 The gain in reducing the probability of no-failure for heating controller comparing to its prototype

Primitive Object	Time, hour				
-	10	10^2	10^3	10^4	10^5
Automata	8,25892738	8,21199097	8,22122646	8,22109243	8,21178650
Groups	21,9252471	21,9258838	21,9256153	21,9228633	21,8953904

During the *second phase* we applied both the "Constant" and "Optimal Reservation" methods to the heating controller shown in Fig. 7.18. The existing model was updated in the process of simulation with the new automata that had not been included into the initial model from the start.

For the "Constant" method the number of automata that had correct data within every group was set to 3. To increase redundancy in group 1, automaton 3 and 4 were added. Automaton 3 was used in order to achieve the required level of redundancy in group 2 as well (Fig. 7.19). Automaton 4 and 5 were also added for group 3 and automaton 3 and 5 were added for group 4. Thus, in the obtained architecture at least three automata in every group have correct output data. Moreover, in group 1 the level of redundancy is higher than required because of the essential logic of such PCA. In case the primitive objects in RBD are the groups of correct output data of automata, the probability of no-failure for the obtained system is the following:

$$P_1 = \left(\left(1-\left(1-p_{g_1}\right)^4\right)\times\left(1-\left(1-p_{g_2}\right)^3\right)\times\left(1-\left(1-p_{g_3}\right)^3\right)\times\left(1-\left(1-p_{g_4}\right)^3\right)\times p_K\right). \quad (7.8)$$

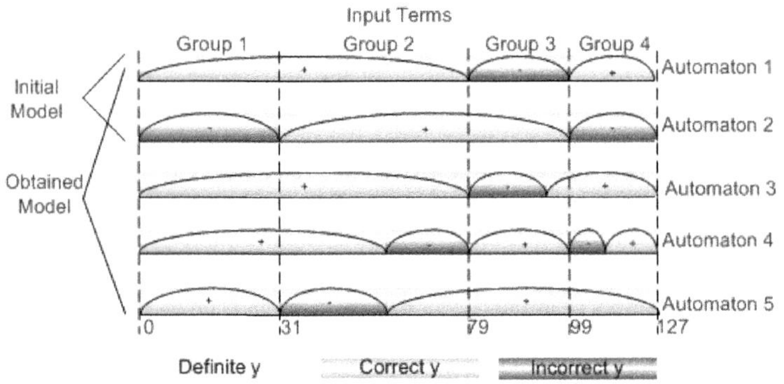

Fig. 7.19 Model of heating controller obtained with the "Constant" method

The degree of reducing the probability of failure for the "Constant" method, if $p_K = 1$ and $p_{g_1} = p_{g_2} = p_{g_3} = p_{g_4} = 0.9$, is the following:

$$W = \frac{1-P_0}{1-P_1} = \frac{1-0.9\times(1-(1-0.9)^2)\times0.9\times0.9}{1-(1-(1-0.9)^4)\times(1-(1-0.9)^3)^3} = \frac{1-0.72171}{1-0.996904} = 84.085690, \quad (7.9)$$

where P_1 is a probability of no-failure for the obtained system; P_0 is a probability of no-failure for the initial system.

So, the reduction of the probability of failure for the system received by the "Constant" method application is 84.085690 if the number of logic cells has grown from 9 to 23.

The analysis of the system architecture represented in Fig. 7.19 shows that there is a switching between the correct and incorrect output data in groups 2, 3 and 4 (for example, automaton 4 within group 2). Such a redundancy has not been taken into consideration in (7.8). Nevertheless, this disadvantage can be avoided by increasing the amount of groups as it is shown in Fig. 7.20. The degree of reducing the probability of failure, while increasing the number of groups from 4 to 7 is the following ($p^*_{g_1} = p^*_{g_2} = p^*_{g_3} = p^*_{g_4} = p^*_{g_5} = p^*_{g_6} = p^*_{g_7} = \sqrt[7]{p^4_g} = \sqrt[7]{0.9^4} \approx 0.941571$):

$$W = \frac{1-P_1}{1-P_2} = \frac{1-(1-(1-0.9)^4)\times(1-(1-0.9)^3)^3}{1-(1-(1-0.941571)^4)^5\times(1-(1-0.941571)^3)^2} = 1.000399, \quad (7.10)$$

where P_2 is a probability of no-failure for the obtained system after a new splitting into groups.

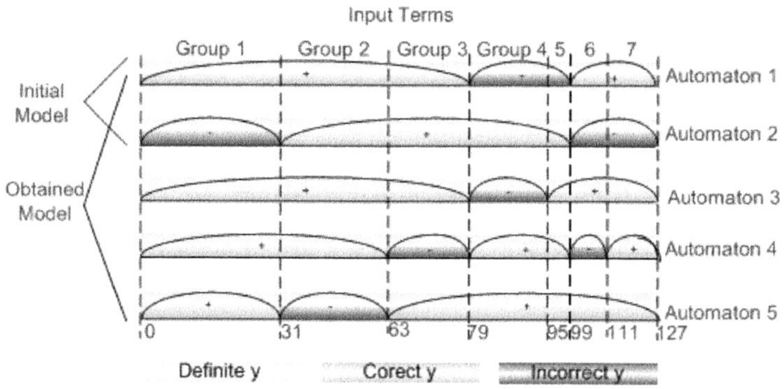

Fig. 7.20 Model of heating controller received with the "Constant" method after increasing the number of groups of output data

Hence, the high level of system reliability can be achieved by precise splitting a set of output data of automata into the groups to allow switching between the automata to be more flexible.

According to the "Optimal Reservation" method based on the algorithm of the shortest descent the resulting model of the heating controller should have the probability of no-failure $P_s^{init}(t) = 0.98$ with the minimum number of allocated logic cells. Two PCA in Fig. 7.18 constitute the minimum functional basis of the heating controller, therefore they can be selected as an initial architecture. 4 logic cells have been allocated for the first automaton and 5 cells for the second one.

Phase 1:

- initial assumption for value of p_i is 0.95 for group 1 and 0.9 for groups 2, 3 and 4 respectively;
- according to (7.5) $k = \frac{\sum_{j=1}^{h} l_j}{n \times h} = \frac{9}{4 \times 2} \approx 1$;

- to meet the condition $\forall i=\overline{1,n}: P_i \geq P_s^{init}(t)$ that requires additional redundancy in groups 1, 3 and 4, automaton 3 (Fig. 7.21) was added into the initial architecture. The probabilities of no-failure for each group are the following: $P_1 = 1-(1-0.95)^2 = 0.9975 > P_s^{init}; P_2 = P_3 = P_4 = 1-(1-0.9)^2 = 0.99 > P_s^{init}$;
- as automaton 3 does not have correct and definite output data within the whole group 4, the value of P_4 is not recalculated;
- the probability of no-failure for the whole system is the following: $P_s = (1-(1-0.95)^2)\times(1-(1-0.9)^2)^3 = 0.967873253 < P_s^{init}$;
- the number of logic cells in the current model is $K_s = K_{automaton\ 1} + K_{automaton\ 2} + K_{automaton\ 3}$=4 +5 + 5 = 14.

Phase 2:

- by adding redundancy to every group (Fig. 7.21) P_i is defined as: $P_1 = 1-(1-0.95)^3 = 0.0999875; P_2 = P_3 = P_4 = 1-(1-0.9)^3 = 0.999$;
- the value of j (7.7) for every group is the following:

$$j_1 = \frac{0.999875-0.9975}{0.999875\times1\times3} = 0.000791766,\ j_2 = j_3 = j_4 = \frac{0.999-0.99}{0.999\times1\times3} = 0.003.$$

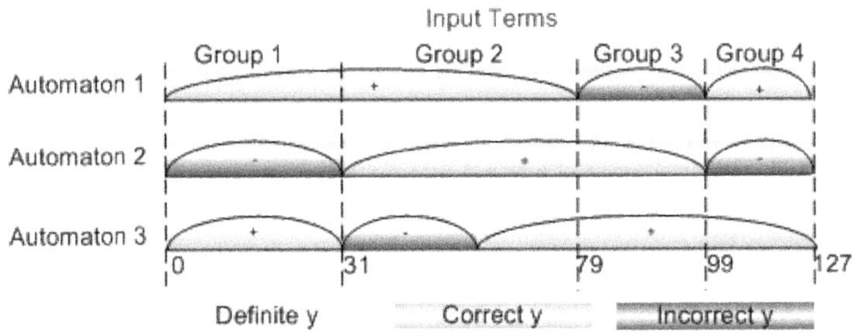

Fig. 7.21. Model of the heating controller received by optimal reservation after phase 1

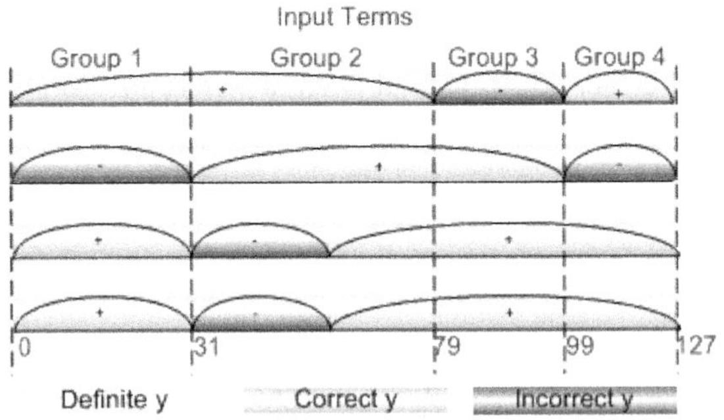

Fig. 7.22 Model of the heating controller received by optimal reservation after phase 2

As the value of j is the maximum for groups 2, 3 and 4, one more automaton was added to the model to provide the required redundancy level in these groups (Fig. 7.22). The probability of no-failure for each group and the whole model is the following:

$$P_1 = 1-(1-0.95)^3 = 0.0999875; P_2 = 1-(1-0.9)^2 = 0.99;$$

$$P_3 = P_4 = 1-(1-0.9)^3 = 0.999; P_s = 0.999875 \times 0.99 \times 0.999^2 = 0.987897487 > P_s^{init}.$$

Resulting model of the heating controller consists of 4 automata and uses 19 logic cells ($K_s = 4 + 5 + 5 + 5 = 19$). The probability of no-failure for the given model equals to 0.987897487 if the probability of no-failure for the primitive objects in group 1 is 0.95 and 0.90 for groups 2, 3 and 4 respectively. Thus, by applying the proposed method, it was proved that the essential logic of PDA and PCA allows flexible digital systems development with the certain properties such as reliability level or system complexity.

During the *third phase* the model in Fig. 7.18 was implemented to FPGA EP1K10TC144-3 (family ACEX 1K) and compared with its prototype. The sub-system that allows switching automata was designed according to Table 7.2 and Fig. 7.18.

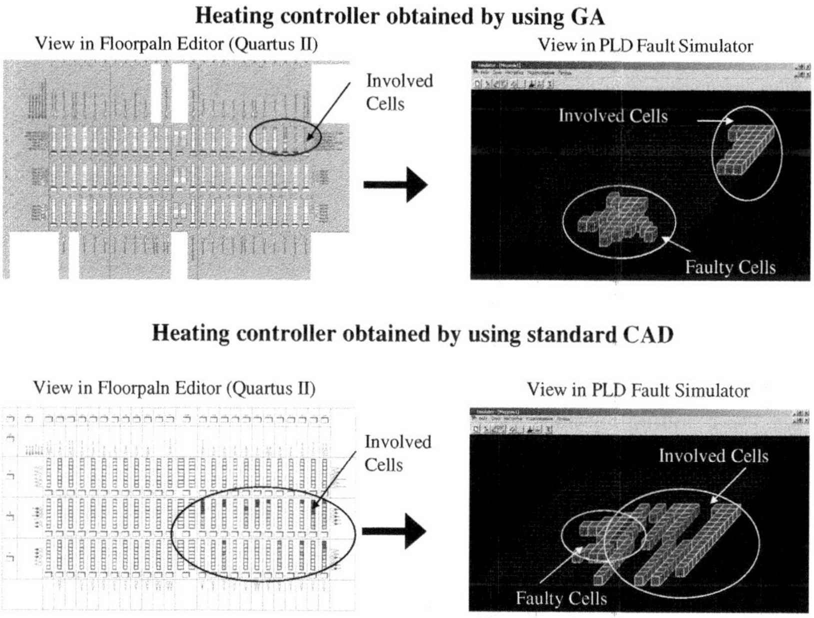

Fig. 7.23 Location of the heating controllers in the FPGA EP1K10TC144-3 and their representation in the PLD fault simulator

Both versions of controller were compared (Fig. 7.23) according to the diversity metric given in (7.1). A fault simulator [32] was used to assess the probability of keeping system operating state for both versions that is calculated as

$$P = \frac{N_1}{N}, \tag{7.11}$$

where N_1 is a number of trials for those the system kept its correct operating during fault simulation; N is a total number of trials.

The experimental results given in Table 7.5 show that the system is able to keep its operating state even though there is a significant number of faulty cells in FPGA. The reason is the compactness of the developed controller: it uses only 27 logic cells in a FPGA, whereas in the prototype the overall number of cells involved is 74.

Table 7.5 The probability of keeping system operating state with the several numbers of faulty cells

Faulty Cells	1 Cell	2 Cells	3 Cells	5% of Chip	10% of Chip	25% of Chip	50% of Chip	75% of Chip
GA-project	0.958	0.951	0.943	0.865	0.810	0.663	0.398	0.080
Prototype	0.870	0.797	0.745	0.480	0.343	0.050	0	0

Both prototype and evolved controller can be gathered to duplex architecture as it is shown in Fig. 7.2 - 7.3 to constitute a multi-version FPGA-based system. Another way is to use the obtained project as a control module for the prototype (Fig. 7.4).

7.7 Conclusions

In this chapter we reviewed the principles of multi-version digital system design. We also introduced the concept that helps to develop a simple and reliable system from unreliable parts. The corner stone of multi-versions design approach is a need to get the resulting versions as different as possible. With the method proposed both classical and non-classical paths were taken, with the latter making use of genetic algorithms. Although analysis showed a significant decrease in the FPGA utilization, on the other hand, applying the method led to the problem of building a reliable system from unreliable parts and sticking to the initial design constraints at the same time. That problem was tackled by introducing an innovative method that made it possible to 1) develop a reliable system from unreliable parts; 2) manage a level of system reliability and complexity; 3) implement efficient switching between different automata, based upon available information about their correctness and definiteness; 4) implement the system on an FPGA chip.

A practical application section proved the feasibility of the selected approach and used methods. A careful study has outlined not only obvious advantages, but also possible implications and pitfalls that method users should be aware of.

References

[1] Avizienis, A., Lapric, J.C.: Dependable Computing: From Concepts to Design Diversity. Proceedings of the IEEE 74(5), 629–638 (1986)

[2] Dobson, J., Randell, B.: Building Reliable Secure Computing Systems out of Unreliable Insecure Components. In: Proceeding of IEEE Symposium on Security and Privacy, pp. 187–193 (1986)

[3] Avizienis, A., Laprie, J.C., Randell, B., Landwehr, C.: Basic Concepts and Taxonomy of Dependable and Secure Computing. IEEE Transactions on Dependable and Secure Computing 1(1), 11–33 (2004)

[4] ITU-T. Terms and Definitions Related to QoS and Network Performance Including Dependability, Recommendations E800, Geneva (1994)

[5] Cotting, M.C., Burken, J.J.: Reconfigurable Control Design for the Full X-33 Flight Envelope. In: Proceedings of AIAA Guidance, Navigation & Control Conference, Montreal, Quebec, Canada, p. 16 (2001)

[6] Saint-Jean, S.B., Torres, R.: HS-Scale: a Hardware-Software Scalable MP-SOC Architecture for embedded Systems. In: IEEE Computer Society Annual Symposium on VLSI (ISVLSI 2007), pp. 21–28 (2007)

[7] Blanke, M., Kinnaert, M., Lunze, J., Staroswiecki, M.: Diagnosis and Fault-tolerant Control, p. 672. Springer, Heidelberg (2006)

[8] Tribble, A.C., Miller, S.P., Lempia, D.L.: Software Safety Analysis of a Flight Guidance System. Rockwell Collins, Inc., 400 Collins Rd, NE Cedar Rapids, IA 52402 USA, `http://shemesh.larc.nasa.gov/fm/papers/Tribble-SW-Safety-FGS-DASC.pdf`

[9] Rushby, J.: Formal Methods and the Certification of critical Systems. Computer Science Laboratory, SRI International menlo Park CA 94025 USA, `http://techreports.larc.nasa.gov/ltrs/PDF/cr4551.pdf`

[10] Ullmann, M., Huebner, M., Grimm, B., Becker, J.: An FPGA run-time system for dynamical on-demand reconfiguration. In: Proceedings of the 18th International Symposium on Parallel and Distributed Processing, p. 135 (2004)

[11] Benini, L., De Micheli, G.: Networks on Chips: A New SoC Paradigm. Computer 35(1), 70–78 (2002)

[12] Strigini, L., Littlewood, B.: A Discussion of Practices for Enhancing Diversity in Software Designs, Centre for Software Reliability. Technical Report LS_DI_TR_04 (2000)

[13] Townend, P., Xu, J., Munro, M.: Building Dependable Software for Critical Applications: Multi-Version Software versus One Good Version. In: Proceedings of the 6th Int. Workshop on Object-Oriented Real-Time Dependable Systems, WORDS 2001, p. 103 (2001)

[14] Kharchenko, V.S., Tarasenko V.V.: Multiversion Design Technologies of On-board Fault-tolerant FPGA Devices. In: Proceedings of MAPLD Conference, Maryland, USA (2001)

[15] Shuqing, W., Jiaping, L., Zipeng, Z., Xiaohui, Y.: Application of Neural Networks and Genetic Algorithm in Knowledge Acquisition of Fuzzy Control System. In: Proceedings of the 6th World Congress on Intelligent Control and Automation, vol. 1, pp. 3886–3890 (2006)

[16] Dunham, B., Fridshal, D., Fridshal, R., North, J.: Design by Natural Selection. In: Synthese, pp. 254–259. D. Reidel Publication Company, Dordrecht (1963)
[17] Thompson, A., Layzell, P., Zebulum, R.: Explorations in Design Space: Unconventional Electronics Design through Artificial Evolution. IEEE Transactions on Evolutionary Computation 3(3) (1999)
[18] Teuscher, C.: Turing's Connectionism. In: An Investigation of Neural Network Architectures. Springer, London (2001)
[19] Shuqing, W., Jiaping, L., Zipeng, Z., Xiaohui, Y.: Application of Neural Networks and Genetic Algorithm in Knowledge Acquisition of Fuzzy Control System. In: Proceedings of the 6th World Congress on Intelligent Control and Automation, vol. 1, pp. 3886–3890 (2006)
[20] Dias, F.M., Antunes, A., Mota, A.M.: Artificial neural networks: a review of commercial hardware. Engineering Applications of Artificial Intelligence 17(8), 945A–952A (2004)
[21] Darwing, C.: The Origin of Species. John Murray, London (1859)
[22] Holland, J.H.: Adaptation in Natural and Artificial Systems. The University of Michigan Press, Ann Arbor (1975)
[23] Savage, M.J.W., Salcic, Z., Coghill, G., Covic, G.: Extended genetic algorithm for codesign optimization of DSP systems in FPGAs. In: Proceedings of IEEE International Conference on Field-Programmable Technology, pp. 291–294 (2004)
[24] Koza, J., Al-Sakran, S., Jones, L.: Cross-Domain Features of Runs of Genetic Programming Used to Evolve Designs for Analog Circuits, Optical Lens Systems, Controllers, Antennas, Mechanical Systems and Quantum Computer Circuits. In: NASA/DoD Conference on Evolvable Hardware, pp. 205–214. IEEE Computer Society Press, Los Alamitos (2005)
[25] Hornby, G., Takamura, S., Yokono, J., Hanagata, O., Yamamoto, T., Fujita, M.: Evolving Robust Gaits with AIBO. In: IEEE International Conference on Robotics and Automation, pp. 3040–3045. IEEE, Los Alamitos (2000)
[26] Hornby, G.: Functional Scalability through Generative Representations: the Evolution of Table Designs. Environment and Planning B: Planning and Design 31(4), 569–587 (2004)
[27] Yakymets, N., Kharchenko, V.: Resource-Oriented Diversification of Fault-Tolerant PLD-Systems. Radio-Electronic and Computer Systems, KhAI 3, 45–50 (2006)
[28] Yakymets, N., Kharchenko, V.: Design of Complex Fault-Tolerant PLD-Based Systems Using Genetic Algorithms. In: Proceedings of IEEE East-West Design and Test Symposium, Yerevan, pp. 429–432 (2007)
[29] Yakymets, N., Kharchenko, V.: Fault-Tolerant Digital Systems Implemented with Partially Definite and Partially Correct Automata. In: Proceedings of the Second International Workshop on Engineering Fault Tolerant Systems (EFTS 2007), Dubrovnik, Croatia (2007)
[30] Kharchenko, V.S., Sklyar, V.V., Volkovoy, A.V.: Multi-version Information Technologies and Development of Dependable Systems out of Undependable Components. In: Proceeding of DepCoS-RELCOMEX Conference, Szklarska Poreba, Poland, pp. 43–50 (2007)
[31] Dem'yanov, V., Malozemov, V.: An introduction to Minimax. Nauka (1972)
[32] Yakymets, N., Ushakov, A.: Certificate of authorship No. 10393 for the computer program "Adjustable generator of cluster faults of logical cells in programmable logic devices" (2004)

8 Synthesis of Compositional Microprogram Control Unit with Dedicated Area of Inputs

Alexander Barkalov[1], Larysa Titarenko[1], Jacek Bieganowski[1], and Alexander Miroshkin[2]

[1] University of Zielona Góra, Institute of Computer Engineering and Electronics
e-mail: {A.Barkalov,L.Titarenko,J.Bieganowski}@iie.uz.zgora.pl
[2] Donetsk National Technical University, Department of Computers,
Apt. 41, 204a, Artema str., 83122 Donetsk, Ukraine
e-mail: arrack@mail.ru

Abstract. The chapter is devoted to CMCU optimization, based on the modification of the microinstruction format. Proposed modifications are intended to eliminate code transformers from the CMCU and reduce the hardware amount of circuits used in the FSM for the microinstruction addressing, as compared with the CMCU basic structure. The reduction of the hardware amount is achieved at the cost of increasing the number of cycles needed for the execution of the control algorithms, and in some cases also at the cost of increasing control memory size.

8.1 Introduction

A Control Unit (CU) is one of the most important parts of any digital system [10]. It is responsible for interplay of all blocks of a system. There are many ways for implementation of a control unit [4]. It can be implemented using the model of a Finite-State-Machine (FSM), and the Moore model is used very often for such a FSM [1]. Use of FSM model permits to set the circuits with the highest possible performance, especially when the single-level models [4] are applied. The second way is the model of a microprogram control unit [3], when a circuit of CU can be implemented using only multiplexors and memory blocks. In this case a control algorithm to be implemented is represented as a microprogram which is kept in a control memory. That model permits to obtain very chap but slow designs. If a control algorithm to be implemented includes long sequences of unconditional jumps, then the model of a Compositional Microprogram Control Unit (CMCU) can be used [3]. Such devices are very useful when a part of a digital system is implemented using Field-Programmable-Gate-Arrays (FPGA) [11]. These chips include look-up table elements, which can be used for implementation of the block of microinstruction addressing and the counter (together with internal flip-flops). The blocks of embedded memory [11, 12] can be used for keeping of a microprogram. Therefore, the CMCU model permits to use all types of blocks of

M. Adamski et al. (Eds.): Design of Digital Systems and Devices, LNEE 79, pp. 193–214.
springerlink.com

an FPGA chip. This model gives the best results for linear control algorithms [1]. Here we propose some new methods for implementation of CMCU. Our proposed methods are based on some modifications of microinstructions in comparison with well-known base methods of CMCU implementation [1, 3, 4].

8.2 Background of CMCU

Let a graph-scheme of algorithm (GSA) Γ be used for representation of a control algorithm [2]. Let $B = \{b_0, b_E\} \cup E_1 \cup E_2$ be a set of GSA vertices and E be a set of arks connected some of these vertices. Here b_0 is an initial (start) vertex of GSA, b_E is a final vertex, E_1 is a set of operator vertices, where $M=|E_1|$, and E_2 is a set of conditional vertices. Each operator vertex $b_q \in E_1$ contains a collection of micro-operations $Y(b_q) \subseteq Y$, where $Y = \{y_1,\ldots,y_N\}$ is a set of data-path microoperations [2]. Each conditional vertex $b_q \in E_2$ includes some logical condition $x_e \in X$, where $X = \{x_1,\ldots,x_L\}$ is a set of logical conditions. A GSA Γ is named linear GSA if the number M exceeds 75% of the total number of its vertices [1].

Let us introduce some definitions used in this chapter.

Definition 1. An operational linear chain (OLC) of GSA Γ is a finite vector of operator vertices $\alpha_g = \langle b_{g_1}, \ldots, b_{g_{Fg}} \rangle$, such that an arc $\langle b_{g_i}, b_{g_{i+1}} \rangle \in E$ corresponds to each pair of adjacent vertices b_{g_i}, $b_{g_{i+1}}$, where i is the component number of vector α_g.

Let D^g be a set of operator vertices, which are components of OLC α_g.

Definition 2. An operator vertex $b_q \in D^g$ is called an input of OLCα_g, if there is an arc $\langle b_t, b_q \rangle \in E$, such that $b_t \notin D^g$.

Definition 3. An input $b_q \in D^g$ is called a main input of OLC α_g, if GSA Γ does not include an arc $\langle b_t, , b_q \rangle \in E$ such that $b_t \in B_1$.

Definition 4. An operator vertex $b_q \in D^g$ is called an output of OLC α_g, if there is an arc $\langle b_t, , b_q \rangle \in E$, where $b_t \notin D^g$.

It follows from the basic properties of GSA [2] that each OLC α_g corresponding to definitions given above should have at least one input and exactly one output. Let I_g^j stand for input j of OLC α_g and O_g for its output. Let inputs of OLC α_g form a set $I(\alpha_g)$.

For GSA Γ we have the following sets:

1. A set of OLC $C=\{\alpha_1, \ldots, \alpha_G\}$, satisfying the following condition

$$\begin{aligned} &D^1 \cup \ldots \cup D^G = B_1; \\ &\left|D^i \cap D^j\right| = 0 \left(i \neq j; i, j \in \{1, \ \ldots, \ G\}\right); \\ &G \rightarrow \min. \end{aligned} \tag{8.1}$$

2. A set of inputs $I(\Gamma)$ of the operational linear chains of GSA Γ:

$$I(\Gamma)=\bigcup_{g=1}^{G} I(\alpha_g). \tag{8.2}$$

3. A set of outputs $O(\Gamma)$ of the operational linear chains of GSA Γ:

$$O(\Gamma)=\{O_1, \ \ldots, \ O_G\}. \tag{8.3}$$

Let the natural microinstruction addressing be executed for microinstructions corresponding to the adjacent components of each OLC $\alpha_g \in C$:

$$A(b_{g_{i+1}})= A(b_{g_i})+1 \ (i=1, \ \ldots, \ F_g-1). \tag{8.4}$$

In expression (8.4) symbol $A(B_{gi})$ stands for the address of microinstruction corresponding to component i of vector $\alpha_g \in C$, where i=1, …, F_g–1.

In this case GSA Γ can be interpreted by compositional microprogram control unit with basic structure of Fig. 8.1 [10]. Let us denote it as unit U_1.

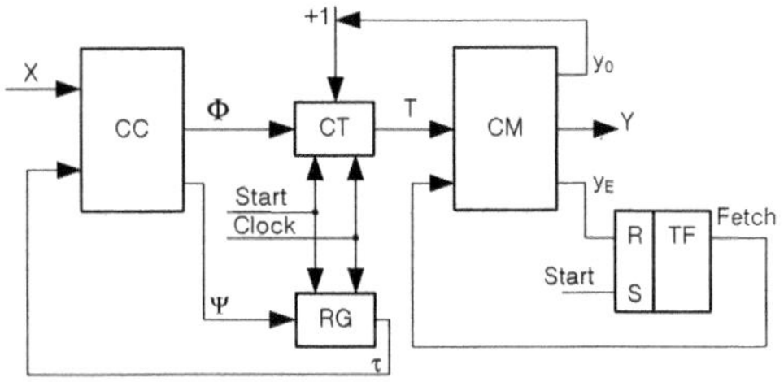

Fig. 8.1 Structural diagram of compositional microprogram control unit with basic structure

In the unit U_1, combinational circuit CC and register RG form a finite state machine S_1, which will be called *microinstruction addressing unit* or FSM S_1. Counter CT, control memory CM and flip-flop TF form microprogram control unit S_2 with natural microinstruction addressing. The unit U_1 operates in the following manner.

The pulse "Start" initializes following actions: the zero code of FSM S_1 initial state is loaded into register RG; start address of microprogram is loaded into counter CT; flip-flop TF is set up (Fetch=1). If Fetch=1, microinstructions can be fetched out of the control memory. Let at time t (t=0, 1, 2, …) the code of state $a_m \in A_1$, where A_1 is a set of FSM S_1 states, be loaded into register RG and address $A(I_g^j)$ of the input j of OLC $\alpha_g \in C$ be loaded into the counter CT. Current microinstruction is read out of CM and its microoperations $y_n \in Y$ initialize some actions of the data-path. If this input is not the output of current OLC $\alpha_g \in C$ ($I_g^j \neq O_g$), additional variable y_0=1 is generated by MCU S_2. If y_0=1, content of register RG is unchangeable and 1 is added to the content of counter CT. It corresponds to a transition between adjacent components of OLC $\alpha_g \in C$. If the output O_g is reached, then y_0=0. In this case circuit CC generates Boolean functions:

$$\Phi = \Phi(\tau, \quad X), \tag{8.5}$$
$$\Psi = \Psi(\tau, \quad X), \tag{8.6}$$

where $\tau=\{\tau_1, \ldots, \tau_{R_1}\}$ is a set of state variables encoding states $\alpha_m \in A_1$. The minimum number of these variables is determined as

$$R_1 = \lceil \log_2 M_1 \rceil, \tag{8.7}$$

where $M_1=|A_1|$. If there is a transition from output O_g to some input under influence of some values of logical conditions, functions (8.5) determine the address of this input $I_i^j \in I(\Gamma)$ which is to be loaded into the counter. Functions (8.7) calculate the code of next state $a_s \in A_1$ to be loaded into RG. Content of both CT and RG is changed by the pulse "Clock". Outputs of the CT, $T=\{T_1,\ldots T_{R_2}\}$ determine next microinstruction address. This set includes

$$R_2 = \lceil \log_2 M_2 \rceil \tag{8.8}$$

variables, where $M_2=|B_2|$. If CT contains the address of microinstruction corresponding to vertex $b_q \in B_1$ such that $\langle b_q, b_E \rangle \in E$, some additional variable $y_E=1$ is generated. If $y_E=1$, the flip-flop TF is cleared. Thus Fetch=0 and microinstruction fetching from the control memory is terminated.

As follows from (8.5), FSM S_1 of unit U_1 implements any multidirectional microprogram transition between output $O_g \in O(\Gamma)$ and input $I_i^j \in I(\Gamma)$ in one cycle of operation. At the same time MCU S_2 implements addressing rule (8.4), used to organize transitions between microinstructions corresponding to adjacent components of OLC $\alpha_g \in C$. Therefore, control memory CM should only keep microoperations $y_n \in Y$ and additional variables y_0, y_E. In other words, an address part is absent in the microinstruction format in case of CMCU U_1. The main disadvantage of CMCU U_1 is the loss of universality, because changes in the interpreted microprogram lead to the redesign of circuit CC. Fortunately, current achievements in semiconductor technology permit to eliminate this drawback.

In this chapter we deal with some modification of CMCU U_1, namely CMCU with common memory [4] denoted here as U_2. It has the following structure (Fig. 8.2)

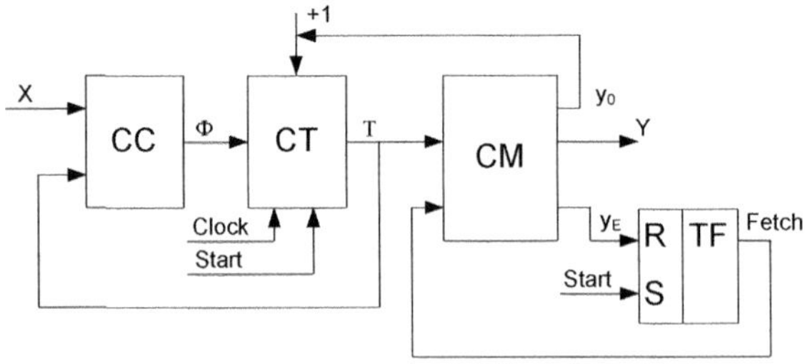

Fig. 8.2 Structural diagram of CMCU U_2

In CMCU U_2, the counter CT is used as a source of the codes for S_1 and addresses for S_2. The Circuit CC implements system

$$\Phi = \Phi(T, X). \tag{8.9}$$

All other blocks of CMCU U_2 execute the same functions as corresponding blocks of CMCU U_1.

8.3 Synthesis of CMCU with Dedicated Area of Inputs

All compositional microprogram control units known from literature have some common feature, namely generation of input addresses by the block CC. This approach can be called hardware address generation, in which the number of outputs in the CC block is equal to R_2 (model U_1 is the only exception). In order to reduce this number, some additional block for address generation is needed (for transformation of object codes). The second approach leads to increasing of the CMCU cycle time, in comparison with its value for CMCU U_1. In case of the CMCU with elementary OLC and code sharing [4], the number of CC outputs is smaller than R_2, but application of these methods can cause either significant increase of the control memory size, in comparison with its minimal value $V_{\min}$, or an increase of the CMCU cycle time. If the increase of time cycle is not desirable, the number of CC outputs cannot be reduced, in comparison with R_2. Let us consider how the number of CC outputs can be reduced in cases when application of code sharing leads to introduction of the address transformer, but performance of the resulting CMCU cannot be worse, than in case of the CMCU U_1. Let us discuss these methods using an example of CMCU U_2. Our discussion is based on results from [5-9].

In case of CMCU U_2, the output addresses of OLC $\alpha_g \in C$ possess the property of randomness. Application of address procedure does not guarantee that, for example, some bit is equal to zero for all input addresses. Situation of this kind would allow to reduce the number of CC block outputs in comparison with R_2. Let the set of OLC inputs $I(\Gamma)$ for GSA Γ include I_0 elements, which can be encoded by only R_3 bits, where

$$R_3 = \lceil \log_2 I_0 \rceil. \tag{8.10}$$

Obviously the following condition is satisfied for the linear graph-schemes of algorithm, where the number of operator vertices exceeds significantly the number of conditional vertices:

$$R_3 < R_2. \tag{8.11}$$

Let the following condition (8.3) be satisfied for GSA Γ:

$$\lceil \log_2 (I_0 + M_2) \rceil = \lceil \log_2 M_2 \rceil, \tag{8.12}$$

where M_2 is the number of operator vertices. Let us choose I_0 cells of control memory to keep OLC inputs and let these cells have addresses from 0 to $(I_0-1)_2$. Let us call this set of cells a dedicated input area (DIA). This fixation of OLC inputs requires execution of unconditional jumps to the real input address, which should be introduced into the special control microinstruction. It leads to some modification of microinstruction formats in comparison with CMCU U_2 [4]. The model of CMCU U_3 with dedicated input area is shown in Fig. 8.3.

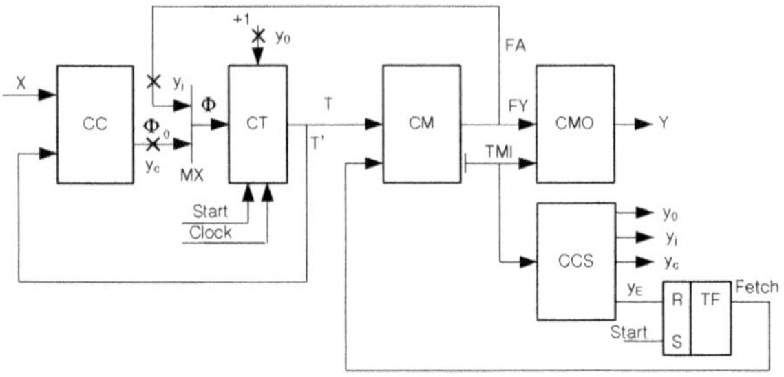

Fig. 8.3 Structural diagram of CMCU U_3

Let us discuss particular qualities of CMCU U_3 in comparison with U_2. In case of the CMCU U_3, there are two formats of microinstructions (Fig. 8.4).

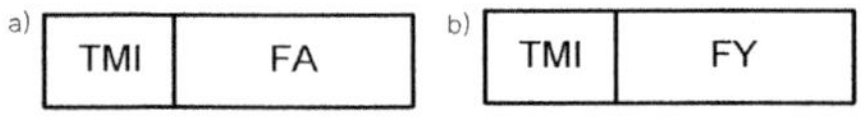

Fig. 8.4 Microinstruction formats for CMCU U_3

The control microinstruction, shown in Fig. 8.4a, contains an address field FA with address for transition from the dedicated input area into the area of microprogram (AMP) containing operational microinstructions. This format includes a field of attribute TMI, with all zeroes (TMI=00). Operational microinstruction (Fig. 8.4b) includes the field TMI and an operational part FY. If TMI=01, this microinstruction corresponds to an OLC output, corresponding in turn to y_C=1. If TMI=10, the microinstruction corresponds to some OLC component, which is not an OLC output. It corresponds to y_0=1. Code TMI=11 indicates that some OLC output connected with vertex b_E is reached. It corresponds to y_E=1. Let us point out that the control microinstruction corresponds to y_j=1.

In case of the control microinstructions, some additional block for generation of microoperations should be used to prevent generation of microoperations $y_n \in Y$ (if y_j=1), because in this case microinstruction would contain information about address of transition only. Multiplexer MX should be used to load into counter CT: either the transition address created by functions Φ_0 (y_C=1), or the address of some cell of the microprogram area, which occupies the field FA of the control microinstruction (y_j=1). Block CCS is used to generate control signals y_0, y_j, y_C, y_E, depending on the content of field TMI.

Compositional microprogram control unit U_3 operates in the following manner. First, zero code is loaded into the counter CT using pulse *Start*, corresponding to the address of main OLC $\alpha_1 \in C$ input, kept in the dedicated input area. At the same time, flip-flop TF is set up and allows microinstruction fetching from the CM control memory (Fetch=1). Current microinstruction is read from the control memory CM and block CCS generates some control signals y_0, y_j, y_C, y_E. If CT contains the address of OLC output, variable y_C=1 is generated together with microoperations $y_n \in Y$. In this case, input memory functions

$$\Phi_0 = \Phi_0(\mathrm{T}, \quad X) \tag{8.13}$$

load the address taken from dedicated input area into the counter CT. The signal y_j is generated and an address from AMP is loaded into CT. If the counter CT contains an address of OLC component corresponding to vertex b_q, such that $\langle b_q, b_E\rangle \notin \mathrm{E}$ and $b_q \neq O_g$, both microoperations $y_n \in \mathrm{Y}(b_q)$ and variable y_0=1 are generated. In consequence, the counter content is incremented and causes transition to the following microinstruction. If the counter CT contains the address of microinstruction corresponding to vertex b_q, such that $\langle b_q, b_E\rangle \in \mathrm{E}$, variable y_E is generated and fetching of microinstructions terminated.

The method of CMCU U_3 synthesis includes the following steps:

1. Transformation of initial GSA.
2. Construction of the OLC set using transformed GSA $\Gamma(U_3)$.
3. Finding addresses for OLC inputs.
4. Microinstruction addressing.
5. Construction of the control memory content.
6. Construction of the transition table of CMCU.
7. Construction of CCS table.
8. Synthesis of CMCU logic circuit using given logical elements.

Let us discuss application of this method for synthesis of the CMCU $U_3(\Gamma_1)$, where the transformed GSA $\Gamma_1(U_3)$ is shown in Fig. 8.5.

Application of addressing procedure to the transformed GSA $\Gamma_1(U_3)$ gives the set $C=\{\alpha_1,\ldots,\alpha_6\}$, where $\alpha_1=\langle b_1,b_2\rangle$, $I_1^1=b_1$, $O_1=b_2$; $\alpha_2=\langle b_3,b_4,b_5\rangle$, $I_2^1=b_3$, $I_2^2=O_2=b_5$; $\alpha_3=\langle b_6,\ldots,b_9\rangle$, $I_3^1=b_6$, $I_3^2=b_8$, $O_3=b_9$; $\alpha_4=\langle b_{10}, b_{11}\rangle$, $I_4^1=b_{10}$, $O_4=b_{11}$; $\alpha_5=\langle b_{12}, b_{13}\rangle$, $I_5^1=b_{12}$, $O_5=b_{13}$; $\alpha_6=\langle b_{14},\ldots,b_{17}\rangle$, $I_6^1=b_{14}$, $O_6=b_{17}$. Thus, we get the set of inputs $I(\Gamma_1)=\{b_1, b_3, b_5, b_6, b_8, b_{10}, b_{12}, b_{14}\}$, and the following values can be found: M_2=17, R_2=5, I_0=8, R_3=3. It means that condition (8.11) holds and application of the method proposed above makes sense. Moreover, because M_2+I_0=25, condition (8.12) is satisfied and this method allows to have smaller number of CC inputs, without increasing the length of microinstruction address, in comparison with CMCU $U_2(\Gamma_1)$.

Addressing of OLC inputs is executed in trivial way, but the address of input I_1^1 should be equal to zero. Let $IA(b_q)$ be the address of input corresponding to vertex $b_q \in B_2$. In case of CMCU $U_3(\Gamma_1)$ these addresses are: $IA(b_1)$=000, $IA(b_3)$=001,… $IA(b_{14})$=111.

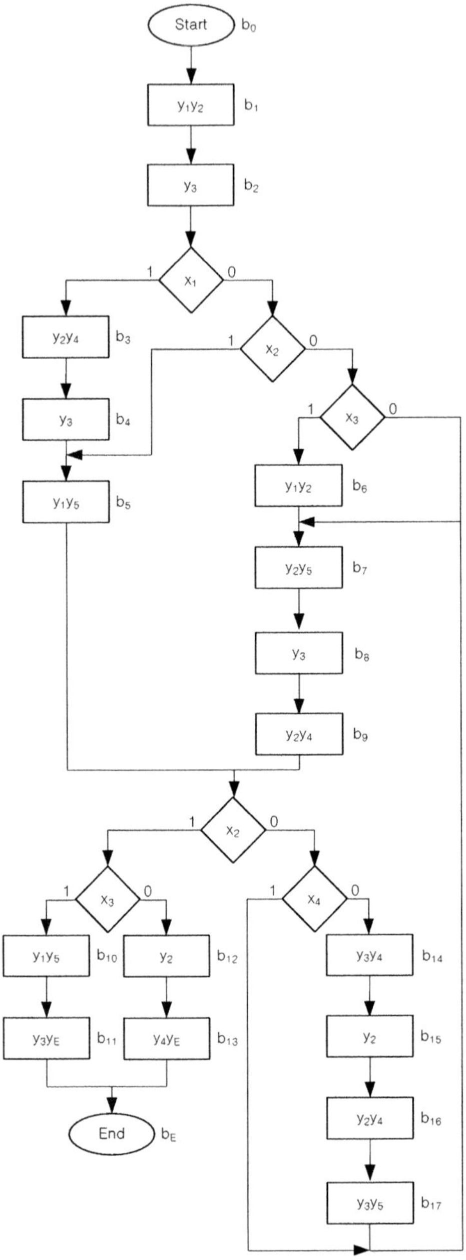

Fig. 8.5 Transformed GSA $\Gamma_1(U_3)$

Application of addressing procedure to GSA $\Gamma_1(U_3)$ results in microinstruction addresses shown in Fig. 8.6.

T_4T_5 \ $T_1T_2T_3$	000	001	010	011	100	101	110	111	
00	b_1	b_3	b_5	b_6	b_8	b_{10}	b_{12}	b_{14}	DIA
01	b_1	b_2	b_3	b_4	b_5	b_{18}	b_7	b_8	AMP
11	b_9	b_{10}	b_{11}	b_{12}	b_{13}	b_{19}	b_{15}	b_{16}	
10	b_{17}	*	*	*	*	*	*	*	

Fig. 8.6 Microinstruction addresses for CMCU $U_3(\Gamma_1)$

First line of the table from Fig. 8.6 corresponds to the dedicated input area and each cell of this line contains an address $IA(b_q)$. The rest of lines corresponds to the area of microprogram AMP and each cell for this part of lines contains an address $A(b_q)$. For example, input I_5^1=b_{12} and its address in DIA is determined as $IA(b_{12})$=001100, whereas its address in AMP is $A(b_{12})$=10011.

Microinstructions to be kept in the control memory are constructed using the following rules:

- any vertex $b_q \in I(\Gamma)$ from DIA corresponds to a control microinstruction of the unconditional jump, where [FA]= $A(b_q)$;
- if vertex $b_q \in D^g$ is not an output of OLC $\alpha_g \in C$, the control memory cell having address $A(b_q)$; should contain operational microinstruction, where $[TMI]=y_0$;
- if vertex $b_q \in D^g$ is connected with final vertex b_E, the control memory cell with address $A(b_q)$ should contain operational microinstruction, where [TMI]=y_E.

Let us denote the construction procedure of the control memory content by symbol P_1. Application of procedure P_1 gives the control memory content shown in Table 8.1.

Let us point out that only 16 cells of the control memory of CMCU $U_3(\Gamma_1)$ are shown in Table 8.1. Two bits are used to encode variables y_0, y_j, y_C, y_E, namely m_1 and m_2. The encoding is executed in such a manner that code 00 corresponds to y_j, code 01 to y_0, code 10 to y_C, and code 11 to y_E. One-hot encoding approach is used to encode microoperations, when the bit capacity R_{CM} of the control memory cell is given by the expression

$$R_{CM} = \max(2+N,\quad 2+\lceil \log_2(I_0+M_2) \rceil). \tag{8.14}$$

In this case R_{CM}=7, which means that fields FA and FY are represented by bits m_3–m_7.

The transition table of CMCU is constructed using the system of transition formulae for outputs of OLC $\alpha_g \in C^1$. In the discussed case we have C^1={α_1, α_2, α_3, α_6} and the following transition formulae:

$$\begin{aligned} &O_1 \to x_1 I_2^1 \vee \bar{x}_1 x_2 I_2^2 \vee \bar{x}_1 \bar{x}_2 x_3 I_3^1 \vee \bar{x}_1 \bar{x}_2 \bar{x}_3 I_3^2; \\ &O_2, O_3 \to x_2 x_3 I_4^1 \vee x_2 \bar{x}_3 I_5^1 \vee \bar{x}_2 x_4 I_3^1 \vee \bar{x}_2 \bar{x}_4 I_6^1; \\ &O_6 \to I_3^2. \end{aligned} \tag{8.15}$$

Table 8.1 Content of control memory for CMCU $U_3(\Gamma_1)$

Address $T_1T_2T_3T_4T_5$	TMI m_1m_2	Content $m_3m_4m_5m_6m_7$	Reference
00000	00	01000	$b_1 \rightarrow A(b_1)$ DIA
00001	00	01010	$b_3 \rightarrow A(b_3)$
00010	00	01100	$b_5 \rightarrow A(b_5)$
00011	00	01101	$b_6 \rightarrow A(b_6)$
00100	00	01111	$b_8 \rightarrow A(b_8)$
00101	00	10001	$b_{10} \rightarrow A(b_{10})$
00110	00	10011	$b_{12} \rightarrow A(b_{12})$
00111	00	10101	$b_{14} \rightarrow A(b_{14})$
01000	01	11000	$b_1 \rightarrow b_2$ AMP
01001	10	00100	$b_2 \rightarrow O_1$
01010	01	01010	$b_3 \rightarrow b_4$
01011	01	00100	$b_4 \rightarrow b_5$
01100	10	10001	$b_5 \rightarrow O_2$
01101	01	11000	$b_6 \rightarrow b_7$
01110	01	01001	$b_7 \rightarrow b_8$
01111	01	00100	$b_8 \rightarrow b_9$

Table 8.2 Transition table for CMCU $U_3(\Gamma_1)$

O_g	$A(O_g)$	I_m^j	$A(I_m^j)$	X_h	Φ_h	h
O_1	01001	I_2^1	00001	x_1	D_5^1	1
		I_2^2	00010	$/x_1x_2$	D_4^1	2
		I_3^1	00011	$/x_1/x_2x_3$	$D_4^1 D_5^1$	3
		I_3^2	00100	$/x_1/x_2/x_3$	D_3^1	4
O_2	01100	I_4^1	00101	x_2x_3	$D_3^1 D_5^1$	5
		I_5^1	00110	x_2/x_3	$D_3^1 D_4^1$	6
		I_3^1	00011	$/x_2x_4$	$D_4^1 D_5^1$	7
		I_6^1	00111	$/x_2/x_4$	$D_3^1 D_4^1 D_5^1$	8
O_3	10000	I_4^1	00101	x_2x_3	$D_3^1 D_5^1$	9
		I_5^1	00110	x_2/x_3	$D_3^1 D_4^1$	10
		I_3^1	00011	$/x_2x_4$	$D_4^1 D_5^1$	11
		I_6^1	00101	$/x_2/x_4$	$D_3^1 D_5^1$	12
O_6	11000	I_3^2	00100	1	D_3^1	13

Transition table of the CMCU $U_3(\Gamma_1)$ corresponds to system (8.15) and includes $H_3(\Gamma_1)$=13 lines (Table 8.2). This table is used to obtain the input memory functions for the flip-flops of counter CT (8.13), as for example:

$$D_3^1 = F_4 \vee F_5 \vee F_6 \vee F_8 \vee F_9 \vee F_{10} \vee F_{12} \vee F_{13} = \bar{T}_1 T_2 \bar{T}_3 T_5 \bar{x}_1 \bar{x}_2 \bar{x}_3 \vee \ldots \vee \bar{T}_1 T_2 \bar{T}_3 \bar{T}_5 .$$

The superscript «1» of function D_3 reflects the fact that D_3 belongs to the set Φ_0. If this superscript is omitted, we obtain $D_3 \in \Phi_0$. It can be found from this formula that the address bit $T_4=0$ for all outputs of OLC, and therefore corresponding variable is absent in system (8.13).

The table for block CCS is constructed in trivial way and in our particular case it is replaced by the Karnaugh map (Fig. 8.7).

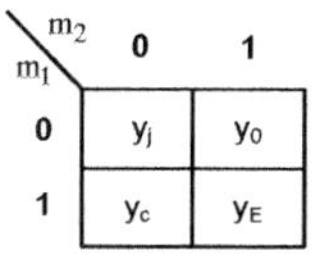

Fig. 8.7 Codes of control variables

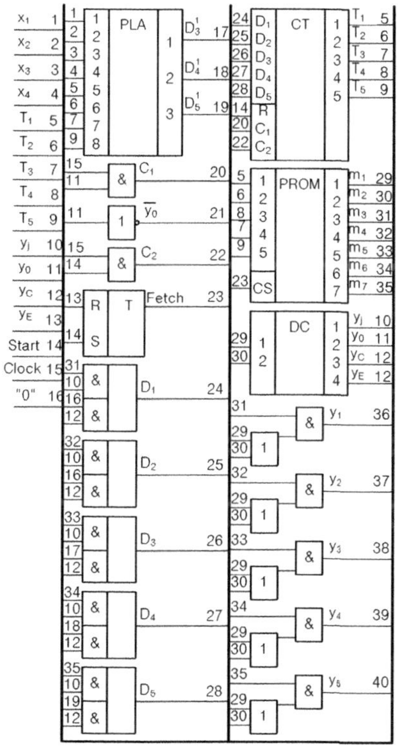

Fig. 8.8 Logic circuit for CMCU $U_3(\Gamma_1)$

Obviously, variables y_0, y_j, y_c, y_E are generated by a decoder with m_1 and m_2 inputs. Logic circuit of CMCU $U_3(\Gamma_1)$ is shown in Fig. 8.8. Here, the two-level block AND-OR implements multiplexer MX, outputs of which correspond to the input memory function of counter CT. The multiplexer is described by the following equations:

$$D_1 = y_j \cdot m_3 \vee 0 \cdot y_c; \quad D_2 = y_j \cdot m_4 \vee 0 \cdot y_c; \quad D_3 = y_j \cdot m_5 \vee D_3^1 \cdot y_c;$$
$$D_4 = y_j \cdot m_6 \vee D_4^1 \cdot y_c; \quad D_5 = y_j \cdot m_7 \vee D_5^1 \cdot y_c. \tag{8.16}$$

It is clear, that system (8.16) can be implemented using FPGA, but the logic circuit shown in Fig. 8.8 reflects main principles of CMCU organization only, without its implementation using modern FPLDs.

The system of microoperations is implemented in the following way. It can be seen from the Karnaugh map (Fig. 8.6), that the operational microinstruction is determined by disjunction $m_1 \vee m_2$. Thus, for example, microoperation y_1 is generated if m_3=1 and $m_1 \vee m_2$=1. This analysis leads to the following system:

$$y_1 = m_3(m_1 \vee m_2); \; y_2 = m_4(m_1 \vee m_2); \; y_3 = m_5(m_1 \vee m_2);$$
$$y_4 = m_6(m_1 \vee m_2); \; y_5 = m_7(m_1 \vee m_2). \tag{8.17}$$

System (8.17) is implemented by the circuit of Fig. 8.6 using AND and OR gates; but can be also implemented with FPGA.

This approach can be applied to obtain some modifications of the CMCU U_3-U_6 models, which are briefly discussed below.

Allocation of the dedicated input area transforms CMCU with special addressing into CMCU U_4, structural diagram of which is the same as the structural diagram of CMCU U_3, but inputs of the block CC of CMCU U_4 are connected with address variables $T' \subseteq T$. The outcome of special microinstruction addressing for CMCU $U_4(\Gamma_1)$ is shown in Fig. 8.9.

T_4T_5 \ $T_1T_2T_3$	000	001	010	011	100	101	110	111
00	b_1	b_8	b_1	b_5	b_9	b_{17}	b_{13}	*
01	b_3	b_{10}	b_2	b_6	b_{14}	b_{10}	*	*
11	b_5	b_{12}	b_3	b_7	b_{15}	b_{11}	*	*
10	b_6	b_{14}	b_4	b_8	b_{16}	b_{12}	*	*

Fig. 8.9 Microinstruction addresses for CMCU $U_4(\Gamma_1)$

In this particular case, output O_1 is determined unambiguously by the generalized interval of a Boolean space 010**, output O_2 by 011**, output O_3 by 100**, and output O_6 by 101**. Therefore, inputs of the block CC for the CMCU $U_4(\Gamma_1)$ are connected with the variables from set $T'=\{T_1, T_2, T_3\}$. It means that the number of CC inputs is smaller, than in case of CMCU $U_3(\Gamma_1)$.

Transformation of the table of Fig. 8.9 into the table shown in Fig.8.10 results in reduction of the number of address variables connected with CC to only two bits.

T_4T_5 \ $T_1T_2T_3$	000	001	010	011	100	101	110	111
00	b_1	b_8	b_{10}	b_{12}	b_1	b_3	b_6	b_{14}
01	b_3	b_{10}	b_{11}	b_{13}	b_2	b_4	b_7	b_{15}
11	b_5	b_{12}	*	*	*	b_5	b_8	b_{16}
10	b_6	b_{14}	*	*	*	*	b_9	b_{17}

Fig. 8.10 New microinstruction addresses for CMCU $U_4(\Gamma_1)$

Analysis of Fig. 8.8 shows that output O_1 is unambiguously determined by the generalized Boolean interval 100**, output O_2 by interval 101**, output O_3 by interval 110**, and output O_6 by interval 111**. We have T_1=1 for all outputs of OLC $\alpha_g \in C^1$, and therefore only variables $T_2,T_3 \in T'$ should be connected with the inputs of CC.

Allocation of the dedicated input area transforms CMCU with optimal addressing into CMCU U_5, with the same structural diagram as in case of CMCU U_3, but the application of optimal encoding of OLC $\alpha_g \in C^1$ components allows to reduce the number of terms in (8.13). The outcome of optimal encoding (more correctly, optimal microinstruction addressing) for CMCU $U_5(\Gamma_1)$ is shown in Fig. 8.11.

T_4T_5 \ $T_1T_2T_3$	000	001	010	011	100	101	110	111
00	b_1	b_8	b_{10}	b_{12}	b_1	b_{14}	b_3	b_6
01	b_3	b_{10}	b_{11}	b_{13}	b_2	b_{15}	b_4	b_7
11	b_5	b_{12}	*	*	*	b_{16}	b_5	b_8
10	b_6	b_{14}	*	*	*	b_{17}	*	b_9

Fig. 8.11 Optimal microinstruction addresses for CMCU $U_5(\Gamma_1)$

In the discussed case, partition $\Pi_C=\{B_1, B_2, B_3\}$ can be formed, where $B_1=\{\alpha_1\}$, $B_2=\{\alpha_2,\alpha_3\}$, $B_3=\{\alpha_6\}$. Analysis of Fig. 8.9 shows that class B_1 is determined by code $K(B_1)$=*10**, class B_2 by code $K(B_2)$=*0***, and class B_3 by code $K(B_3)$=*11**. In this case all address assignments corresponding to the components of OLC $\alpha_g \in C^1$ are treated as insignificant and are used for optimization of the codes of classes.

Let us transform system (8.6) into the form:

$$\begin{aligned} &B_1 \rightarrow x_1 I_2^1 \vee \bar{x}_1 x_2 I_2^2 \vee \bar{x}_1 \bar{x}_2 x_3 I_3^1 \vee \bar{x}_1 \bar{x}_2 \bar{x}_3 I_3^2; \\ &B_2 \rightarrow x_2 x_3 I_4^1 \vee x_2 \bar{x}_3 I_5^1 \vee \bar{x}_2 x_4 I_3^1 \vee \bar{x}_2 \bar{x}_4 I_6^1; \\ &B_3 \rightarrow I_3^2. \end{aligned} \tag{8.18}$$

System (8.18) corresponds to the transition table of CMCU $U_5(\Gamma_1)$ with $H_5(\Gamma_1)$=9 lines (Table 8.3).

This table is used to construct system (8.13). For example, the following equation can be derived from Table 8.3: $D_3^1 = F_4 \vee F_5 \vee F_6 \vee F_8 \vee F_9 = T_2 \bar{T}_3 \bar{x}_1 \bar{x}_2 \bar{x}_3 \vee \ldots \vee T_1 T_2$. Comparison of equations for the function D_3^1 of CMCU $H_3(\Gamma_1)$ and $H_5(\Gamma_1)$ shows that in the second case the number of terms is 1.6 times smaller and the number of literals reduced by two elements.

Table 8.3 Transition table for CMCU $U_5(\Gamma_1)$

B_i	$K(B_i)$	I_m^j	$A(I_m^j)$	X_h	Φ_h	h
B_1	*10**	I_2^1	00001	x_1	D_5^1	1
		I_2^2	00010	$/x_1x_2$	D_4^1	2
		I_3^1	00011	$/x_1/x_2x_3$	$D_4^1 D_5^1$	3
		I_3^2	00100	$/x_1/x_2/x_3$	D_3^1	4
B_2	*1***	I_4^1	00101	x_2x_3	$D_3^1 D_5^1$	5
		I_5^1	00110	x_2/x_3	$D_3^1 D_4^1$	6
		I_3^1	00011	$/x_2x_4$	$D_4^1 D_5^1$	7
		I_6^1	00111	$/x_2/x_4$	$D_3^1 D_4^1 D_5^1$	8
B_3	*11**	I_3^2	00100	1	D_3^1	9

Allocation of the dedicated input area in case of CMCU with transformation of addresses turns it into the CMCU U_6, with the structural diagram of Fig. 8.12. In this case block CC generates functions (8.13)

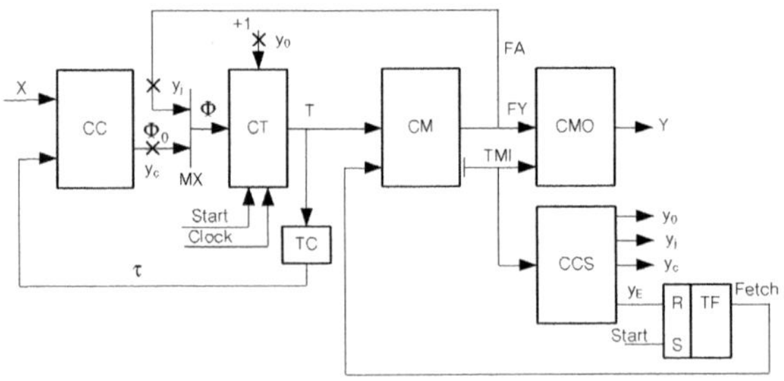

Fig. 8.12 Structural diagram of CMCU U_6

All other blocks of both CMCU U_5 and U_3 implement similar functions. Synthesis method used for CMCU U_6 can be interpreted as a modification of the synthesis method applied for CMCU U_3 and includes some additional steps such as construction of partition Π_C of the set C^1, encoding of classes $B_i \in \Pi_C$, and construction of the table for block TC.

Let us consider an example of CMCU $U_6(\Gamma_1)$ synthesis. The microinstruction addresses for the CMCU are given in Fig. 8.8. As it was shown earlier, we can get the partition $\Pi_C=\{B_1, B_2, B_3\}$, where $B_1=\{\alpha_1\}$, $B_2=\{\alpha_2, \alpha_3\}$, $B_3=\{\alpha_6\}$. It is sufficient to have two variables from set $\tau=\{\tau_1, \tau_2\}$ to encode classes $B_i \in \Pi_C$. Let us use the codes: $K(B_1)=01$, $K(B_2)=00$, $K(B_3)=10$. Transition table for CMCU $U_6(\Gamma_1)$ is constructed using the system of transition formulae (8.18) and includes $H_6(\Gamma_1)=9$ lines (Table 8.4).

This table is used to construct equations (8.13). We find, for example, the equation:

$$D_3^1 = F_4 \vee F_5 \vee F_6 \vee F_8 \vee F_9 = \bar{\tau}_1 \tau_2 \bar{x}_1 \bar{x}_2 \bar{x}_3 \vee \bar{\tau}_1 \bar{\tau}_2 x_2 x_3 \vee \ldots \vee \tau_1 \tau_2 . \quad (8.19)$$

Table 8.4 Transition table for CMCU $U_6(\Gamma_1)$

B_i	$K(B_i)$	I_m^j	$A(I_m^j)$	X_h	Φ_h	h
B_1	01	I_2^1	00001	x_1	D_5^1	1
		I_2^2	00010	$/x_1x_2$	D_4^1	2
		I_3^1	00011	$/x_1/x_2x_3$	$D_4^1 D_5^1$	3
		I_3^2	00100	$/x_1/x_2/x_3$	D_3^1	4
B_2	00	I_4^1	00101	x_2x_3	$D_3^1 D_5^1$	5
		I_5^1	00110	x_2/x_3	$D_3^1 D_4^1$	6
		I_3^1	00011	$/x_2x_4$	$D_4^1 D_5^1$	7
		I_6^1	00111	$/x_2/x_4$	$D_3^1 D_4^1 D_5^1$	8
B_3	10	I_3^2	00100	1	D_3^1	9

The corresponding table of code transformer TC is constructed in a traditional way and shown in Table 8.5. This table is used to construct functions $\tau=\tau(T)$, which in our case have the form: $\tau_1=T_1T_2T_3$, $\tau_2 = T_1\overline{T}_2\overline{T}_3$. They are used to the synthesis of CMCU $U_6(\Gamma_1)$ logic circuit, which is executed as in case of CMCU $U_3(\Gamma_1)$.

Table 8.5 Table of code transformer for CMCU $U_6(\Gamma_1)$

a_g	$C(O_g)$	B_i	$K(B_i)$	τ_g	g
a_1	100**	B_1	01	τ_2	1
a_2	101**	B_2	00	–	2
a_3	110**	B_2	00	–	3
a_6	111**	B_3	10	τ_1	4

Allocation of the dedicated input area turns CMCU with transformation of GSA into CMCU U_7, having the same structural diagram as CMCU U_3.

The main disadvantage of this approach is the higher number of algorithm execution cycles, due to the existence of control microinstructions. Besides, some additional chip resources are needed to implement the system of microoperations, even in case of hot-one microoperation encoding The following method can be used to eliminate this disadvantage.

8.4 Optimization of Compositional Microprogram Control Unit with the Dedicated Input Area

Let control microinstruction have the following format (Fig. 8.13).

TMI	FY	FA

Fig. 8.13 Format of control microinstruction

In this case, following actions can be executed during one cycle of CMCU operation: generation of microoperations using the code from field FY and generation of transition address using the code from field FA. Both the format of operational microinstruction and principle of allocation for the first I_0 cells of the control memory for input addresses of OLC $\alpha_g \in C$ are also used here.

Application of such control microinstructions transforms the CMCU U_2 into CMCU U_8 (Fig. 8.14).

Compositional microprogram control unit U_8 operates as follows. The input address of OLC $\alpha_l \in C$ is loaded into the counter CT using pulse *Start*. At the same time the flip-flop TF is set up. Current microinstruction is fetched from the control memory CM and its field TMI is transformed into the control signals y_0, y_j, y_c, y_E. If signal y_0=1 is generated simultaneously with microoperations $y_n \in Y$, the content of CT is incremented and corresponds to the transition inside current OLC. If signal y_j=1 is generated, it corresponds to a transition from the dedicated input area DIA into the area of microprogram AMP. In this case, the transition from some input of OLC $\alpha_g \in C$ to next component is executed. If signal y_c=1, it corresponds to the transition from the output of OLC $\alpha_g \in C$ and the content of counter CT is determined by functions Φ_0. If signal y_E=1, the algorithm execution should be finished. In this case, flip-flop TF is reset and the fetching of microinstructions from control memory is terminated.

Microoperations $y_n \in Y$ are represented by some code in the fixed field FY and therefore the block CMO is absent in case of the hot-one encoding of microoperations (Fig. 8.14). This approach has one serious disadvantage, namely the field FA is not used by microinstructions from the microprogram area AMP. This disadvantage can be partly eliminated due to the partition of control memory CM into two parts (Fig. 8.15). The part CM_1 includes FA field only and therefore information fetching is executed using the leftmost address bits from set T_j, where $|T_j|=R_3$. Both the operational part of microinstructions and the field TMI are kept in the part CM_2, which is addressed using the whole address.

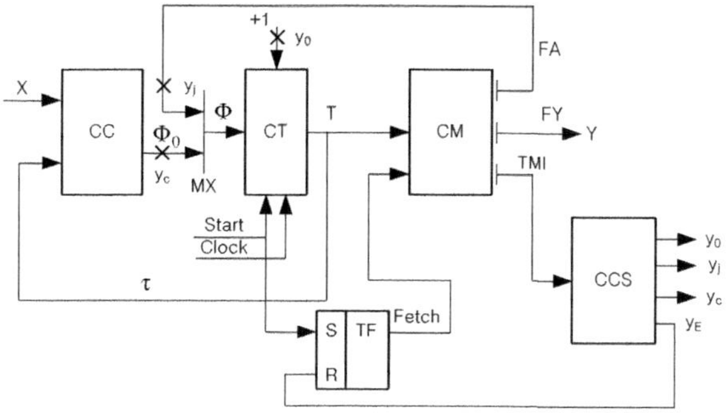

Fig. 8.14 Structural diagram of CMCU U_8

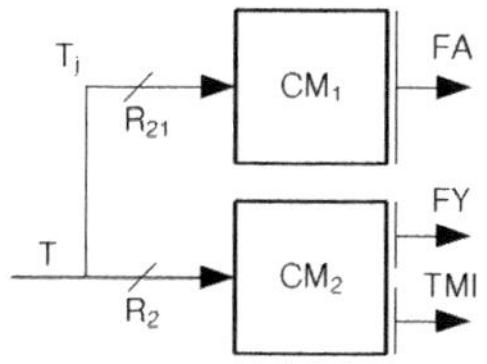

Fig. 8.15 Structural diagram of control memory for CMCU U_8

Obviously, fetching of the transition address is executed for all microinstructions, regardless of their type. This address is used only for particular microinstructions, when $y_j=1$.

The synthesis method used for CMCU U_8 includes the following steps:

1. Transformation of the initial GSA Γ.
2. Construction of OLC set C for the transformed GSA $\Gamma(U_8)$.
3. Addressing of inputs for OLC $\alpha_g \in C$.
4. Addressing of microinstructions.
5. Construction of the control memory content for CM_1.
6. Construction of the control memory content for CM_2.
7. Construction of the CMCU transition table.
8. Construction of the table for block CCS.
9. Synthesis of CMCU logic circuit for given elements.

Let us discuss the application of this method for synthesis of the CMCU $U_8(\Gamma_1)$, where the transformed GSA $\Gamma_1(U_8)$ is the same as the one shown in Fig. 8.5. Outcomes of the first three synthesis steps are the same for CMCU $U_3(\Gamma_1)$ and $U_8(\Gamma_1)$. Thus, the following set of inputs can be found: $I(\Gamma_1)=\{b_1, b_3, b_5, b_6, b_8, b_{10}, b_{12}, b_{14}\}$. In our case the inputs have the addresses: $IA(b_1)=000,\ldots,IA(b_{14})=111$.

Microinstruction addressing is executed as follows. First, all main inputs are removed from OLC $\alpha_g \in C$, as the first stage of addressing. Standard addressing procedure is then applied to the transformed OLC $\alpha_g \in C$, as the second stage of addressing. This is the same procedure as the one used in case of the CMCU U_1.

In this example, removing the main inputs results in the OLC set $C=\{\alpha_1,\ldots,\alpha_6\}$, where $\alpha_1=\langle b_2\rangle$, $\alpha_2=\langle b_4, b_5\rangle$, $\alpha_3=\langle b_7, b_8\ b_9,\rangle$, $\alpha_4=\langle b_{11}\rangle$, $\alpha_5=\langle b_{13}\rangle$, $\alpha_6=\langle b_{15}, b_{16}\ b_{17},\rangle$. Addressing the microprogram area AMP starts from the address, which exceeds by 1 the last address taken from the dedicated input area DIA. Resulting microinstruction addresses of the CMCU $U_8(\Gamma_1)$ are shown in Fig. 8.16.

T_4T_5 \ $T_1T_2T_3$	000	001	010	011	100	101	110	111
00	b_1	b_8	b_2	b_8	b_{15}	*	*	*
01	b_3	b_{10}	b_4	b_9	b_{16}	*	*	*
11	b_5	b_{12}	b_5	b_{11}	b_{17}	*	*	*
10	b_6	b_{14}	b_7	b_{13}	*	*	*	*
	DIA		AMP					

Fig. 8.16 Microinstruction addresses for CMCU U_8

The control memory content of DIA area can be found using the following rules:

- if vertex $b_q \neq O_g$ (g=1,…,G), field TMI of the memory cell with address $IA(b_q)$ contains code of variable y_j, its field FY contains microoperations $y_n \in Y(b_q)$, and its field FA contains address $A(b_t)$, where $\langle b_q, b_E\rangle \in E$;
- if vertex $b_q = O_g$ (g=1,…,G) and $\langle b_q, b_E\rangle \notin E$, field TMI of the memory cell with address $IA(b_q)$ contains code of variable y_j, its field FY contains microoperations $y_n \in Y(b_q)$,, and its field FA contains the transition address;
- if vertex b_q is connected with the final vertex b_E, field TMI of the memory cell with address $IA(b_q)$ contains code of variable y_E, its field FY contains microoperations $y_n \in Y(b_q)$, and its field FA is empty.

In our example, content of the control memory CM_1 is shown in Table 8.6. In this case $T_j=\{T_3, T_4, T_5\}$.

In this table, bits α_1–α_5 represent the transition address and form the field FA.

Construction of the control memory content for AMP area is executed in the same way as in case of CMCU U_3. Content of the control memory CM_2 includes both areas DIA and AMP; shown in Table 8.6.

As in the previous case, bits m_1, m_2 represent TMI codes, where code 00 corresponds to y_j, code 01 to y_0, code 10 to y_c, and code 11 to y_E. Bits m_3–m_7 contain hot-one code of the collection of microoperations $y_n \in Y(b_q)$, where $q=1,\ldots,M_2$. Let us point out that for the vertex b_5 from AMP area, the field $FY=\varnothing$. It corresponds to an idle cycle of the data-path.

Table 8.6 Content of the control memory CM_1 for CMCU $U_8(\Gamma)$

Address $T_3T_4T_5$	Content $a_1a_2a_3a_4a_5$	Comment
000	01000	$b_1 \to A(b_2)$
001	01001	$b_3 \to A(b_4)$
010	01010	$b_5 \to O_2A(b_5)$
011	01011	$b_6 \to A(b_7)$
100	01101	$b_8 \to A(b_9)$
101	01110	$b_{10} \to A(b_{11})$
110	01111	$b_{12} \to A(b_{13})$
111	10000	$b_{14} \to A(b_{15})$

Transition table for CMCU U_8 is constructed by analogy with the construction of the corresponding table for CMCU U_3. In the discussed case it is the same as in case of CMCU $U_3(\Gamma_1)$, represented by Table 8.2. Equations for function (8.13) for both CMCUs are the same, but the system of formulae for multiplexer MX (8.16) is transformed due to presence of the block CM_1. In our case this system is transformed into the form:

$$\begin{aligned}
D_1 &= y_j \cdot a_1 \vee 0 \cdot y_{c;} \\
D_2 &= y_j \cdot a_2 \vee 0 \cdot y_{c;} \\
D_3 &= y_j \cdot a_3 \vee D_3^1 \cdot y_{c;} \\
D_4 &= y_j \cdot a_4 \vee D_4^1 \cdot y_{c;} \\
D_5 &= y_j \cdot a_5 \vee D_5^1 \cdot y,
\end{aligned} \tag{8.20}$$

where variables α_1 - α_5 represent the bits of FA field.

Table 8.7 Content of control memory CM_2 for CMCU $U_8(\Gamma_1)$

Address $T_1T_2T_3T_4T_5$	TMI m_1m_2	Content $m_3m_4m_5m_6m_7$	Comment
00000	00	11000	b_1 I_1^1 DIA
00001	00	01010	b_3 I_2^1
00010	00	10001	b_5 I_2^2
00011	00	11000	b_6 I_3^1
00100	00	00100	b_8 I_3^2
00101	00	10001	b_{10} I_4^1
00110	00	01000	b_{12} I_5^1
00111	00	00110	b_{14} I_6^1
01000	10	00100	b_1 O_1 AMP
01001	01	00100	$b_4 \to b_5$
01010	10	00000	b_5 O_2

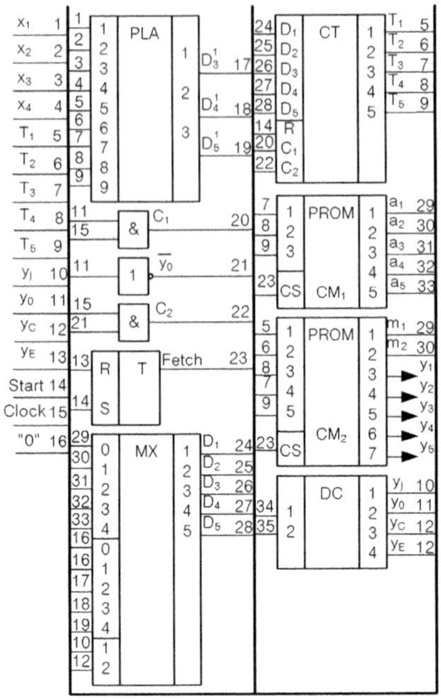

Fig. 8.17 Logic circuit of CMCU $U_8(\Gamma_1)$

Logic circuit of CMCU $U_8(\Gamma_1)$ is shown in Fig. 8.17. In this case OLC $\alpha_g \in C^1$ have the following output addresses: $A(O_1)$=01000, $A(O_2)$=01010, $A(O_3)$=01101, $A(O_6)$=10010. Because each address bit has both direct and complementary values, the inputs of block CC are connected to all R_2=5 feedback variables. Multiplexer MX is shown here as a block replacing the set of logical elements «AND-OR» from Fig. 8.8. The control memory of CMCU is divided into two blocks (CM_1 and CM_2), contents of which are determined by corresponding tables.

Application of this approach transforms the CMCU with special addressing of MIs into CMCU U_9 (special microinstruction addressing and allocation of the dedicated input area), CMCU with optimal addressing of MIs into CMCU U_{10} (optimal microinstruction addressing and allocation of dedicated input area), and CMCU with transformation of GSA into CMCU U_{11} (transformation of the initial GSA and allocation of the dedicated input area). The structural diagram s for these CMCUs are the same as for CMCU U_8, and their synthesis methods are some extensions of methods used for the basic models of CMCU, obtained by adding the steps in which construction of tables for blocks CM_1 and CM_2 is performed.

Use of the control microinstructions having format from Fig. 8.13 transforms CMCU with address transformer into CMCU U_{11} with the structural diagram shown in Fig. 8.18. All blocks of CMCU U_{10} play the same roles as the blocks of corresponding basic models of CMCU with code transformer and CMCU U_8. This synthesis method combines the methods used earlier for CMCU with code transformer and U_8.

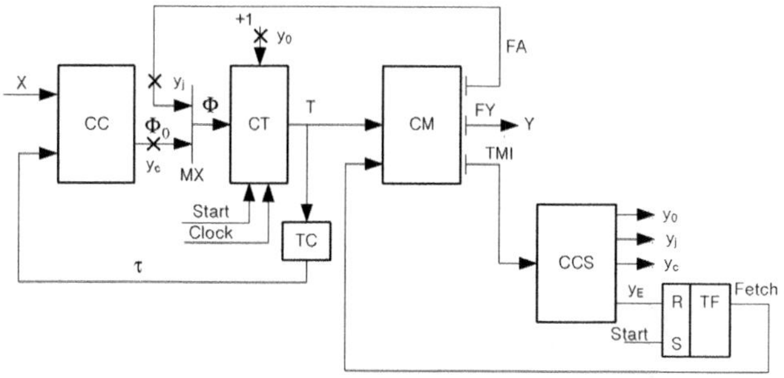

Fig. 8.18 Structural diagram of CMCU U_{11}

Using the control microinstruction format of Fig. 8.13, instead of the format given in Fig. 8.4, allows to reduce the number of microinstructions in the control memory from I_0 to IO_0, where IO_0 is the number of OLC inputs, which are also the outputs of the same OLC. It means that condition (8.12) is transformed into the following one:

$$\lceil \log_2(IO_0 + M_2) \rceil = \lceil \log_2 M_2 \rceil. \tag{8.21}$$

Because $IO_0 < I_0$, the probability of satisfying condition (8.21) and hardware amount of logic CMCU circuit can be reduced, due to allocation of the dedicated input area DIA.

8.5 Conclusions

Some new models of compositional microprogram control units are proposed in this chapter. All proposed methods are targeted on reduction of hardware amount in the addressing circuit of CMCU. Reduction is achieved due to modification of microinstruction formats in use.

All proposed models are based on introduction of the special area of inputs, called as dedicated area of inputs. I permits to diminish the number of bits generated by the block of microinstuction addressing. Such an approach can be used for optimization of the CMCU with common memory [3, 4], where the same bits are used for addressing o microinstruction and for this address generation. The second task can be solved due to use of additional code transformer. The code transformer cab be eliminated due to use of an additional address field in the microinstruction format.

Here we only show the possible models of CMCU. To prove an effectiveness of these models we are going to create the special CAD tools and investigate the hardware amount for well-known and proposed methods of the synthesis of CMCU.

References

[1] Adamski, M., Barkalov, A.: Architectual and Sequential Synthesis of Digital Devices. University of Zielona Góra Press, Zielona Góra (2006)
[2] Baranov, S.I.: Logic synthesis of Control Automata. Kluwer Academic Publishers, Dordrecht (1994)

[3] Barkalov, A., Palagin, A.: Synthesis of Microprogram Control Units. IC NAS of Ukraine, Kiev (1997) (in Russian)
[4] Barkalov, A., Węgrzyn, M.: Design of control units with programmable logic. University of Zielona Góra Press, Zielona Góra (2006)
[5] Barkalov, A., Titarenko, L., Bieganowski, J.: Synthesis of control unit with modified operational linear chains. PAK 5, 5–17 (2007)
[6] Barkalov, A., Titarenko, L., Bieganowski, J.: Synthesis of control unit with modified microinstructions. In: Proc. of the Inter. Conf. Mixed Design of Integrated Circuits and Systems MIXDES 2007, pp. 157–160 (2007)
[7] Barkalov, A., Titarenko, L., Bieganowski, J.: Synthesis of control unit with modified system of microinstructions. In: Proc. of IEEE East-West Design & Test Symposium EWDTS 2007, pp. 545–549 (2007)
[8] Barkalov, A., Titarenko, L., Bieganowski, J.: Design of control unit with modified operational linear chains. In: Proc. of the Inter. Conf. TCSET 2008, Lviv-Slavsko, Ukraine, pp. 259–262 (2008)
[9] Barkalov, A., Titarentko, L., Kołopieńczyk, M.: Optimization of control memory size of control unit with codes sharing. In: Proc. of the IX Inter. Conf. CADSM 2007, Lviv-Polana, Ukraine (2007)
[10] Gajski, D.: Principles of Digital Design. Prentice Hall, New York (1997)
[11] Maxfield, C.: Design Warrior's Guide to FPGAs. Academic Press Inc., Orlando (2004)
[12] Xilinx Inc. (2010), `http://www.xilinx.com`

9 PeNLogic – System for Concurrent Logic Controllers Design

Marek Węgrzyn and Agnieszka Węgrzyn

University of Zielona Góra, Institute of Computer Engineering and Electronics,
ul. Podgórna 50, 65-246 Zielona Góra, Poland

Abstract. In the paper the CAD system dedicated for modeling, verification, and synthesis of concurrent controllers modeled by interpreted Petri net is presented. Petri net model can be prepared as graph or as textual form. Controllers specified by Petri nets can be analyzed and implemented using method suitable for such model. For verification of Petri net another part of system is used. Moreover, the results of verification are decomposition of net into several communicating state machines (as finite state machines, FSMs). After verification it is possible to transform Petri net model into HDLs model (VHDL and Verilog) and alternatively into EDIF or XNF netlist format. Such prepared models are also simulated and synthesized using other academic or commercial CAD systems. The system has been developing at University of Zielona Góra. Development of new methods of modeling, verification and synthesis has been contributed to make an attempt the new integrated version of the system. In addition, using of Java environment gives opportunity to work out the system that is platform independent.

9.1 Introduction

The specific application often dictates the system design requirements, such as modularity and flexibility. In general, the design procedure involves the integration of analytical and graphical descriptions. Graphical descriptions, such as control-interpreted Petri nets [11], SFC (Sequential Function Chart) [8] and Grafcet [7], provide established techniques for proper system designs. They have helped industrial engineers to understand the system behavior and performance over many years.

A behavioral representation describes the system's functionality independently of its implementation. It treats a system as a black box, and defines how the black box responds to any combination of input values. The process of designing a system proceeds from a behavioral specification (SFC diagram or Petri net) to a programmable logic implementation (FPGA, Field Programmable Gate Array). A design in purely behavioral form, like a Petri net, is converted into an intermediate rule-based description, void of any technology-specific implementation details. The final FPGA implementation is generated by automatic synthesis using CAD

M. Adamski et al. (Eds.): Design of Digital Systems and Devices, LNEE 79, pp. 215–228.
springerlink.com

tools, instead of manual, tedious design process. The textual format serves as a bridge with some related university tools.

In the paper the integrated environment for design of concurrent logic controllers, called PeNLogic, is presented.

9.2 PeNLogic System

The PeNLogic system has been developing for many years. The first version as a set of different application was presented in [4,15]. However, development of new methods of modeling, verification and synthesis has been contributed [2] to make an attempt the new integrated version of the system. In addition, using of Java environment gives opportunity to work out the system that is platform independent.

The core of the PeNLogic system is Petri net models of concurrent logic controllers. Petri net can be prepared as graph or as textual form. Controllers specified by Petri nets can be analyzed and implemented using method suitable for Petri nets [1]. For verification of Petri net another part of system is used. This part bases on formal method of Petri net analysis, which is widely described in [13]. Moreover, the results of verification are decomposition of net into several communicating state machines (as finite state machines, FSMs) [19,21]. After verification it is possible to transform Petri net model into HDLs model (VHDL [12,22] and Verilog [10,18]) and alternatively into EDIF or XNF netlist format [17]. Such prepared models are also simulated and synthesized using other academic [3,4,20] or commercial CAD systems.

Fig. 9.1 shows a general schema of the PeNLogic system.

9.2.1 Petri Net Modeling of Concurrent Controllers

As opposed to sequential automaton, concurrent automaton can be in one or more internal state at the same time. Maximal sets of concurrently occurring local states are defined by global state of automaton. Any subset of concurrent local states is called partial state. In concurrent automata local relation are considered that relates internal partial states (current and next) and suitable input and output states of automata. Interpreted Petri net is one of the forms for representing of concurrent automaton.

On the other hand, Petri nets as a graphical tool provides a unified design methodology for specifying discrete-event systems. They can be applied in various stages of the design implementation from hierarchical system description to its physical realization. A Petri net is used as a tool for the modeling and analysis of digital circuits, especially concurrent controllers [1,3,5,12,13].

In PeNLogic system it is possible to specify Petri net in textual formats [16]: PNSF2, PNSF3 and CCPNML format and as a graph (figure). For preparing of graphical form of hierarchical, colored, interpreted Petri net, Petri net Editor was implemented. Using this part of the system there is generated from a net graph into PNSF2, PNSF3 and CCPNML formats. Moreover, a net graph and each type of format is stored in database using relational model.

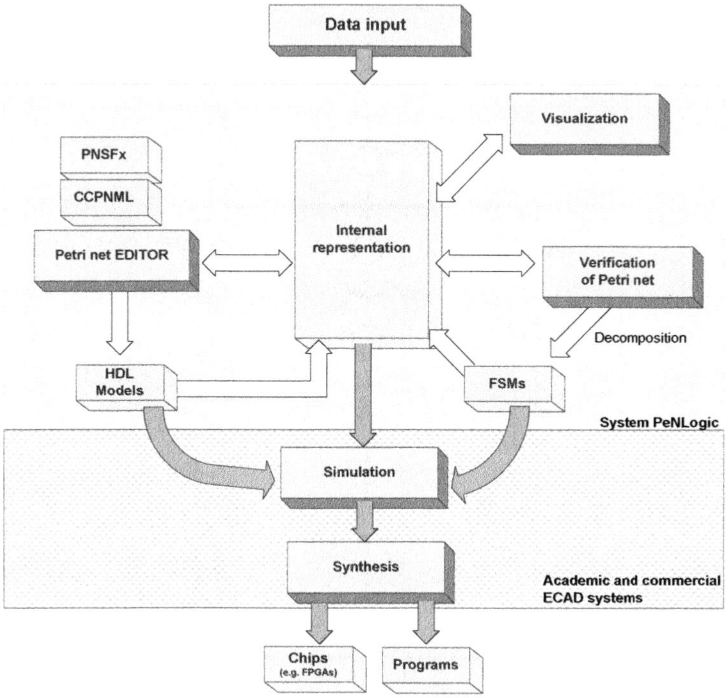

Fig. 9.1 Data flow in PeNLogic system

The proposed method, which is applied in PeNLogic system, will be presented on an example [17].

An Example

In this section, the specification of *drill station* controller is given, to be later related with control interpreted Petri net. In the example, several operations may occur simultaneously. A work piece is loaded in one station, clamped and drilled in the second one and finally at the third station is tested for depth. Some actions can occur independently of others, while other actions can take place after all stations have completed their individual programs.

The Petri net model which describes the behavior of the process controller is presented in Fig. 9.2. Parallel parts (subnets) of the net between transitions *t2* and *t3* represent the programs related to the concurrent actions at the particular three stations. The production cycle can be automatically repeated depending on signal R or finished and started again. Fig 9.3 shows PNSF2 specification of the controller.

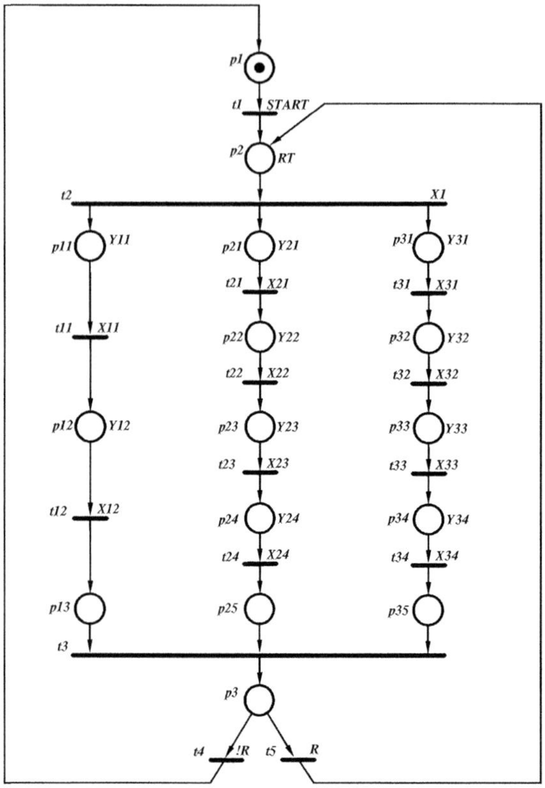

Fig. 9.2 Petri-net-based model of control system

9.2.2 Analysis of Petri Net

Digital systems, especially concurrent controllers, can be analyzed by Petri nets [11,13]. So there are many advanced methods of Petri net analysis developed. The model is formally verified based on well-known Petri net theory. The main tasks of Petri net analysis are checking some properties of the net, i.e. liveness, boundedness, persistence etc. There exists several methods for analysis of Petri nets. However, most of those methods check properties and answer a question, whether the nest is or not live, bounded, etc. without presenting exactly places in defect [11,13].

Calculation of deadlocks and traps is one of the important analysis tasks, because a good system should not contain events, which can never occur. Testing the liveness of Petri net depend on finding deadlocks and traps and researching dependence between theirs.

```
.clock CLK
.inputs START R X11 X12 X21 X22 X23 X24 X31 X32 X33 X34
.comb_outputs RT Y11 Y12 Y21 Y22 Y23 Y24 Y31 Y32 Y33 Y34
.part Controller
.places p1 p2 p3 p11 p12 p13
.places p21 p22 p23 p24 p25 p31 p32 p33 p34 p35
.transitions t1 t2 t3 t4
.transitions t11 t12  t21 t22 t23 t24 t31 t32 t33 t34
.net
 t1: p1 * START |- p2;
 t2: p2 * X1 |- p11 * p21 * p31;
 t3: p13 * p25 * p35 |- p3;
 t4: p3 * !R |- p1;
 t5: p3 *  R |- p2;
 t11: p11 * X11 |- p12;
 t12: p12 * X12 |- p13;
...
.MooreOutputs
 p2 |- RT;
 p11 |- Y11;
 p12 |- Y12;
...
.marking p1
.end
```

Fig. 9.3 PNSF2 model (a part)

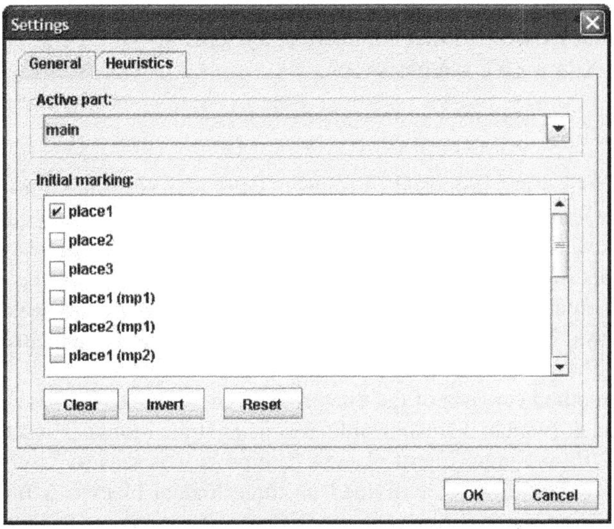

Fig. 9.4 Dialog box for verification settings

In considered approach, for checking liveness of Petri net, Thelen method is used [9]. This method allows efficient calculation of prime implicants of a Boolean function represented in such form. In presented system the Thelen method for the mentioned logical equations is applied; it allows obtaining sets of deadlocks and traps in form of ternary vectors. The method is based on generating and searching tree for conjunction normal form (CNF). Such approach is very time consuming, because number of deadlocks and traps in Petri nets increase exponentially with the number of places and transitions [11]. For this purpose, parallel extension of the algorithm was proposed. In addition, for time reduction, some heuristic method was developed. In Figure 9.4 an example of PeNLogic dialog box for verification settings is presented. There is possible to change an initial marking of a Petri net and to change heuristic methods, which are used during calculation.

Basing on algorithm for finding all deadlocks and traps in Petri net and checking dependencies between sets of deadlocks and traps, it is possible to answer the question if Petri net is bounded and live. The results of presented verification method are answer on question if analyzed net can be decomposed. If net can be decomposed, then it is got vectors representing decomposed net. Such vector corresponds to one automaton. Using those vectors in the next step there is generated KISS format for FSMs [21].

9.2.3 HDL Modeling, Simulation and Synthesis

In the PeNLogic system there are two modules for transforming Petri net models into HDLs model, i.e. into VHDL and Verilog. The goals of HDL modeling is preparing model for simulation and alternative way of synthesis. For both, simulation and synthesis well-known commercial systems are used, e.g. Aldec A-HDL simulator or Xilinx XST synthesis tool.

VHDL Modeling

Several methods were proposed to transfer Petri nets specifications into VHDL-based for performance and reliability analysis. VHDL syntax can support the intermediate level models. It makes possible to describe the highest level of the system and its interfaces first, then refining to greater detail. In the system there were implemented five different methods [22]. In this chapter, there are presented only two ways. The first one is based on rule description and uses sequential description. The second one uses structural VHDL description.

The first method consists of the one process. In this process following variables are declared: *T* (which is responsible for firing transitions), *P* (actual marking places) and *NP* (next marking of places). The process is sensitive to rising edge of a clock signal. The process is divided on three logical blocks. A function of the first one is firing of transitions. A transition is fired when its input places are marked and input signals are activated. These conditions are checked by *IF-THEN-ELSE* instruction. The second block carries out adding tokens to transition output places. A place has a token when its input transitions fired or the place was marked and its output transitions did not fire. This condition is also checked by *IF-THEN-ELSE* instructions. If there are not places marked then initial marking is arbitrary set. The last statement in this block is assignment *NP* to *P*. The last block sets outputs. If a Moore machine is designed, then outputs are assigned to places. If a Mealy machine is designed, then outputs are assigned to transitions.

This description is natural and can be manually written by designer. Very important feature of this model is that such a description is synthesizable (e.g. Xilinx XST).

The second method is based on structural description. The Petri net nodes, places and transitions are represented using two VHDL entities (two kinds of objects). Arcs of Petri net are represented by signals in VHDL code. A transition fires when its input predicate is true. A predicate is a logical equation assigned to the appropriate signals related to the transition. Setting outputs is done in the same manner as in the first method. A designer creates his own library of classes of elementary building blocks of the control system, then combines these objects (from particular class) to build larger parts. Since there exist a documented executable description of the sub-system in VHDL, the upgrades or refinements requires only a replacement of the modified elements.

In such model, a sequential process is separated from the combinatorial process. In specification with two processes, the first process has in its sensitivity list a clock signal (*CLK*) that synchronizes the system, and a reset signal (*RESET*) to set-up automata into initial state. The set-up of current state and next state is done in the first process. The second process has in its sensitivity list the current state (represented the local place). The automaton outputs are set-up in the second process. Such code is easy readable, because outputs changes are separated from changes of the states sequence.

Verilog Modeling

In PeNLogic system there is also possible to transform model of Petri net into Verilog model [18]. The presented model (Fig.9.5) is divided on two parts. The first one reflects some logical functions of Petri nets, e.g. conditions for transitions firing and functions for controller outputs depended on active local states. For this purpose, statements assign are used. The second part represents the structure of the net as a set of processes (statements always) related to separate places. Additional signal reset is used for setting almost all places into state '0', except those places that are initially marked (initial marking of the net).

9.2.4 Petri Nets Decomposition

Concurrent Logic Controller can be presented using a concurrent view of the modeled system behavior [1,5,20]. The logic controller model should retain the natural partitioning of the behavior imposed by the designer. The functionality very often is represented as a set of concurrent blocks of a manageable size that communicate using few signals. On other hand, hierarchical Petri nets give possibility for modeling of designed systems in more compact form, even for complex systems [17].

```
module CARTS( CLOCK, reset, M1, M2, A, B, C, D, E, L2, L1, R2, R1 );
      input CLOCK, reset, M1, M2, A, B, C, D, E;
      output L2, L1, R2, R1;

      assign L1 = P5 | P6;
      assign L2 = P12 | P13;
      assign R1 = P2 | P4;
      assign R2 = P9 | P11;

      wire PRED0;
      assign PRED0 = ~D;

      wire  T1, T2, T3, T4, T5, T6, T7, T8, T9, T10, T11, T12;
      assign T1 = P1 & M1;
      assign T2 = P2 & D;
...
      assign T9 = P10 & P7 & PRED0;
      assign T10 = P11 & B;
      assign T11 = P12 & E;
      assign T12 = P13 & C;

      reg  P1,P2,P3,P4,P5,P6,P7,P8,P9,P10,P11,P12,P13;

      always @(posedge CLOCK)
      begin
            if (reset) P1 <= 1;
            else P1 <= ( P1 & ~( T1 )) | ( T6 );
      end
      always @(posedge CLOCK)
      begin
            if (reset) P2 <= 0;
            else P2 <= ( P2 & ~( T2 )) | ( T1 );
      end
...
      always @(posedge CLOCK)
      begin
            if (reset) P7 <= 1;
            else P7 <= (P7 & ~(T3 |T9)) |(T5|T11);
      end
endmodule
```

Fig. 9.5 A part of an example of Verilog model

During decomposition a Petri net is divided into a set of subnets. These subnets have to satisfy some restriction, e.g. a subnet must include only places which are sequential to each other, or cannot contain multi-input or multi-output transitions [5]. For explanation decomposition method of Petri net into automata subnet, basic information is presented. Two places p_i and p_j are concurrent, if exist such marking of the net, in which these two places p_i and p_j are marked simultaneously. In decomposed Petri nets none places can be concurrent.

Decomposition of Petri nets can be based on coloring of Petri nets. The decomposition method described in [1] is based on coloring of Petri net. In the method, attributes are introduced to each place of the net a minimum number of colors in such way, that two concurrent nodes should have different colors. A Colored Petri net model carries information about belonging to sequential process. Each automaton (state machine, SM) component is represented by P-invariants. There are several methods for determination of P-invariants in Petri net, e.g. [13].

The initial partitioning generates interacting Finite State Machines (FSM) with separate state registers [1,17]. They form a conservative interpreted SM-colored Petri net. A Petri net can be expressed graphically, then decomposed [21] into Linked State Machines (LSMs), and finally translated into an equivalent set of Verilog models.

The generic architecture of interactive FSMs is a set of interconnected FSMs which exchange data (local internal state signals or output signals) through input and output ports. Each component is characterized by input and output ports that connect it with other components and external controller ports. The set of communicating FSMs is called a concurrent state machine (Concurrent Controller) if two or more FSMs are not exclusively active.

Decomposition into SM-Components

The Petri net from Fig. 9.2 have a sequential subnet (with places *p1*, *p2* and *p3*), and concurrent subnet (remaining places), where three places are marked simultaneously. Hence, the net can be decomposed onto three conservative nets (Fig. 9.6) - State Machine components. In addition, to two subnets an additional place should be added (Fig. 9.6.b and Fig. 9.6.c).

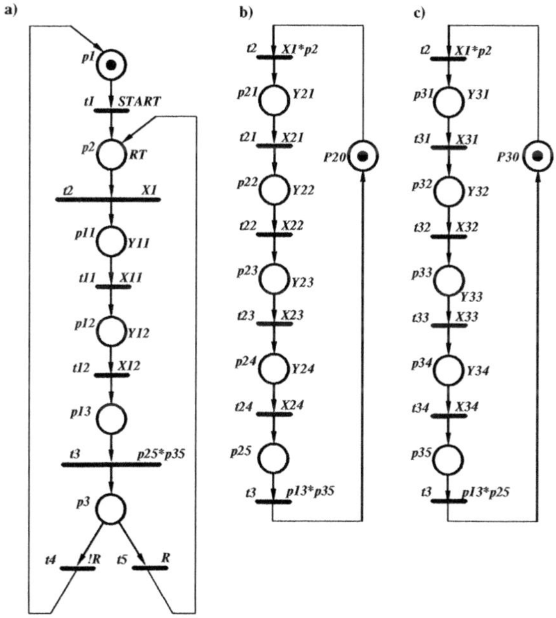

Fig. 9.6 The decomposed Petri net

In the first SM component (Fig. 9.6.a), the condition for the transition *t3* is a logic *AND* function of signals *p25* and *p35* that are the outputs from the appropriate places, i.e. they can be represented by the appropriate flip-flops at one-hot encoding. In the second subnet (Fig. 9.6.b), the additional place is called *P20*. The condition for the transition *t2* is a logical *AND* of the input signal *X1* and the signal from the place *p2*, and for transition *t3* is a logic *AND* of signals *p25* and *p35*. Similar situation is in case of the third subnet (Fig. 9.6.c).

Verilog Modeling and Synthesis

Several methods were proposed to transfer Petri nets specifications into VHDL for performance and reliability analysis. But in the literature Verilog-based modeling of Petri nets is not well-known [10,18]. Verilog syntax can support the intermediate level models. It makes it possible to describe the highest level of the system and its interfaces first, then to refer to greater details.

```
parameter [5:0] p1  =  6'b000001,
                p2  =  6'b000010,
                p11 =  6'b000100,
                p12 =  6'b001000,
                p13 =  6'b010000,
                p3  =  6'b100000;

always @(posedge CLK, posedge RESET)
begin
   if(RESET == 1)
      PLACE_A <= p1;
   else
   if(CLK == 1)
   begin
      case (PLACE_A)
         p1:  if (START==1) PLACE_A <= p2;
         p2:  if (X1==1) PLACE_A <= p11;
         p11: if (X11==1) PLACE_A <= p12;
         p12: if (X12==1) PLACE_A <= p13;
         p13: if (p25==1 && p35==1)
                         PLACE_A <= p3;
         p3:  if (R==0) PLACE_A <= p1;
              else PLACE_A <= p2;
         default PLACE_A <= p1;
      endcase
   end
end

always @(PLACE_A)
begin
      case (PLACE_A)
         p2:      Y <= 3'b001;
         p11:     Y <= 3'b010;
         p12:     Y <= 3'b100;
         default Y <= 3'b000;
   endcase
end
```

Fig. 9.7 Verilog model of the SM-component from Fig. 9.6.a

The Verilog language describes a digital system as a set of modules. Each of these modules has an interface to other modules as well as a description of its contents. The basic Verilog statement for describing a process is the `always` statement. The `always` continuously repeats its statement, never exiting or stopping. A behavioral model may contain one or more `always` statements. The `initial` statement is similar to the `always` statement except that it is executed only once. The `initial` provides a means of initiating input waveforms and other simulation variables before the actual description begins its simulation. The `initial` and `always` statements are the basic constructs for describing concurrency.

Because of the fact, that Verilog can model concurrency, for example using structure `always`, Petri nets can be effectively described by Verilog language. However, in the presented method synchronization of the concurrent processes is realized by the additional signals as transition conditions. Implementation of decomposed Petri nets onto set of SM-components can be carried out in different methods [6]. In the presented paper a modeling of automata in Verilog is focused.

Fig. 9.7 shows a part of Verilog model with using of two processes (two `always` statements). As an example, there is presented the SM-component from Fig. 9.6.a, only.

For place encoding one-hot method has been applied. Such approach is recommended for using FPGA devices as a final implementation technology. In the model, the sequential process is separated from the combinatorial process. In specification with two processes, the first process has in its sensitivity list a clock signal (*CLK*) that synchronizes the system, and a reset signal (*RESET*) to set-up automata into initial state. The set-up of current state and next state is done in the first process. The second process has in its sensitivity list the current state (represented the local place). The automaton outputs are set-up in the second process. Such code is easy readable, because outputs changes are separated from changes of the states sequence. IN the example, the outputs *Y12*, *Y11* and *RT* correspond to related bits of a output vector `Y[2:0]`, respectively.

The design process can be greatly simplified by means of FPGA compilers, which are recently available, like Xilinx ISE (supporting VHDL and Verilog). The obtained models of the subnets were simulated in Aldec A-HDL and MTI ModelSim. The effective simulation allows the SM-components to be rapidly modified, tested and debugged before the device is programmed. If a design change is needed, it is a simple matter to reedit the original specification. Since the Verilog model of environment (*testbench*) is described in modular, and parameterized fashion it is not necessary to construct its behavior description from extended, configuration graph. Testing unit can be simply build from the previously verified library modules by merging selected sub-blocks. The final, exhaustive testing also does not make any problem because the size of considered Petri net models is rather small. Then the models were synthesized by both Xilinx XST and Mentor Graphics Precision. One of the most flexible and high-density devices is Xilinx Virtex or Spartan FPGAs [23], which are used for prototyping of Reprogrammable Logic Controllers (RLCs). The bitstreams are generated by Xilinx ISE.

9.2.5 Direct Mapping into Netlist

This section describes a sub-system for the synthesis of logic controllers described Petri nets. As a final technology programmable logic is considered, e.g. Xilinx FPGA and CPLD devices (the library of present available devices is shown in Fig. 9.8). The interpreted Petri-net model (or equivalent SFC) is translated into a rule-based specification that is composed of the discrete local state symbols, input signal symbols and output signal symbols of the controller The textual format of Petri net (for example, PNSF, PNSF2, CCPNML) used as an entry format [16]. Then, the textual Petri net description is translated directly into the netlist. Presently, there are applied two formats, EDIF and XNF (Xilinx proprietary format). The next steps of the design implementation are realized by the vendor CAD/CAE system. For Xilinx devices it is ISE system. Functional and timing simulations are performed also by the vendor CAD/CAE system.

The presented method of synthesis of controllers is based on creating a one-to-one direct mapping between the Petri net and the hardware realization [17]. This mapping of Petri net into FPGA is based on the correspondence between a transition and a simple combinational circuit and the correspondence between a place and a clearly defined subset of the state register. Dealing with concurrency the designer is confronted with some problems that will not arise in the logic synthesis of sequential systems. One of them is the concurrent local state encoding.

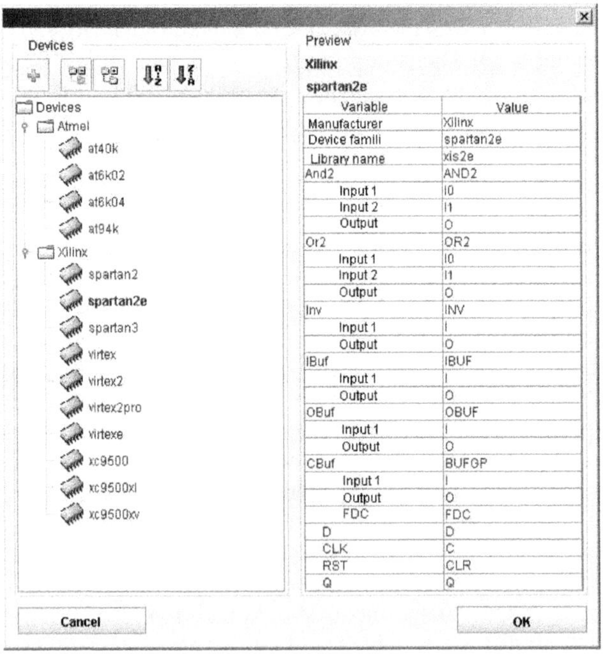

Fig. 9.8 Information on examples of the library content

The logic decision rules, which exactly reflect the description in textual Petri net specification, are transformed from the transition-oriented form (`T1: P1* M1 -> P2;`) into the place-oriented description (`P1 <= (P1 & ~(T1)) | (T6)`). This approach is similar to the Verilog model presented in Fig. 9.7. For place encoding concurrent one-hot method has been applied. Such approach is recommended for using FPGA devices as a final implementation technology. The concurrent one-hot encoding of Petri net is treated as the simplest case of more general mapping. The one-hot method produces fast designs with a simple combinational part, especially for implementations in FPGA [2,3,17]. It is not possible to assume that all flip flops, except those representing active local states that are set to 1, are set to 0 since several places can be marked simultaneously.

9.3 Conclusions

In the paper the academic CAD system dedicated for concurrent controllers was presented. For specification of controllers Petri nets are used. The system contains analyzes module based on checking of some Petri nets properties. In addition, the results of verification are decomposition of net into several communicating state machines. There are also possible to simulate the controllers using different HDL models produced by the presented system. The hardware representations of the controllers are obtained either as the direct mapping from the system, or by synthesis results of prepared HDL models. Moreover, the PeNLogic system was extended on module for visualization of some controlled parts [14].

The system is still under developing. Nowadays, there are considered additional modules as simulation of Petri net and transformation of decomposed FSMs into HDLs models. During preparation are Verilog and VHDL models of standard functions and functional blocks (according to the standard IEC 61131-3 [8]). For sake of simplicity there is also developing the new data-flow diagram based on Tcl/Tk languages. Other works are directed on implementation of web-based version of the system.

References

[1] Adamski, M.: Parallel Controller Implementation using Standard PLD Software. In: Moore, W.R., Luk, W. (eds.) FPGAs, Abingdon EE & CS, England, pp. 296–304 (1991)

[2] Adamski, M., Węgrzyn, M.: Design of Reconfigurable Logic Controllers from Petri Net-based specifications. In: The 4th IFAC Workshop on Discrete-Event System Design, DESDes 2009, Gandia, pp. 233–238 (2009)

[3] Adamski, M., Karatkevich, A., Wegrzyn, M. (eds.): Design of Embedded Control Systems. Springer, New York (2005)

[4] Adamski, M., Węgrzyn, M., Wolański, P.: Simulating and synthesising of reconfigurable logic controllers using VHDL. In: Proc. of the 42 IWK, Ilmenau, Germany, vol. 1, pp. 522–527 (1997)

[5] Biliński, K.: Application of Petri Nets in parallel controllers design. Ph.D. Thesis, University of Bristol, Electrical and Electronic Engineering Department (1996)

[6] Chmielewski, S., Węgrzyn, M.: Modelling and synthesis of automata in HDLs. In: Romaniuk, R.S. (ed.) Proceedings of SPIE, Photonics Applications in Astronomy, Communications, Industry, and High-Energy Physics Experiments, vol. 6347 (2006); 63470J1-13
[7] David, R., Alla, H.: Petri Nets and Grafcet. Prentice Hall, New York (1992)
[8] International Electrotechnical Commission, International standard IEC 61131-3, Programmable Controllers, Part 3: Programming Languages (1992)
[9] Mathony, H.J.: Universal logic design algorithm and its application the synthesis of two-level switching circuits. IEE Proceedings, Pt.E - Computers and Digital Techniques 136(3), 171–177 (1989)
[10] Minns, P., Elliott, I.: FSM based Digital Design using Verilog HDL. John Wiley & Sons, Ltd., Chichester (2008)
[11] Murata, T.: Petri Nets: Properties, Analysis and Applications. Proceedings of the IEEE 77(4), 541–580 (1989)
[12] Pardey, J., Bolton, M.: Logic Synthesis of Synchronous Parallel Controllers. In: Proceedings of the 1991 IEEE International Conference on Computer Design on VLSI in Computer & Processors, pp. 454–457 (1991)
[13] Węgrzyn, A.: Symbolic Analysis of Logical Control Devices using Selected Methods of Petri Net Analysis. Ph.D. Thesis, Warsaw University of Technology (2003) (in Polish)
[14] Węgrzyn, A., Jóźwiak, I.: Visualization of control process by means of Petri nets and database. In: Romaniuk, R.S. (ed.) Proceedings of SPIE, Photonics Applications in Astronomy, Communications, Industry, and High-Energy Physics Experiments, vol. 6347 (2006); 63472Q1-8
[15] Węgrzyn, A., Węgrzyn, M.: PeNCAD System - From Modeling to Synthesis of Concurrent Controllers. In: The 5th IEEE East-West Design & Test International Symposium, Yerevan, pp. 384–389 (2007)
[16] Węgrzyn, A., Węgrzyn, M.: On Textual Specification of Petri Nets for Control Algorithms. In: The 5th IEEE East-West Design & Test International Symposium, Yerevan, pp. 603–607 (2007)
[17] Węgrzyn, M.: Hierarchical implementation of Logic controllers by means of Petri nets and FPGAs. Ph.D. Thesis, Warsaw University of Technology (1998) (in Polish)
[18] Węgrzyn, M.: Implementation of Safety Critical Logic Controller by means of FPGA. Annual Reviews in Control 27, 55–61 (2003)
[19] Węgrzyn, M.: Petri Net Decomposition Approach for Partial Reconfiguration of Logic Controllers. In: Adamski, M., Gomes, L., Węgrzyn, M., Łabiak, G. (eds.) Discrete-Event System Design 2006, A Proceedings volume from the IFAC Workshop, DESDes 2006, pp. 323–328. University of Zielona Góra Press, Rydzyna (2006)
[20] Węgrzyn, M., Adamski, M., Monteiro, J.L.: The Application of Reconfigurable Logic to Controller Design. Control Engineering Practice 6, 879–887 (1998)
[21] Węgrzyn, M., Węgrzyn, A.: Implementation of Concurrent Logic Controllers based on Decomposition into State Machine Components. Radio-Electronics and Informatics 3, 44–47 (2006)
[22] Wolański, P.: VHDL Register Transfer Level modeling of digital systems by means of Petri nets. Ph.D. Thesis, Warsaw University of Technology (1998) (in Polish)
[23] Xilinx Inc. (2010), http://www.xilinx.com

Part III
Testing, Modeling and Signal Processing

10 Methods of Signals Processing in Radio Access Networks

Vladimir Popovskiy

Kharkov National University of Radioelectronics, 14. Lenin Ave., 61166 Kharkov, Ukraine
e-mail: tkc@kture.kharkov.ua

Abstract. This article considers optimum stochastic methods of radio signal processing, including parameter assessment tasks and management of parameters of receiving and transmitting devices. Such tasks are formed by state variable methods using Kalman-Bucy optimum recursive procedures. It's recommended to solve the management problems basing on the division theorem. The article gives analysis of steadiness and efficiency of state and management assessment procedures in steady-state and unsteady-state conditions. It gives recommendations regarding the choice of parameters and efficiency of processing devices taking into account statistics of signals and constraints attributable to certain telecommunication technologies. It analyzes a proposal, within which recursive procedures are used efficiently. The main tasks are united on Multiple-input/Multiple-output (MIMO) principle and are aimed at solving the access problems in mobile communication networks, Wi-Fi and Wi-Max systems, etc. Such tasks include: space-time encoding, multipath effect reduction, radio link power improvement, interference effect reduction, adaptation to channel parameter changes and current signal interference situation, possible repeated use of frequencies.

10.1 General Information

Important component of any information telecommunication system is the access network. Access network can be built in any physical environment: wire, wireless (radio), and optical. In this regard wireless environment is the most critical, where transformations of signals and noise levels can prove considerable, and signals and interference often have random character, especially in cellular networks, Wi-Fi and Wi-Max [1].

At the access level users of wire and wireless local networks have possibility of direct access to network services. There are many tasks of system-wide orientation at this level. They are as follows: providing multiplicity of access, electromagnetic compatibility, increasing interference resistance, transmission capacity, etc. Many of these tasks result in the tasks of signals processing. This is particularly what this section deals with.

M. Adamski et al. (Eds.): Design of Digital Systems and Devices, LNEE 79, pp. 231–261.
springerlink.com

Processing supposes implementation of corresponding transformations of signals with the purpose of achieving particular useful properties. As applied to OSI standard, information signals processing takes place on the first physical and partially on the second channel levels of a seven - level model. At higher levels, other types of signals are processed: signals of signaling, control, etc.

A signal in sufficiently general view can be represented through parameters amplitudes $A(t)$ and phases $\varphi(t)$ as follows:

$$x(t) = A(t)e^{-j\varphi(t)}, \tag{10.1}$$

where vector $x(t)$ determines direction of currents or potentials, besides in radio it shows spatial and polarization descriptions of this signal. Unfortunately, in practice, the signal alone does not exist (10.1). As a rule, this signal is to be perceived by means of measurements, observations or other objective realizations. Therefore, a signal often appears jointly with existing noises, additive $v(t)$ and multiplicative $H(t)$ interference:

$$y(t) = H(t) \otimes x(t) + v(t) + n(t), \tag{10.2}$$

where $H(t) \otimes x(t)$ - is convolution operation; noises $n(t)$ are the sum of real physical noises always occurring in electric circuits and communication channels, equivalent noises, generated by measurement errors, noises of quantum and other different factors in total representing a sample from Gaussian white noise.

Obviously, correlation (10.2) may be represented in a frequency field:

$$Y(\omega) = H(\omega)X(\omega) + U(\omega) + N(\omega), \tag{10.3}$$

where all components are Fourier transforms of corresponding functions. Solution of processing tasks in a frequency domain (10.3) appears easier, especially in the conditions, where reflecting dynamics of the state different processes is not required. However, in case the task requires presentation of dynamics or nonstationarity, it is necessary to solve problems in temporary domain. Methods of solution for such tasks are often interpreted as methods of state variables [2.3]. The state of an object $x(t)$ (including random object) is represented by differential or difference equation of the type:

$$\frac{dx(t)}{dt} = \Phi(x,t,\xi) \tag{10.4}$$

where $\xi(t)$ - is a random component, Gaussian white noise which is generating (forming) random process $x(t)$. Based on obtained observations (10.2), (10.3) signals processing is carried out, reduced as a rule, to recursive estimation of condition $\hat{x}(t)$. Estimation of $\hat{x}(t)$ in some cases may be considered as processing terminal stage, in other cases the estimation is used for the tasks of controlling the condition (10.4) or for observation (10.2).

This section mainly uses temporal representations of states in space [3, 4].

Another specific feature of tasks in this section is that treatment of signals in spatial and polarization regions is investigated while comparing corresponding solutions in a traditional frequency-time domain. It is obvious, that expansion of region of feasibility is instrumental in the improvement of quality of these solutions. Therefore, in the conditions when a frequency-time processing resource becomes scarce, efforts of developers focus more on a spatially-polarization resource that has not been employed until recently. New types of processing appear: space-time processing, space-frequency-time processing, space-time encoding, space-time access, etc. [4,5,6].

To solve tasks of processing with the active use of space-polarization parameters it is necessary to get realization of the electromagnetic field in two or more points in space in the form of (10.1), (10.2) or (10.3). Thus, these expressions can represent signals on the output of antennas orthogonal by polarization or mutually spaced [5].

Efficiency of processing tasks using space-polarization parameters in a number of cases exceeds efficiency of processing tasks of similar solutions in the domain of frequency-time parameters. In addition, transition to space-polarization field allows saving an expensive frequency resource. We will consider the above tasks in more detail.

10.2 Specific Features of Radio Access at Physical Level

10.2.1 General Description of Physical Processes at Radio Access

Telecommunication systems with radio access provide an important physical signal conversion: a signal on transmitting side in the form of conduction currents is transformed into electromagnetic field (bias current). There is a reverse transformation on a receiving side. All these transformations occur in antenna. In full-duplex lines, transmitting antenna is often simultaneously a receiving antenna (Fig. 10.1).

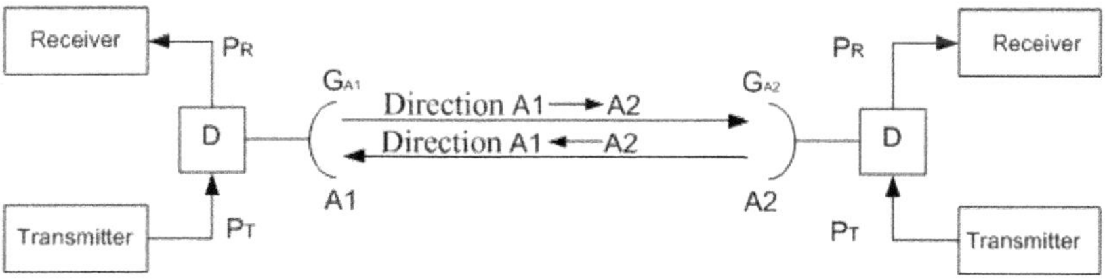

Fig. 10.1 Structure of full-duplex radio communication line

Antenna gain factor G_A depends upon both signal operating frequency $f = c/\lambda$ and aperture effective area S_{ef} :

$$G_A = 4\pi \frac{S_{ef}}{\lambda^2}, \tag{10.5}$$

where $S_{ef} = S \cdot K_{us}$, $K_{us} = 0{,}5...0{,}8$ is a coefficient of antenna area utilization S.

Expression (10.5) determines the maximum level of reception (transmission) of signal in the main lobe of directivity pattern (DP). In other directions $G(\theta, \varphi)$ this level is lower. Fig. 10.2. represents an example of antenna DP.

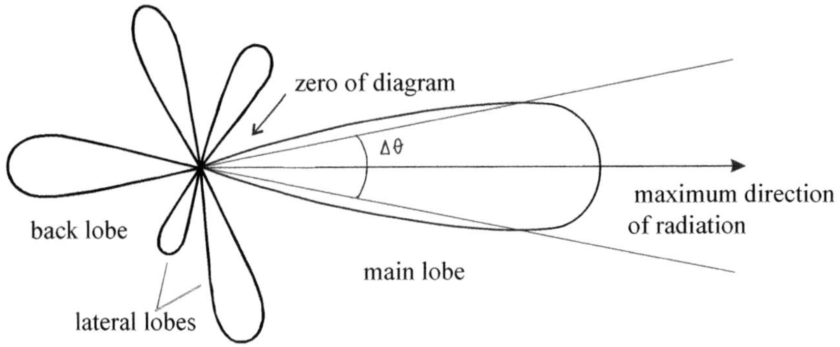

Fig. 10.2 Antenna directivity pattern

The figure shows, that it is possible to select the field of the main lobe in DP, where its width $\Delta\theta$, field of lateral and back lobes are determined. Usually access is provided due to mutual orientation (positioning) of main lobes of transmitting and receiving antennas one upon another. In doing so necessary energetic in radio line, determined by equation of transmission is provided:

$$P_R = P_T + G_T + G_R + W_{fr} + W_{add}. \text{ [dB]} \tag{10.6}$$

where P_T, P_R - are powers of transmitting and receiving signals accordingly; G_T, G_R are gain factors of transmitting and receiving antennas accordingly; $W_{fr} = \lg(\lambda/4\pi R)^2$ is signal attenuation in free space at a distance R; W_{add} is additional multiplier, allowing to take into account attenuation due to actions of different physical mechanisms in the course of distribution of radio waves, including the effect of signals fading.

The solution of radio access by several (many) users located in different points of space, for which directions are determined by angles θ_i, i=1,2,…,N is a more complex task. The easiest way to provide such access is to use an omnidirectional antenna with circular DP. This is what they do in cellular communication both on subscriber station and on base side. Simplicity of decision is achieved due to the losses of energy in radio line, because non directional antenna has a low gain factor G_A.

However, there is another, more rational solution for multiple access. Thus, the MIMO technology suggests using multi-beam DP of antennas of cellular communication base station, antennas of access point, antennas of communications transponder. Fig. 10.3 represents the example of such multi-beam antenna DP. Beams in the direction of subscriber stations can be set only for the period of communication session. In this case, due to the increase of the gain factor in each of θ_i - directions energy of radio line improves. At the same time, and that is more important, there is a possibility of simultaneous communication with i-subscribers at the same frequency f_0. In other words, it is possible to economize the spectrum of the used operating frequencies by *i-times*, or to increase base station capacity by *i-times* accordingly. Such method of multiple access is called space-time access or frequency re-use method. We will consider this method in more detail, as well as other methods of using space-polarization physical parameters to increase high-quality indexes of communication at the level of access.

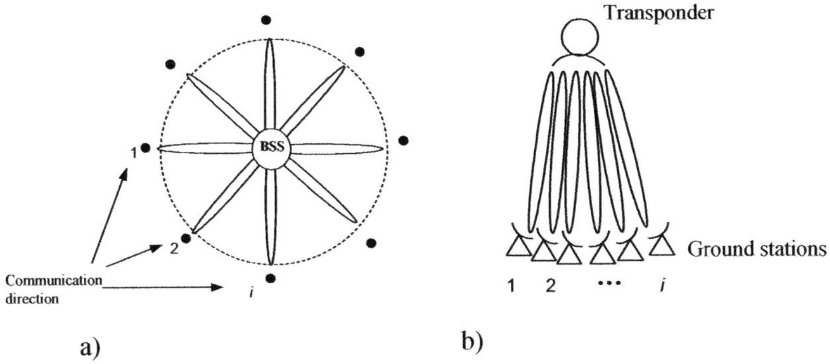

Fig. 10.3 Examples of directivity patterns of multibeam antennas for space-time access a) on cellular communication base station (BSS) with the orientation to i -direction communication; b) on satellite system transponder with the orientation to i -earth stations.

10.2.2 Space-Time Access Method

There exists a sufficiently simple algorithm of functioning of BSS with space-time access. Procedure of the station resource allocation includes four basic stages:

1. Acquiring a request from subscriber. This request is received by non-directional or near-omnidirectional antenna. At the same time direction the of request signal arrival is determined. For greater transmission reliability, the request signal is usually of broadband structure by frequency or by time.
2. One of narrow beams of antenna directivity pattern is set in the direction of the calling subscriber.
3. Omnidirectional antennas of base stations within the roaming limits carry out the search of destination user. After its response, narrow MBA beam of certain BSS is also set in this direction.

4. Communication between subscribers is provided, and, at the end of session, the beams are removed or reoriented to other subscribers.

Fig. 10.4. represents a diagram of organization of BSS communication directions with subscribers i and j in space-time access.

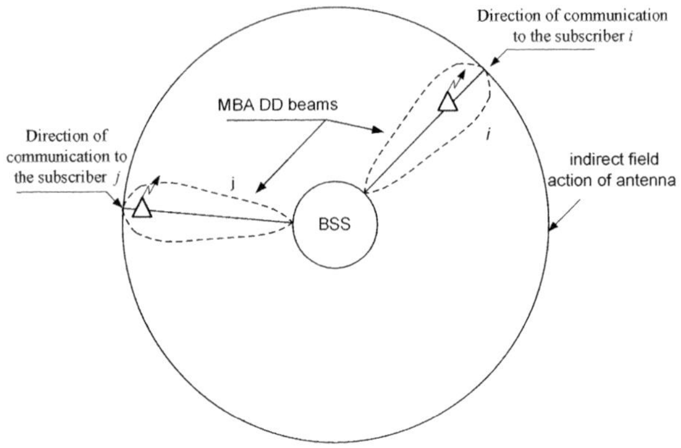

Fig. 10.4 Diagram of organization of BSS communication between subscribers i and j in space-time access

Multibeam antenna DP can be created using different methods. It is known [5,7] from the theory of antennas that DP $G(\theta)$ in a distant area (when sizes of aperture $d << R$ are distances to the view point) and distribution of electromagnetic field by aperture $f(x)$ are bound by Fourier transform:

$$G(\theta)=\int_{\alpha} g(\theta,x) f(x) e^{-j\cdot r} dx, \tag{10.7}$$

where $g(\theta,x)$ is multiplier of directivity of element dx,
$r = x\cdot\sin\theta + y(x)\cdot\cos\theta$ is path-length difference.

Distribution function $f(x)$ is complex; it is realized by selecting amplitudes and phases of currents in aperture points.

The easiest way is to implement an aperture in the form of discrete antenna elements, each of which has its function of radiation $\psi_i(u)$, $u = \pi d \sin\theta / \lambda$. Such discrete antenna is called an antenna array (AA).

Resulting DP of AA forming many beams is represented as

$$G(u)=\sum_{i=1}^{N} \mathrm{w}_i \psi_i(u), \tag{10.8}$$

where w_i is complex weight coefficient.

Controlling the beams of such antenna is carried out by means of changing parameters of weight coefficients w_i, included into the circuit of each AA element.

Analyzed mechanism of creation of required DP (10.8) is universal enough in the sense, that all spatial–polarization tasks in AA are executed after the same pattern: the MBA creation, spaced reception, adaptive reception, adaptive compensation of interference, adaptive antenna arrays (AAA) operation, fazed AA (FAA), multiple access, all tasks united under MIMO brand. Fig. 10.5. displays the example of such antenna array.

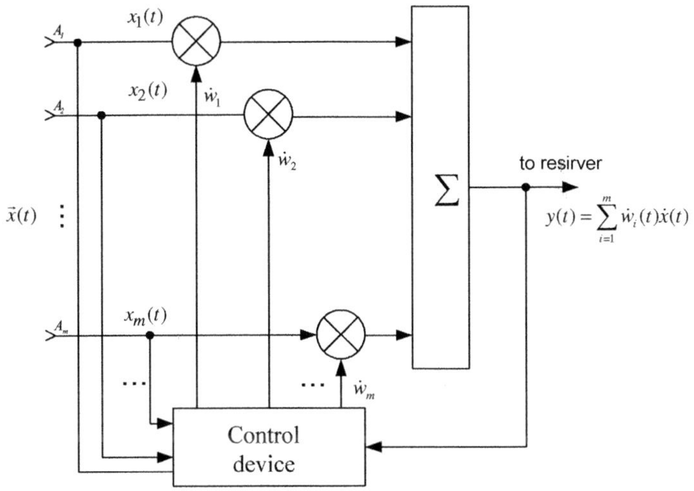

Fig. 10.5 General structure of weighting adjustment of plural signals $S_i(t)$ with a complex coefficient $w_i(t)$

Topical variety of AA and multibeam antennas are hybrid structures: reflector antennas with AA elements in the focus, each forming its own beam.

The task of determining subscriber signal direction of arrival can be solved by any of the known procedures [6], ESPRIT, Mvsic, etc. Alongside with that recursive procedure of spatial spectrum estimation can be more constructive. The procedure consists of three basic recursive components:

1) dedicating spatial window by control vector:

$$\begin{aligned}\mathrm{V}(\theta)=[&\exp(\sin(\theta/180)\cdot\pi),\exp(j\sin(\theta/180)\cdot\pi),...,\\&\exp(j(n-1)\cdot\sin(\theta/180)\cdot\pi),...\exp(j(N-1)\cdot\sin(\theta/180)\cdot\pi)\end{aligned}, \tag{10.9}$$

where n is the number of antenna element, $n=\overline{0,N-1}$; N is quantity of antenna elements; θ is the angle (direction of spatial window);

2) suppressing signals coming from all undedicated directions. This procedure can be realized with the use of Widrow, Applebaum [4,5], Kalman-Bucy [7] algorithms or other known algorithms of controlling vectors of AA weight coefficients. Thus, modification of Applebaum algorithm for suppressing signals beyond dedicated window is given by:

$$\mathrm{w}(k+1)=\mathrm{w}(k)-2\mu(k)\left[\mathrm{X}(k)\mathrm{w}^{\mathrm{T}}(k)\mathrm{X}(k)-\mathrm{V}(\theta)\right], \qquad (10.10)$$

where $\mu(k)$ is a step constant, $\mathrm{X}(k)$ is vector of signals;

3) 3) recursive definition of mark of matching the amplitude and space phase in a dedicated window

$$P(k+1,\theta)=P(k,\theta)+\mathrm{w}(\theta)^T\left(\mathrm{X}(k)\cdot\mathrm{X}(k)^T\right)\mathrm{w}(\theta). \qquad (10.11)$$

By the beginning of the third phase, the transitional processes of the second phase must be completed. In so doing there is a mark of a signal getting to protective window.

Further narrow beams of BSS antenna are set in the direction of the marked signals. Fig. 10.6 provides block diagram of space-time access algorithm, where a separate beam in M–devices of the WCV control is formed for each of M–subscriber.

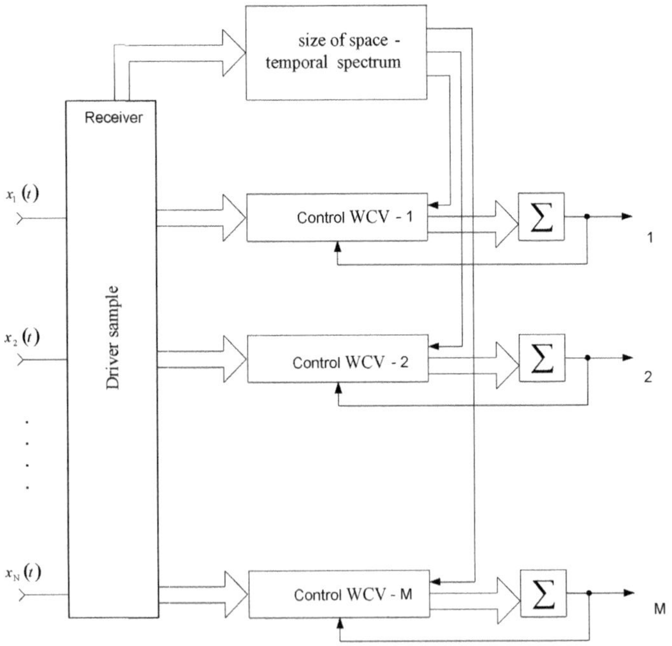

Fig. 10.6 Block diagram of space-time access algorithm

10.2.3 Polarization in Access Tasks

Tasks of access with the use of polarization parameters of signals and antennas have special values. Polarization is determined by the imaginary figure, which is drawn by the tip of electric field tension vector of the signal, radiated by the proper antenna on a plane perpendicular to direction of distribution. Therefore, whip type linear antenna radiates linear polarization (horizontal, vertical or with inclination). It is possible to create circular or elliptic polarization by more complex antennas.

For any concrete signal polarization (linear, circular, elliptic) there is another one which is orthogonal to it. Pairs of orthogonal polarizations are represented by Fig. 10.7.

Antenna can be matched with a signal by polarizations and then maximal accepted signal is separated (reception of signals matched by polarization). Another extreme case: a signal can appear orthogonal with regard to antenna polarization (e.g. antenna is horizontally polarized, and a signal has vertical linear polarization).

In general case some angle $\gamma = 2\gamma_{\Pi}$ appears between antenna polarization and signal polarization. Thus, if in the reception point the field level, measured by Pointing vector, makes Π, maximum power of signal received by antenna with an effective area S_{ef} makes $P_R = \Pi S_{ef}$. This power however depends upon both the said angle γ_{Π} , and the degree of polarization of this signal m_{Π} :

$$P_{R\,ex} = 0.5P_R(1 - m_{\Pi}) + m_{\Pi} P_R \cos^2 \frac{\gamma_{\Pi}}{2} . \qquad (10.12)$$

It follows from formula (10.12), that with non-polarized, randomly polarized signal, when $m_{\Pi} = 0$, on the output of antenna the level of this signal $P_{R\_ex} = 0.5P_R$ and this level does not depend upon the type of antenna polarization.

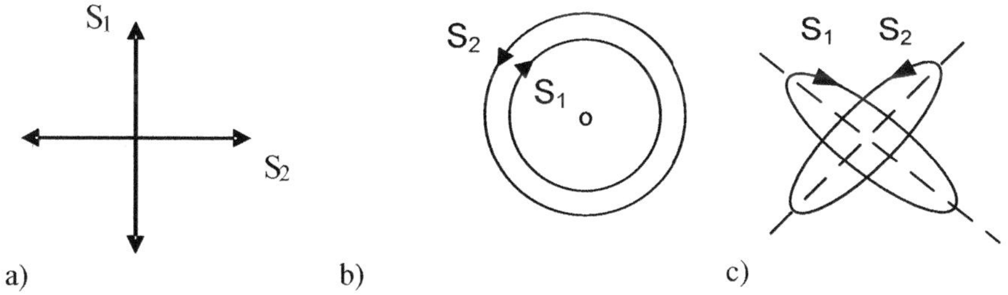

Fig. 10.7 Various polarization bases of the orthogonal signals S1 and S2 a) linear basis; b) circular basis; c) elliptic basis

In other words: in case of any antenna, it is possible to receive only half of non-polarized signal level only. In another extreme case, when a signal is fully polarized, $m_{\Pi} = 1$, it is possible to select antenna polarization so, that all possible received signal power is allocated (with $\cos^2 \frac{\gamma_{\Pi}}{2} = 1$, when $\gamma_{\Pi} = 0$). That is with $\gamma_{\Pi} = 0$ we have the reception matched by polarization. A function is thus maximized (10.12):

$$\sum_{i=1}^{N} w_i \psi_i(u) \underset{w_i}{\rightarrow} \max, \quad N = 2. \tag{10.13}$$

Apparently, there are conditions, when a function (10.13) acquires zero value:

$$\sum_{i=1}^{N} w_i \psi_i(u) \underset{w_i}{\rightarrow} 0, \quad N = 2. \tag{10.14}$$

Conditions (10.14) are characterized by the presence of «zero» polarizations orthogonal in relation to signals. Such orthogonalization is used for interference suppression, for their polarization rejection. Thus, polarization processing of signals comes to estimation of polarization parameters of signals (their amplitudes and phases or quadrature component) proper control of polarization basis of receiving antenna, matched or orthogonal signal, depending upon the task.

Besides the tasks of polarization matching with useful signal and tasks of polarization rejection of interference, polarization is used to solve other important tasks, including [7]:

- tasks of frequency re-use, when independent information streams are transmitted on two orthogonal polarizations, that allows to double productivity of access element;
- tasks of polarization modulation and demodulation of signals, when, for example, "1" is transmitted by vertical polarization, and "0" is transmitted by horizontal one. Such decision is rather constructive, as the process of modulation is not carried out in the transmitter radio circuit but directly in antenna;
- tasks of diversity polarization reception. Such task is especially effective in the multibeam radio channels of cellular channels types and trunking communications. In this case the given diversity reception can be realized directly in the subscriber station, because due to the compactness of two orthogonally polarized antennas placed in one electric center it is possible to maintain existing dimensions of the mobile station;
- tasks of adaptive reception by polarization, when polarization changes of parameters of signals or interference are accordingly traced by polarization of the receiving station.

The last-mentioned tasks of adaptive reception are of special importance. We will discuss them in more detail.

10.2.4 Adaptation in the Tasks of Access

It is possible to define adaptation as a process of optimization of the corresponding algorithm to variable and random in time external effects. As applied to the tasks of access such external effects are:

- interfering effects from other radioelectronic systems, sources of artificial or natural origin;
- effects from environment of radio signals propagation, causing random changes of parameters of received signals, their fading;
- random nature and instability of traffic, requiring the proper reaction from the side of technology of control at the level of access with the purpose of providing necessary level of LMS service. Tasks of adaptive processing also come down to the tasks of estimation of corresponding signals and interference parameters, to control of the state or surveillance.

For detailed consideration and substantiation of adaptation methods it is necessary to attract a rather comprehensive and serious mathematical apparatus of theory of management, Markov processes, methods of state variables, etc. [2…7]. However, taking into account the trend of this edition, we will concentrate only on the results and examples of application of adaptive methods. We will indicate only the necessity of substantiation of adaptive methods by criteria of procedure stability, provision of observability, manageability, identifiability and adaptability of suggested algorithms.

10.2.5 Suppression (Rejection) of Interference. Adaptive Antenna Arrays and Adaptive Interference Cancellers

Priority in development of idea of adaptive antenna arrays (AAA) belongs to B.Widrow. Many authors worked to develop his ideas [5,6,7 etc.].

The process of interference suppression on a receiving side by AAA lies in forming «zeros» of DP so that they meet the condition (10.14)

$$\sum_{i=1}^{N} w_i(t) S_i^{(v)}(t) \xrightarrow[\dot{w}_i]{} 0, \quad N = 1,2,...,N \,, \tag{10.15}$$

where $S_i^{(v)}(t)$ are interference signals received by N antenna elements. It is assumed that $S^{(v)}(t)$ is included in the accepted realization $x(t)$ additively:

$$x(t) = S^{(i)}(t) + S^{(v)}(t) + S^{(n)}(t)\,. \tag{10.16}$$

where indices "*i*", "*v*" and "*n*" belong accordingly: to the information signal, interference signal and noise.

Fig. 10.5 provides a general diagram, which enables solving a problem (10.15). The solution of this task is achieved under proper choice (estimation) of weight coefficients vector $w(t)$, by controlling this vector. In the theory of control, such task is interpreted as observation control task.

According to (10.15) and Diagram 10.5, first multiplication of $w_i\, x_i$ is carried out, and then their addition on general register. Dimension of WCV is determined by dimension of AAA. This dimension can be from 2 and to a rather large number: tens and hundreds. In practice, the AAA dimension usually does not exceed units, rarely tens. It is important to note that the number of antenna elements N determines the possible number of suppressed interference. There may be no more, than $N-1$ interference. Antenna elements may be spaced apart to distances $d > \lambda/2$. If AAA is implemented with the use of two orthogonal polarized antenna elements, they are located in one electric center and AAA is able to suppress only one interference.

The WCV control unit, where this adaptive procedure is realized, is the optimum algorithm of this WCV estimation. Historically the first and simple enough is Widrow algorithm [4,5]:

$$\frac{d\hat{w}(t)}{dt} = \mu\left[x(t)\hat{w}(t) - y_э(t)\right]x(t), \qquad (10.17)$$

where $y_э(t)$ is a certain standard signal desirable for the reception and identical in structure to a useful signal $S^{(i)}(t)$; $\mu \le 1$ is convergence coefficient of this gradient procedure, $\hat{w}(t)$ is the WCV estimation, optimum by the criterion of minimum of mean-square difference of received signal $S^{(v)}(t)$ from standard $y_э(t)$.

Algorithm (10.17) is usually implemented in a discrete form:

$$\hat{w}(k+1) = \hat{w}(k) + \mu(k+1)\left[x(k)\hat{w}(k) - y_э(k)\right]x(k), \qquad (10.18)$$

where context μ and $y_э$ is the same as in (10.18), k is the discretization interval. Investigations show that noise suppression level may reach 20…30 dB and greater. This level is often quite enough for providing steady communication in a radio line directed to the point of access, BSS or transponder.

Another adaptive algorithm for suppressing noise, also suggested by Widrow, is the adaptive noise compensator (ANC). In order to implement it, it is necessary to create a reference reception channel, free of useful information signal, when $S^{(v)}(t) \to 0$:

$$x_{on}(t) = S^{(v)}(t) + S^{(n)}(t). \qquad (10.19)$$

With known direction of arrival or polarization of useful signal, it is easily to obtain such channel by orthogonal antenna receiving base against useful signal. Algorithm of ANC is represented in Fig. 10.8.

Analytical representation of algorithm of WCV estimation of ANC differs slightly from (10.17):

$$\frac{d\hat{w}(t)}{dt} = \mu\left[x_{on}(t)\hat{w}(t) - x_{oc}(t)\right]x_{on}(t) \qquad (10.20)$$

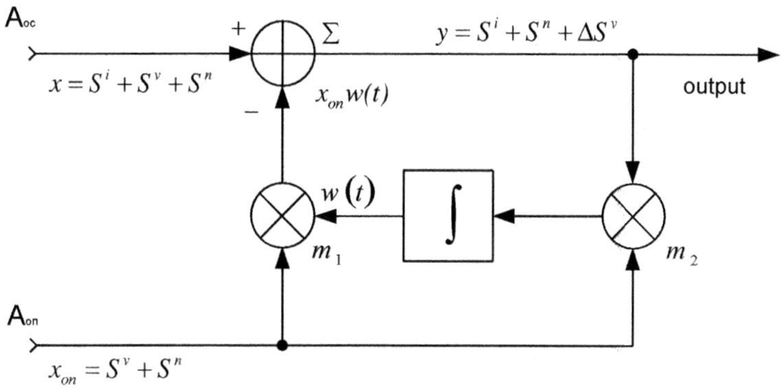

Fig. 10.8 Structure of algorithm for WCV control in the adaptive noise compensator

Its functioning can be described as follows. Difference in square brackets will be realized on the output of summing unit Σ. This difference is multiplied by $x_{on}(t)$ and being multiplied by μ it represents a derivative of WCV. The integral of this derivative is required WCV $\hat{w}(t)$, which is used in multiplier m_1. Another interpretation of algorithm is possible. In summing unit Σ there is subtraction (compensation) of the weighed interference $x_{on}(t) w(t)$. Excess from this deduction $\Delta S^{(v)}$ is multiplied by $x_{on}(t)$ in m_2 and the result is integrated. We obtain as a result:

$$\int y x_{on} dt = \int S^{(i)} x_{on} dt + \int \Delta S^{(v)} x_{on} dt + \int S^{(h)} x_{on} dt. \qquad (10.21)$$

Due to uncorrelatedness of subintegral functions, the first and third intervals in the right part are on average equal to zero. The second integral is significant, because $\Delta S^{(v)}$ and x_{on} contain common correlated components. Result of the second integral is a control signal, influencing the formation of WCV $\hat{w}(t)$. This influence will exist until it is not minimized by $\Delta S^{(v)}$, i.e. until interference in the main reception channel is compensated to minimum.

It should be noted that based of Widrow algorithms more effective, fast algorithms [2…7] synthesized on the basis of Markovian theory of filtration are developed.

It can be shown that algorithms (10.17), (10.18) and (10.20) are optimal for a situation when phase front of interfering signal $S^{(v)}$ is flat and not fluctuant. In radio access line, this front is distorted due to multipath propagation, spatial displacement of transmitter and/or receiver antennas. It requires selecting the model of the WCV condition in the form of

$$dw(t)/dt = F(t)w(t) + G(t)\xi(t), \qquad (10.22)$$

where $F(t), G(t)$ are matrices of the condition and generation, $\xi(t)$ is generating Gaussian white noise with the spectral density of power $N_{\xi}(t)$.

Equation of condition (10.22) allows using formalism of Kalman-Bucy filters [2,3,7] for the estimation of WCV:

$$d\hat{w}(t)/dt = F(t)\hat{w}(t) + P(t)X(t)N_{v}^{-1}\left[X(t)\hat{w}(t) - y_{э}(t)\right], \qquad (10.23)$$

where $X(t)$ is the adopted realization (10.16), $P(t)$ is a posteriori dispersion of WCV error estimation, determined from Rikkati equation

$$dP(t)/dt = F(t)P(t) + P(t)F^{T}(t) - P(t)X(t)N_{v}^{-1}X(t)P(t) + G^{T}(t)N_{\xi}G(t). \qquad (10.24)$$

Principle difference of algorithm (10.23) from classic FCH is that not a signal but optimum weight coefficient is subject to estimation. Thus equation (10.24) appeared dependent upon the received signal $X(t)$. Fig.10.9 represents the AAA algorithm functioning in accordance with (10.23).

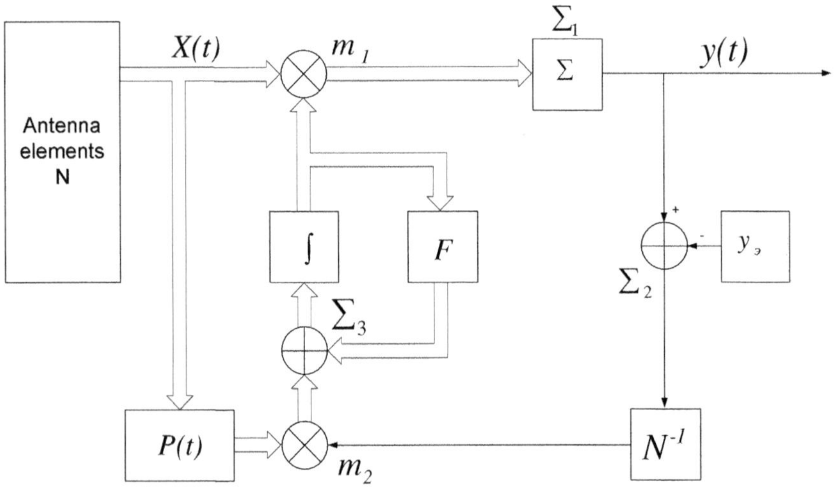

Fig. 10.9 Structural diagram of adaptive antenna array functioning according FCB algorithm

As the adaptive noise compensators operate in the same signal-noise situations, for the WCV of ANC state equations (10.22), (10.23) and (10.24) are also true with replacement of $X(t)$ by $X_{on}(t)$, which is determined from (10.19). Fig. 10.11 provides ANC structural diagram. It is characteristic that algorithm ANC represented by Fig.10.8 is extended to l-channels of basic reception (on l-element AA).

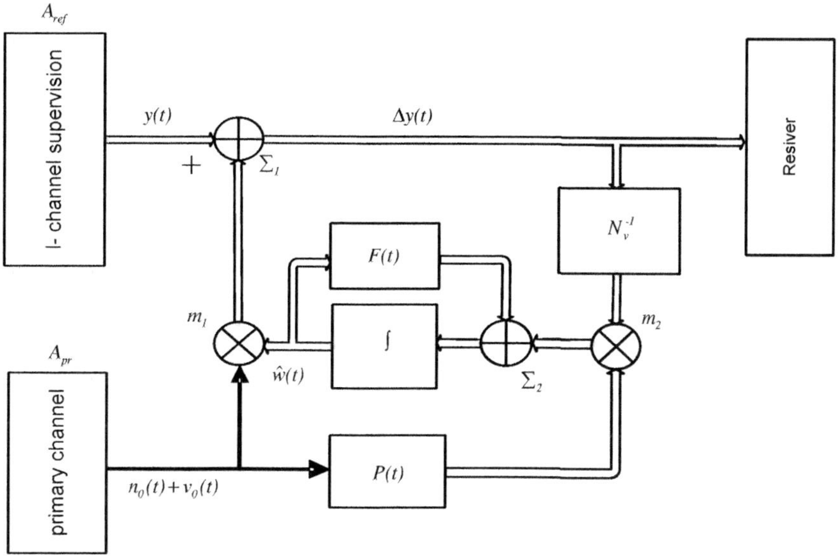

Fig. 10.10 Structural diagram of multichannel adaptive noise compensator

ANC algorithm functions as follows. Noise from reference channel after multiplier m_1, where it acquires corresponding to each i-th channel bias and scale $\hat{w}_i(t)$, is subtracted in summing unit Σ_1. The results of deduction are supplied to the input of measuring device or receiver and concurrently, after corresponding gain by N_v^{-1}, - to multiplier m_2. To another input of the multiplier the weighed value of signal from reference channel y_{pr} is supplied. Value of feedback is determined by multiplying by N_n^{-1}, which is set inversely proportional to watch noises spectral density in each reception channels. Integrator, where value $F(t)$ determines permanent integrations, carries out operation of statistical averaging of multiplication results. Voltage resulting from interaction of noise component from reference channel $v_o(t)$ and that part of uncompensated excess of noise in i-th reception channel, contained in $\Delta y(t)$, which is correlated with $v_o(t)$, appears on the output of integrator. It is obvious, that useful signal $s_i(t)$ due to uncorrelatedness with $v_o(t)$ is not supplied to compensation input. The very voltage from the integrator output, fed to one of multiplier inputs m_1, is the estimated WCV $\hat{w}(t)$. It follows from comparison of the considered ANC structural diagram and Widrow diagram (Fig. 10.8), that both diagrams are similar. At the same time, ANC in Fig.10.10 takes into account the presence of noises in reference channel $N_o(t)$, random changes of noise parameters $v(t)$ and $v_o(t)$, possibility of their correlation and mutual correlation with the help of matrix $F(t)$ elements. Function

$P(t)N_o(t)N_v^{-1}$ (10.22), also stipulating stability of procedure acts as a step coefficient determining the speed of convergence. Thus, an algorithm (10.22) is a certain modernization of Widrow algorithms, since it is optimal for more general conditions and more complicated signal-interference situation.

10.2.6 Control of Multipath Effect in Access Radio Lines

There are rapid attenuations of received signals, caused by multipath effect, re-emission from moving objects in access radio lines of cellular, trunked, pager communication, in Wi-Fi technologies, Wi-Max because of moving of communication objects, motion of surrounding subjects and people. In spite of actions taken (encoding with interlace, floating reception threshold, etc) there are signal dropouts, brief or long losses of communication. The periods of these disappearances make from fractions of a second to several seconds. Experience suggests that the statistical structure of such signals fading is different in different points of space, and their polarization changes at random. This fact makes it possible in order to increase reliability of communication to recommend space and/or polarization diversity methods of reception.

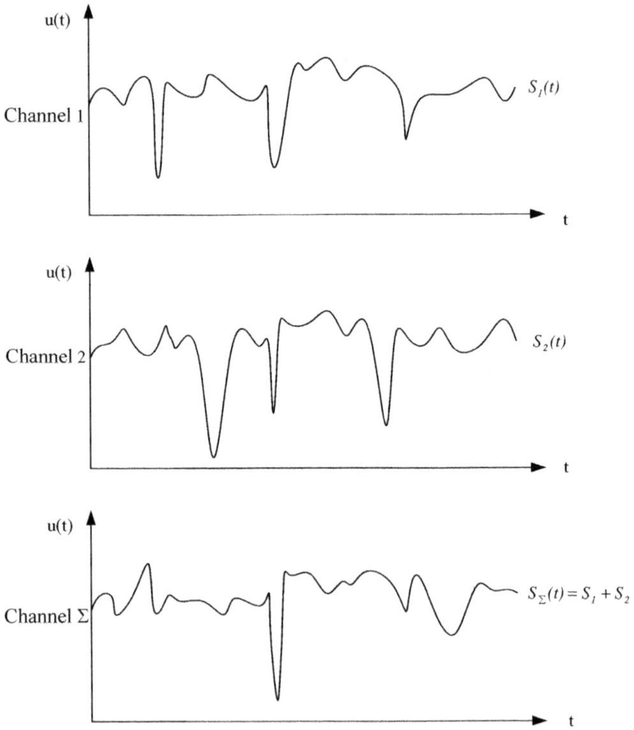

Fig. 10.11 Fragments of amplitudes of the signals received in diversity channels 1 and 2, and total amplitude

From a communication theory [1,2,3] it has been known that, diversity reception provides the greater effect, than independent signals in diversity channels. Fig.10.11. represents the example of fadeout signals.

Picture shows that if there is deep fading of signal amplitude to almost a zero level in separate channels 1 and 2, in a sum channel this fading is considerably less noticeable. With the increase of number of diversity branches, we manage not only to minimize the impact of fading but also to increase the level of regular composition.

Thus, there are three tasks. Task 1 is the discovery of several (two or more) independent representations of received signal. Two signals can be used for this purpose, received by orthogonally polarized antennas, two or more signals, received in different points of space, spaced to a distance $d \geq (10...100)\lambda$ or signals diverted by frequency to an interval $\Delta f \geq 1...1{,}5$ MHz. Signal-arrival-angle diversity method is only mentioned in references, but there is no information on its implementation. Time diversity is used; however, in interactive technologies it is not effective due to excessive delays. Fig.10.12 represents different variants of spatial diversity by transmission (a), on reception (b), on reception and transmission simultaneously (c).

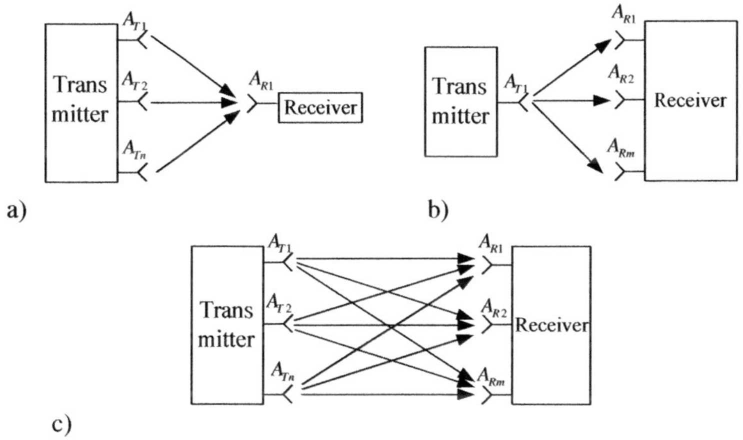

Fig. 10.12 Alternative diagrams of multiple reception of signals: a) at n - antenna on transmission; b) at m - antenna on reception; c) at n - antenna on transmission and m - on reception

Task 2 is addition of diversity signals. It can be solved with provision for (at phase locked addition) or without provision for the phase carrier frequency (predetector or postdetector addition). Phase locked addition is more effective. In this rate, when adding 2 signals with voltage U_1 and U_2, differing by phase to angle φ we obtain total power:

$$P_{\Sigma} \approx (U_1 + U_2)^2 = U_1^2 + U_2^2 + 2U_1U_2\cos\varphi$$

As is evident, with phase locked addition, when $\cos\varphi=0$, we get $P_\Sigma \approx P_1 + P_2 + 2P = 4P$. Not counting the phase, the third addend must be dropped and $P_\Sigma \approx 2P$. In spite of power loss by 2 times postdetector addition is used where it is necessary to maximally simplify circuit design. Phase - locked addition requires the additional circuit design decisions on synchronization of high-frequency signals components.

Task 3 is determining weight coefficients of the composed signals, because their value influences the signal/noise ratio of the resulting signal. Therefore, it is possible to sum up all signals with the same weight, for example, $w_i = 1$ for all branches of diversity (linear addition). In this case, those branches, where a signal is small, will contribute only to the growth of noise level. It is possible at every moment of time to choose one branch only, where useful signal is maximum (auto choice), but branches will be cast aside, where signals are less and they could give positive contribution to the grand total.

It is possible to show that the best and optimal is the addition with weight w_i, proportional to the level of useful signal in the same i-th channel (quadratic addition). In this case (10.9) appears as

$$\sum_{i=1}^{N} w_i \psi_i(t) = \sum_{i=1}^{N} w_i S_i(t) = \sum_{i=1}^{N} S_i^{(v)}(t) S_i(t), \tag{10.25}$$

where $w_i = S_i^{(v)}(t)$.

A number of methods of solution of the above three tasks are known. We will mention the most popular one, where solutions of the second and the third tasks are united into one general task. Fig.10.13 represents general diagram of this solution.

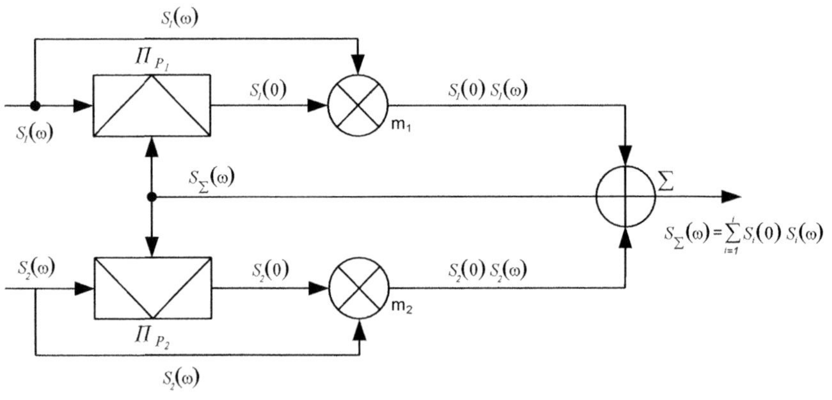

Fig. 10.13 Algorithm of quadratic coherent addition of diversity signals $S_1(\omega)$ and $S_2(\omega)$ without considering quadrature component

Weighing of signals $S_1(t)$ and $S_2(t)$ takes place in multipliers m_1 and m_2, on the second entrance of which signal $S(0)$ converted to zero frequency is supplied from the outputs of converters Πp_1 or Πp_2. In-phase operation is achieved by general signal $S_\Sigma(\omega)$ from the output of summing unit Σ, which controls the phase of converter reference signal.

Fig.10.14 shows graphs of probability of erroneous reception of single character (BER) from signal level ratio and noise $h^2 = P_s / P_n$ for different reception ratio.

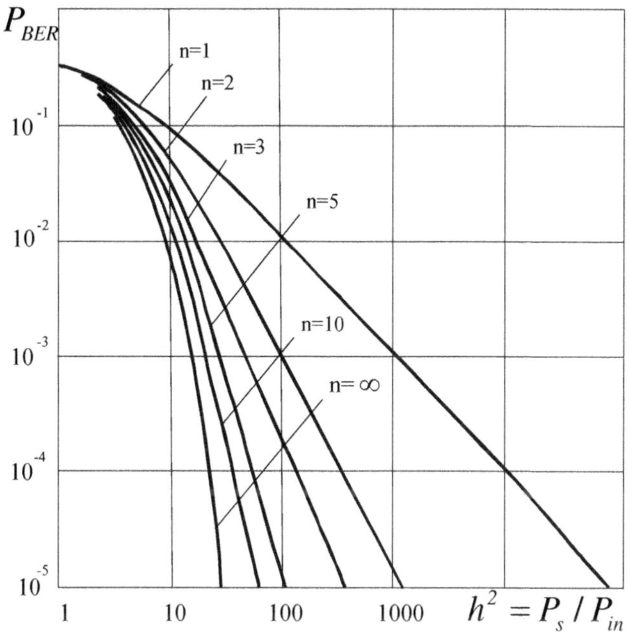

Fig. 10.14 Diagram of signals reception anti-jamming ability at different number of diversity branches

It follows from the diagram that with the increase of order of diversity n the anti-jamming ability becomes better, however transition from single $(n=1)$ to double $(n=2)$ reception provides the greatest contribution, and all subsequent additions of diversity branches provide progressively smaller positive contribution.

Fig.10.15 shows graphs P_{ber} for the reception of signals in the two diversity branches with different correlation coefficient r. It is clear from the diagram that diversity reception of uncorrelated signals $(r = 0)$ provides considerable advantage. Still in channels with high correlation $(r = 0.5)$, this advantage is yet considerable. It is obvious that in order to increase the efficiency of diversity reception it is necessary to achieve reduction of statistical coupling in diversity branches.

In practice, control of WCV, included into (10.25), is carried out during decomposition of $S_1(t) = A(t)\cos(\omega t + \varphi(t))$ into quadrature components $S_i(t) = S_s \sin\omega t + S_c \cos\omega t$ with subsequent estimation of these elements $\hat{S}_s, \hat{S}_c$, that results in the value $A(t) = S(0) = \sqrt{\hat{S}_s^2 + \hat{S}_c^2}$. The resultant value $S(0)$ is further used as a control signal in multipliers m_1 and m_2 (see Fig.10.13). Division into two procedures: stochastic estimation $\hat{S}_s, \hat{S}_c$ and deterministic control does not preclude properties of procedure optimality, since the conditions of separation theorem are correct here [1,2,3].

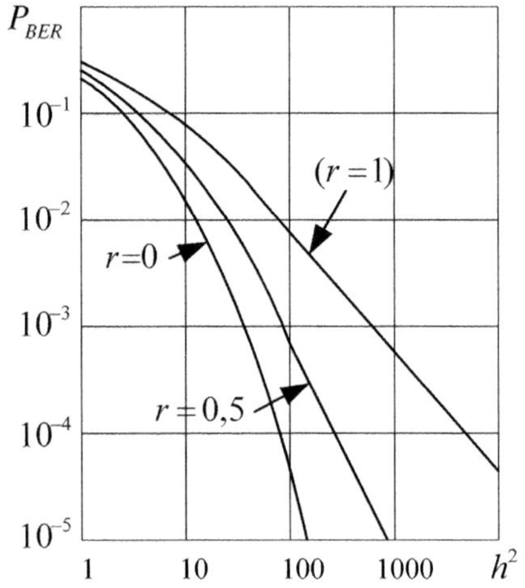

Fig. 10.15 Curves P_{BER} for the reception of signals in two diversity branches with different correlation coefficient of r

The use of broadband signals (BBS) is a variety of diversity reception method, among which the DSSS methods (Direct Sequence Spread Spectrum) and FHSS signals (Frequency Hopping Spread Spectrum) are most widely used. These signals are used in Wi-Fi systems, especially they are recommended in IEEE 802.11 standards etc, where multipath effect takes place.

10.2.7 Space-Time Coding

Methods considered in (10.25) with the use of n-antennas are those, that minimize the destructive impact of multipath effect. However, there are technological solutions, which are invariant in relation to multipath effect, or even this phenomenon is used as positive. Such technologies, within MIMO concept, include space-time encoding.

Core of space-time coding method consists in the fact that n-independent information signals can be transmitted simultaneously on the same operating frequency f_p by n transmitting antennas (Fig.10.16). Each of these signals $S_i^{(T)}$, $i=\overline{1,n}$ on the way to n-receiving to antennas is subject to corresponding fading h_{ij},

$$\begin{cases} S_1^{(R)} = S_1^{(T)}\mathrm{h}_{11} + S_2^{(T)}\mathrm{h}_{21} + \ldots + S_n^{(T)}\mathrm{h}_{n1}, \\ S_2^{(R)} = S_1^{(T)}\mathrm{h}_{12} + S_2^{(T)}\mathrm{h}_{22} + \ldots + S_n^{(T)}\mathrm{h}_{n2}, \\ \cdot\ \cdot\ \cdot\ \cdot\ \cdot\ \cdot\ \cdot\ \cdot\ \cdot\ \cdot\ \cdot\ \cdot\ \cdot\ \cdot \\ S_n^{(R)} = S_1^{(T)}\mathrm{h}_{1n} + S_2^{(T)}\mathrm{h}_{2n} + \ldots + S_n^{(T)}\mathrm{h}_{nn}. \end{cases} \tag{10.26}$$

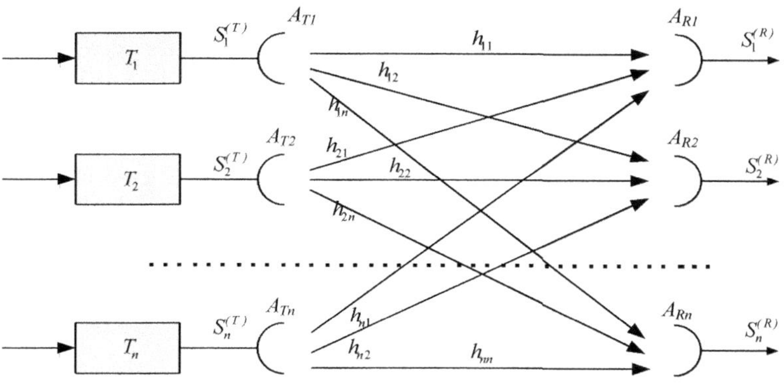

Fig. 10.16 Diagram of n information signals transmission via multibeam radio channel with space-time coding

Thus (10.26) is the system of n-linear equations with n-independent unknown $S_i^{(T)}$. The system is heterogeneous, as absolute terms $S_i^{(R)} \neq 0$ and it must be consistent in order to have a solution and must be defined so that this solution is unambiguous. It is obvious that all these properties of the system (10.26) depend on the combination of ratios h_{ij}, and matrix determinant of these ratios must not be equal to zero:

$$\|\mathrm{H}\| \neq 0\,. \tag{10.27}$$

Such conditions (10.27) are met in the absence of linear dependence between equations of the systems (10.26), when none of equations is a linear combination of the others. These conditions are met in practice due to different values h_{ij} in each of $n \times n$ directions of radio signals distribution. At the same time, on a receiving side there is permanent control of the above conditions and, in case of failure to meet them, on transmitting side, required linear independence between equations (10.26) is achieved with the help of corresponding selection of levels of signals $S_i^{(T)}$. Thus, signals $S_i^{(R)}$ splitting occurs with the help of unique space-time code.

In order to provide the solution of the system (10.26) and division of signals $S_i^{(T)}$, the value of coefficients h_{ij} must be measured and estimated in each of i, j directions. For this purpose a test signal with the known parameters is periodically transmitted in radio line, to provide capability to measure and estimate coefficients h_{ij}, which keep their value for a certain short time interval $\Delta\tau_n$, which is an index step. As a result, a procedure of recursive evaluation $\hat{\mathrm{h}}_{ij}(k) = \Phi(\hat{\mathrm{h}}_{ij}(k-1))$ occurs. Obtained estimations $\hat{\mathrm{h}}_{ij}$ are further used for the resolution of the system (10.26). There are two more important conditions, required for reliable operation of space-time encoding algorithm:

- In a radio channel between a transmitter and receiver there must be sufficiently large allowance of high-frequency level of useful signal, that gives possibility of reliable high-quality reception at the relatively large (-10...-30) dB signals fading.
- Fading of multipath signals is called rapid, however it is possible to specify short time intervals, where parameters of signals which are used in n-antenna elements may be considered practically fixed. Intervals that are sensibly shorter than fading correlation intervals may be considered as intervals with quasi-permanent parameters:

$$\Delta\tau_n \leq 0{,}1\tau_{\mathrm{cor}} \tag{10.28}$$

Obviously, it is possible to carry out transmission not only using the method of space-time encoding but also using polarization-time encoding, where $S_i^{(T)}$ represents signals which have the proper polarization structure.

Space-time encoding is most effectively used in Wi-Fi office systems, however, with proper modernizations it is possible to use it in bigger Wi-Max systems, in cellular and other systems.

10.3 Recommendations on Practical Use of Signal Processing Algorithms

10.3.1 Formalization of Kalman-Bucy Algorithm

The above-mentioned tasks are solved on physical level during radio access due to estimation of corresponding parameters of signals $\hat{x}_i$. The FKB formalism in a discrete form results in the solution of the following equations in problem space [2,3]:

- constitutive equations:

$$x(k+1) = F(k+1,k)x(k) + G(k)\xi(k), \ k = 1,2...; \tag{10.29}$$

- personal equation:

$$y(k) = H(k)x(k) + n(k)\,. \tag{10.30}$$

In the equation, in the same way as in (10.30), not only white Gaussian noises $n(k)$ but also interferences (colored noises) $v(k)$ can be taken into account. The removal of the latter is the AAA and ANC function:

- Filter equation:

$$\hat{x}(k) = F(k,k-1)\hat{x}(k-1) + K(k)\left[y(k) - H(k)F(k,k-1)\hat{x}(k-1)\right], \quad (10.31)$$

where $K(k) = P_{\tilde{x}}(k)H^T(k)N_n^{-1}(k)$ is the scaling function determining the FKB optimum from positions of exactness of estimation and speed of its convergence to the steady state;

- Dispersion of a posteriori error of estimation $\tilde{x} = x - \hat{x}$

$$P_{\tilde{x}}(k+1/k) = F(k+1,k)P_{\tilde{x}}(k)F^T(k+1,k) + G(k)N_{\xi}(k)G^T(k); \quad (10.32)$$

- A priori dispersion $\tilde{x}$

$$P_{\tilde{x}}(k) = \left[I - K(k)H(k)\right]P_{\tilde{x}}(k/k-1); \quad (10.33)$$

Direct use of the algorithm (10.29)...(10.33) for estimation of a particular parameter often results in a failure. There are several reasons for this:

a) incorrect selection of filter parameters F, G, N_n, N_{ξ}, when parameters of the selected model (10.29), (10.30) disagree with the real signal-to-noise environment;
b) incorrect selection of discretization step $\Delta\tau$ in relation to correlation interval τ_{cor};
c) disagreement of discreteness of measurements entry y(k) and discreteness of the estimation procedure (10.31);
d) technological features of estimation procedure, etc.

10.3.2 Recommendations on Planning of Estimation Algorithms

In the tasks of designing of processing devices, it is expedient to carry out preliminary mathematical computer-aided modeling. The structural diagram represented by Fig. 10.17. will be useful for this purpose.

For conducting the experiment, it is necessary to select adequate arrays of generation noises $\xi(k)$ and observation noises $n(k)$. Size of the array must be large enough, and sample characteristic of power spectral density (PSD) must be most uniform in the frequency band (deviations exceeding ±(25...30) % are not desirable). Relation of $\xi(k)$ noise PSD to $n(k)$ noise PSD can be interpreted as a relation of levels of useful signal and noise in the observation channel $h^2 = N_{\xi}/N_n$, and if $N_n - 1$ here, this relation becomes numerically equal to N_{ξ}.

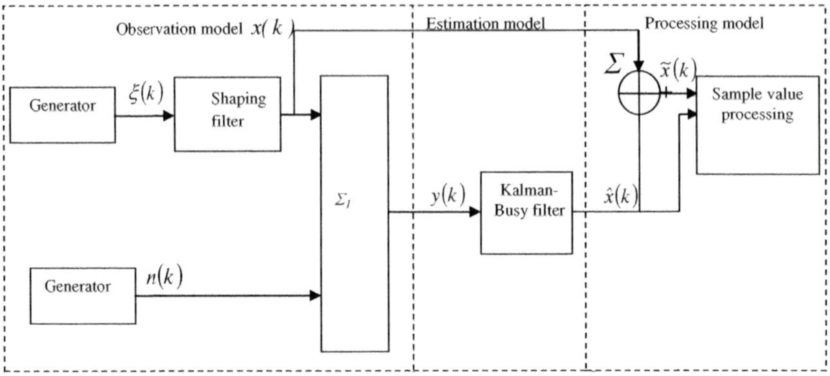

Fig. 10.17 Structural diagram of computer-aided experiment

As a result of the experiment, different situations with convergence to the steady state of procedure (10.31) are possible which occurs on condition that a posteriori dispersion is $P_{\tilde{x}}(k/k-1) \to const$. At the most, the steady state of filter occurs already on 4 to 10 steps of discretization. Fig. 10.18 shows dependence of convergence time of FKB on the size of discretization step.

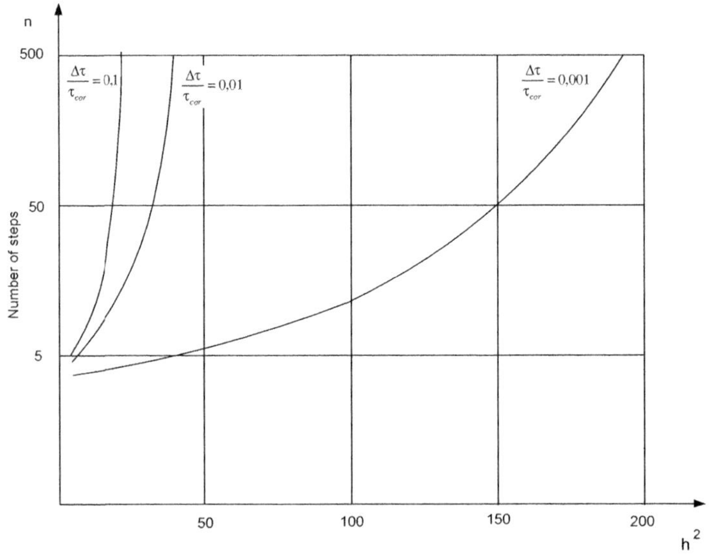

Fig. 10.18 Diagram of FKB convergence depending on the selected parameter h^2 at various sizes of discretization step $\Delta\tau / \tau_{cor}$

The diagram implies that time of convergence to the steady state increases with the increase of the estimated signal h^2 level. This time can increase to infinity, which is the indication of procedure divergence. The mode of relatively large steps of discretization (at $\Delta\tau/\tau_{cor}=0{,}1$) is especially critical. Here, steady convergence is observed at the ratio $h^2 \le 20$ only.

Fig. 10.19 shows dependence of convergence time of FKB with deviation of model parameters from real situation (at mismatching of signal strength $N_{\xi}^{(F)}$ set in a filter in relation to model $N_{\xi}^{(M)}$ level)

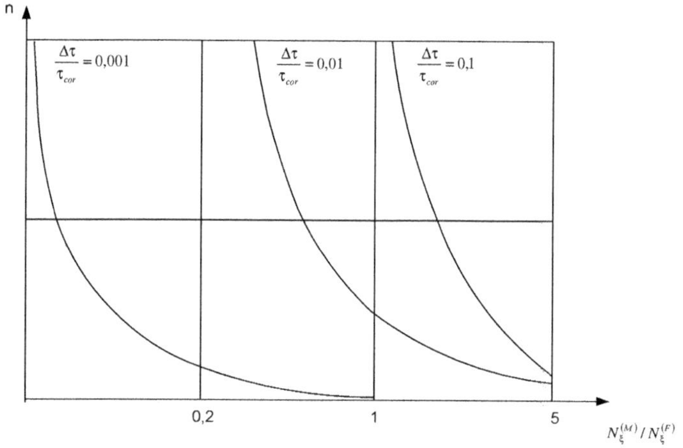

Fig. 10.19 Diagram of convergence time of filter depending on the relation of model PSD $N_{\xi}^{(M)}$ to filter PSD $N_{\xi}^{(F)}$

As it is evident from the graphs, the sensitivity of estimation procedure to deviation of parameters of the selected model to any side is not symmetric. Moreover, selecting understated value of signal strength does not only worsen the convergence but also improves it that enables an important practical conclusion: it is necessary to understate the signal-to-noise ratio in estimation procedure, which will be instrumental in the improvement of convergence, and quality of estimation in this case (a posteriori dispersion of estimation error) does not practically change. The other method of convergence improvement is the reduction of discretization interval, which evidently helps to make a more exact estimation.

Research of sample values of a posteriori dispersion of estimation error $\tilde{x}$ obtained by the method of ensemble averaging with $i \ge 10...20$ of independent sample arrays $x_i(k)$ can be very informative. Thus, sample estimate of dispersion $\sigma_{\tilde{x}}^2=(i-1)^{-1}\sum_{i=1}^{i}\tilde{x}^2$ at $h^2=10$ for the step of $\Delta\tau/\tau_{cor}=0{,}01$ proved to be equal to $\sigma_{\tilde{x}}^2=0{,}3608$, for $\Delta\tau/\tau_{cor}=0{,}001$ its value was $\sigma_{\tilde{x}}^2=0{,}1133$. Time of convergence to the steady state was 30…40 intervals of discretization.

Time of convergence is of great practical importance, because when different nonstationary effects occur it is the time of transient mode. More detailed effect of nonstability can be explored by presenting the equation (10.30) in the form of:

$$y(k) = H(k)x(k) + C(k)\sin(l_k + \frac{\Delta\tau}{\tau_{cor}}) + n(t), \tag{10.34}$$

where $C(k)$ determines amplitude of instability components, l determines value of change of smooth period of sinusoidal nonstability changes.

The program of model formation according to algorithm (10.29) and calculation of estimation according to algorithm (10.31) is represented below.

10.3.3 Program of Estimation Calculation with the Help of FKB

```
clear
   N=1000;
   t1=1:N;
   D=1;
   T=1;
   T0=10;
   w=randn(size(t1));
   v=randn(size(t1));
   x(1)=0;
   F=exp(-T/T0);
   G=sqrt(D*F*(1-F));

   H=1;
for k=1:N-1
   x(k+1)=F*x(k)+G*w(k);
end
z=H*x+v;
figure(1)
plot(t1,z);
Vw=2*ones(size(t1));
Vv=2*ones(size(t1));

Vx(1)=1;
V1x(1)=1;
x1(1)=F*0;
K(1)=1;
er3(1)=1;

for k=2:N
   K(k)=V1x(k-1)*H'*Vv(k)^(-1);
   Vx(k)=F*V1x(k-1)*F'+G*Vw(k)*G';
   V1x(k)=(1-K(k)*H)*Vx(k);
   x1(k)=F*x1(k-1)+K(k)*(z(k-1)-H*F*x1(k-1));
er3(k)=z(k-1)-H*F*x1(k-1);
end
er=(x-x1);
er1m=mean(er)
er1=mean((er-er1m).^2)
```

10.3.4 Recommendations for Designing Adaptive Noise Compensators

Being more general ANC and AAA (10.23) allow taking into account more complex signal to noise environment, and their quality is determined by the extent of noise suppression $v(k)$ and improvement of the ratio of signal P_S level to the sum of interferences and noise on the ANC or AAA output:

$$h^2 = P_S /(P_v + P_n) . \tag{10.35}$$

In the process of ANC development, it is necessary to consider the absence of useful signal in the reference channel, and the level of interference v_0 should exceed the corresponding value in the primary channel. To explain this, we will show results of the computer-aided experiment.

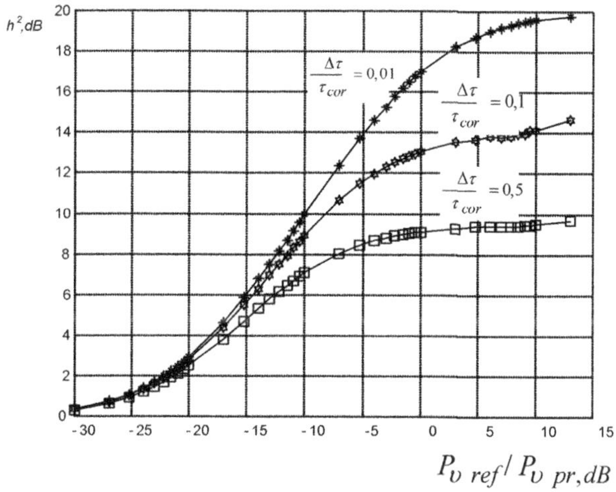

Fig. 10.20 Diagrams of ANC efficiency depending on the relation of noise levels in reference and primary channels

To obtain numerical values let us set the following relations between levels of useful signals and noise in the primary channel: $P_s / P_n = 20\ dB$, $P_v / P_n = 20\ dB$. Let us change interference power v_o in the reference channel in relation to interference power in the primary channel from — *30* to *15 dB*. As a result of computer-aided experiment, graphs of h^2 depending on the relation of powers of interference in the reference channel to power of interference in the primary channel $P_{v.pr}$ are obtained (Fig.10.20). It is evident from the graphs that with the increase of interference power in the reference channel $P_{v.ref}$ efficiency increases smoothly, and at $P_{v.ref} / P_{v.pr} \geq 6...10\ dB$ it achieves maximum values. This fact is of practical importance: when selecting parameters of the reference channel, it is important that the level of interference $P_{v.ref}$ in it exceeds the corresponding level $P_{v.pr}$ in the primary channel. From the obtained graphs, we can make a conclusion that the ANC operation efficiency is substantially influenced by discretization interval: efficiency grows with more frequent discretization.

The other important requirement to reference channel is the minimum level of desired signal, as its presence results in losses of the ANC efficiency because useful signal $v(t)$ is also compensated together with noise. Fig. 10.21 shows diagrams of ANC efficiency losses depending on the level of desired signal in the reference channel in relation to the level of signal in primary channel $P_{s\ pr} / P_{s\ ref}$.

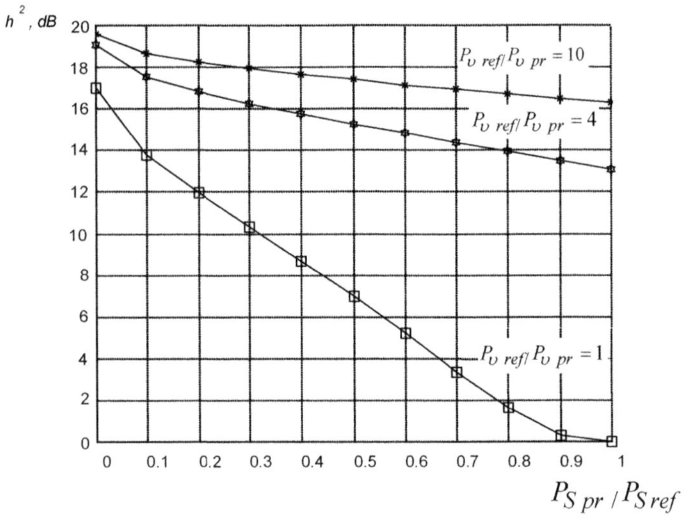

Fig. 10.21 Diagrams of ANC efficiency losses in the presence of useful signal in reference channel with a relative level $P_{s\ ref} / P_{s\ pr}$

It is characteristic that in case of exceeding noise level in the reference channel as compared to the primary one (two overhead curves in Fig.10.21) losses of the ANC efficiency are insignificant. However, if these levels are equal (third curve) losses are considerably essential.

10.3.5 Recommendations for Planning Adaptive Antenna Arrays

By analogy with ANC, where development of a reference channel free of useful signal is required, a similar channel on output Σ_2 is created in AAA after deduction of reference signal $y_э$ (Fig. 10.9). Analysis shows that incomplete matching of structure of reference and useful signals results in incomplete compensation of this signal and accordingly to decline of the AAA efficiency, similar to that in ANC.

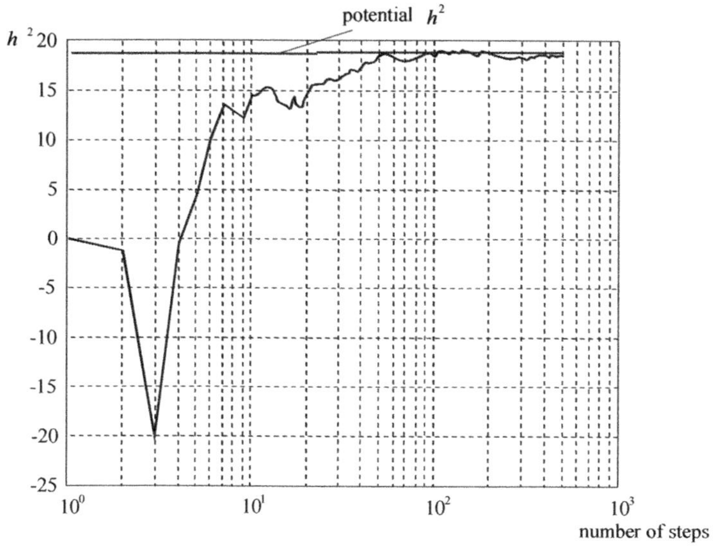

Fig. 10.22 Graphs of transient mode for four-element AAA

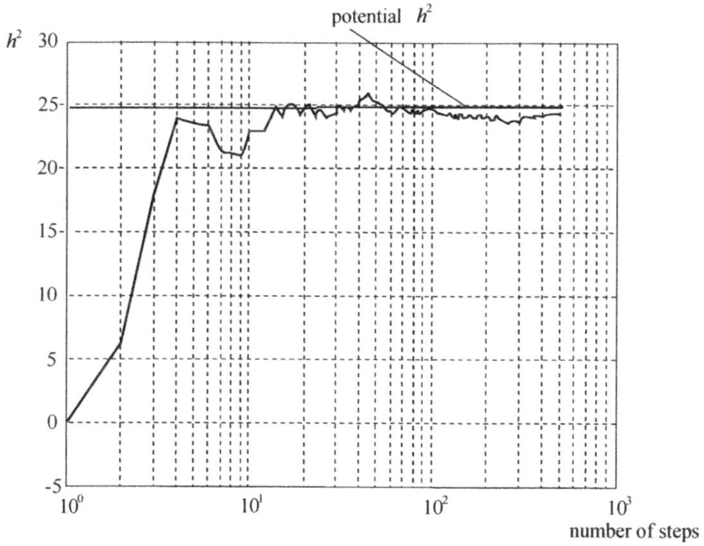

Fig. 10.23 Graph of transient mode for eight-element AAA

Simulation modeling of AAR algorithm showed that convergence time of procedure (10.23) to the steady state is situated within the limits of 10…20 discretization intervals. Fig.10.22 and Fig. 10.23 show diagrams of transient modes of adaptation algorithms, built on condition of equality of levels of useful signals and interferences $P_s = P_v$ and with relation of levels P_s / P_v =20 dB in each receiving channel. As it is evident from the diagrams, potential of the algorithm efficiency increases with the increase of the number of AAA antenna elements.

Attention should be drawn to one more feature of ANC and AAA algorithms. It is known that hysteresis phenomena are characteristic for the systems of White type and they become apparent in AAA and ANC control algorithms. WCV control linkage backlash appears because of the post-tuning drift, retaining WCV in a position, which conforms to current signal-interference environment. In this case, post-tuning drift is proportional to gain factor in control circuit. Indicated drifts are characteristic also for the systems of phase-locked loop and automatic frequency control. Dependence of h^2 on the size of the backlash area, delineated in relative units is shown in the diagram (Fig. 10.24). Diagrams show that with expansion of backlash area ANC efficiency gains ambiguous nature and with the dimensions of the area greater than $0{,}04\Delta w$ the mode of bifurcation and chaotic behavior is observed.

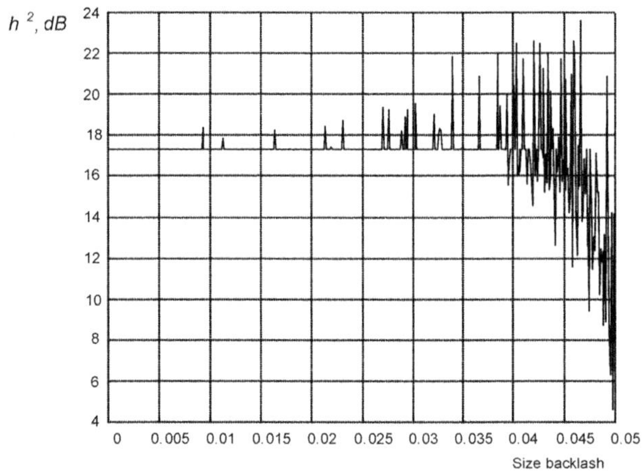

Fig. 10.24 Diagrams of efficiency losses of adaptive algorithm depending on WCV adjusting characteristic backlash

10.4 Conclusions

In recent years more firms, dealing with manufacture of telecommunications equipment, scientists and specialists draw attention to the necessity of the use of space-polarization resources to provide quality solution of access tasks in different technologies which agree upon concept of creation of the fixed-mobile systems (FMS).

Consideration of various space-polarization methods in access tasks implies that these methods are transformed easily enough from one into another. In other words using 2-dimensional or *n*-dimensional spatial or polarization basis on receiving side or on both sides simultaneously there is a possibility to solve such important problems as providing access of electromagnetic compatibility, filling in deficit of frequency resource, increase of interference immunity, authenticity and reliability of transmission etc.

References

[1] Lyons, R.G.: Understanding Digital Signal Processing, vol. (4). Prentice hall, Englewood Cliffs (2004)
[2] Van Tress, H.L.: Detection, Estimation and Modulation Theory. Pt I. John Wiley and Sons, Inc., NY (1967)
[3] Snyder, D.: The state – variable approach to continuous estimation with applications to analog communication theory. The MIT Press, Cambridge (1971)
[4] Widrow, B., et al.: Adaptive Antenna Systems. Proc. IEEE 55, 2143–2159 (1967)
[5] Monzingo, R.A., Miller, W.: Introduction to Adaptive Arrays. John Wiley & Sons Inc., NewYork (1980)
[6] Marpl-jr, S.L.: Digital spectrum analysis and its application / Transl. from English. Mir, Moscow (1990)
[7] Popovskij, V.V., Rodimov, A.P.: Statistical theory of polarization – timing processing of signals / M.: Radio i svjaz (1986)

11 Recursive Code Scales for Moving Converters

Alexander Ojiganov

Saint-Petersburg State University of Information Technologies, Mechanics and Optics; Department of Computing Techniques, Russia

Abstract. The methods of construction of recursive code scales (RCS), and also algorithms of placing on a scale of reading out elements (RE) are considered, results of research of correcting possibilities of such scales are shown. RCS for synthesis of drawing of an information path of a scale of sequence have received the name pseudo-random (PRCS) and composite code scales (CCS). Offered scales can be applied as the coded element in moving converters. Recursive scales at the expense of use in them of only one information code path more technologically traditional scales, code paths (CP) which are carried out, as a rule, in an ordinary binary code or in the Gray code. Thus, RCS allows only at the expense of redundancy introduction on number of reading out elements without use of additional control paths to form the codes which are correcting and (or) finding out errors of reading.

11.1 Pseudo-Random Code Scales

11.1.1 Pseudo-Random Code Scales for Converters of Angular Movings

The method of construction circular PRCS is based on use of the theory of M-sequences [1]. Let us adhere to the standard terminology.

1. The M-sequence is a sequence of binary symbols $\boldsymbol{a}$ the lengths, $M = 2^n - 1$ received on a certain rule which is set by a primitive polynomial $h\ (x)$ degrees n.
2. The primitive polynomial $h\ (x)$ represents a polynomial which is not resulted over field Galua GF (2), i.e.

$$h(x) = \sum_{i=0}^{n} h_i x^i \quad , \tag{11.1}$$

where h_0=h_n=1, and h_i=0,1 at 0 <i <n.

M. Adamski et al. (Eds.): Design of Digital Systems and Devices, LNEE 79, pp. 263–288.
springerlink.com

3. M-sequence symbols a_{n+j} satisfy to a recursive parity

$$a_{n+j} = \mathop{\Xi}_{i=0}^{n-1} a_{i+j} h_i \quad , \mathrm{j} = 0,1...,\tag{11.2}$$

where the sign Ξ denotes summation on the module two, and indexes at M-sequence symbols undertake on module M. Initial values of symbols of M-sequence $a_0 a_1 ... a_{n-1}$ can get out any way, except for a zero combination. For definiteness to these symbols we will give values $a_0 = ... = a_{n-2} = 0$, $a_{n-1} = 1$.

4. M-sequences belong to the class of cyclic codes and can be set by means of a generating polynomial

$$g(x) = (x^M + 1) / h(x),\tag{11.3}$$

where $h(x)$ it is defined according to expression (11.1), $M=2^n-1$.

For each M-sequence of length M exists exactly M various cyclic shifts which can be executed by multiplication of a generating polynomial $g(x)$ on x^j where $j=0,1...,$ $M-1$.

It is known [2], that for any value n exists precisely $\varphi(M = 2^n - 1) / n$ various polynomials $h(x)$, being not resulted and primitive. The function named $\varphi(M)$ function of Euler, represents the number of positive integers smaller or equal M and mutually simple from M. The number $\varphi(M)$ very quickly grows with growth n, number of polynomials of degree n, generating sequences of the maximum length, also quickly increases with growth n.

In Table 11.1 polynomials $h(x)$ to degree 16 inclusive which have the minimum number of nonzero factors h_i are resulted and can be used for generation of corresponding M-sequences of length $M = 2^n - 1$[1].

Let's formulate a method of construction of n-digit circular PRCS [3,4].

1. Depending on demanded word length of a scale n from table 11.1 the polynomial $h(x)$ gets out.
2. Using a recursive parity (11.2) the sequence $\boldsymbol{a}$ is generated.
3. Elementary sites of a scale are carried out according to M-sequence symbols $\boldsymbol{a}$ where individual symbols of sequence correspond to active, and zero symbols to passive sites of an information path. For definiteness sequence symbols are displayed on an information code path in a direction of movement of an clock hand as it should be $a_0 a_1 ... a_{M-1}$.
4. Placing on a scale of reading out elements is carried out. As PRCS are under construction according to M-sequence symbols, it is possible to define by cyclic shifts a placing order on a scale n RE. Differently, m RE (m=1,2..., n) j_m -th cyclic shift $x^{jm} g(x)$ M-sequences is put in conformity. Then the polynomial defining an order of placing on a scale n RE, looks like

$$r(x) = \sum_{m=1}^{n} x^{j_m} \quad ,\tag{11.4}$$

where $j_m \in 0,1..., M\text{-}1$. Having put j_1=0, according to (11.4) we will have the second, the third, ..., the n-th RE displaced (for definiteness in a direction of movement of an hour hand) concerning the first RE on $j_2, j_3...,$ j_n elementary sites of an information path of a scale accordingly.

Table 11.1 Primitive polynomials

n	h (x)	$M = 2^n - 1$	n	h (x)	$M = 2^n - 1$
1	x+1	1	9	$x^9 + x^4 + 1$	511
2	$x^2 + x + 1$	3	10	$x^{10} + x^3 + 1$	1023
3	$x^3 + x + 1$	7	11	$x^{11} + x^2+1$	2047
4	$x^4 + x + 1$	15	12	$x^{12} + x^7 + x^4 + x^3 + 1$	4095
5	$x^5 + x^2 + 1$	31	13	$x^{13} + x^4 + x^3 + x + 1$	8191
6	$x^6 + x + 1$	63	14	$x^{14} + x^{12} + x^{11} + x + 1$	16383
7	$x^7 + x + 1$	127	15	$x^{15} + x + 1$	32787
8	$x^8 + x^6 + x^5 + x + 1$	255	16	$x^{16} + x^5 + x^3 + x^2 + 1$	65535

In the given subsection placing RE along path PRCS with step to one quantum is considered. Possibility of other variants of placing will be shown in 11.3.

In the converters of moving based on a method of reading, the code scale at a full turn should provide reception of number of the various code combinations equal to number of quanta of moving. We will show that it is true for converters of moving on the basis of circular PRCS. For this purpose we will prove the following statements.

Statement 11.1 Circular PRCS allows to receive exactly M various n-digit code combinations corresponding to sequence from M of quanta of moving.

Proof. We will consider a fragment of M-sequence from n consecutive symbols. It corresponds to some code combination $a_j a_{j+1}... a_{j+n-1}$, reproduced from information path PRCS in the reading out knot from n elements. Reading out elements on PRCS are located with step to one quantum, position of the coded element could be any. After scale moving on k quanta $(k < M)$ from an information path of a scale in the reading out knot the n-digit code combination $a_{j+k}a_{j+k+1}$ will be reproduced... $a_{j+k+n-1}$. A condition of equality of these code combinations is the following one:

$$a_j a_{j+1} \ldots a_{j+n-1} = a_{j+k} a_{j+k+1} \ldots a_{j+k+n-1}. \qquad (11.5)$$

It means, that the M-sequence period is equaled k.

It contradicts property of M-sequence, on which its period $M = 2^n - 1$ [1]. Hence, these code combinations should be various. As the number of symbols of M-sequence is equal M to each moving PRCS on one quantum there corresponds the n-digit code combination and it will be equal M, as was to be shown.

The statement 11.1 is defining as testifies to basic possibility of construction of converters of moving on the basis of PRCS with one information code path.

Statement 11.2. Resolution circular PRCS is defined by a parity:

$$\delta = 360^0 / M \quad (11.6)$$

The proof is obvious, as PRCS has a code path with number of quanta and $M = 2^n - 1$ allows receiving at a full turn of scale M various n-digit code combinations.

Let us explain the method stated above, and also statements 11.1 and 11.2, on an example of construction three-digit circular PRCS with placing RE according to a polynomial $r(x) = 1 + x + x^2$. Such a scale is shown in Fig. 11.1.

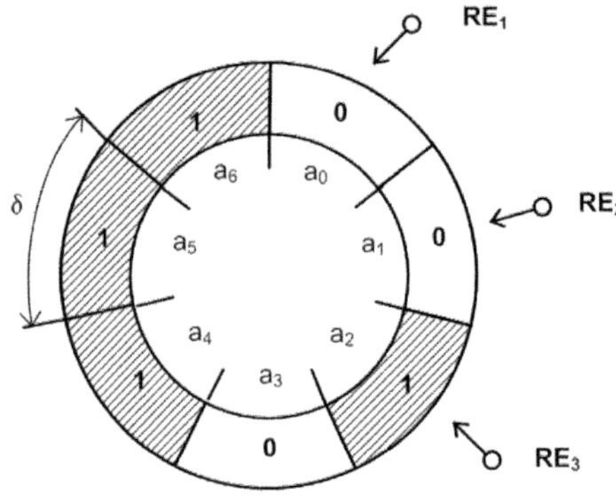

Fig. 11.1 Three-digit PRCS

In an example for simplicity it is accepted n=3 and corresponding the primitive not resulted polynomial is chosen from table 11.1, $h(x) = x^3 + x + 1$, where h_0=h_3=1, h_1=1, h_2=0. Here length of M-sequence M=2^3-1=7, and M-sequence $\boldsymbol{a}$=$a_0a_1a_2a_3a_4a_5a_6$=0010111. At initial values of M-sequence a_0=a_1=0, a_2=1, sequence symbols a_3, a_4, a_5 and a_6 are received according to a recursive parity (11.2) which in the given example looks like, $a_{3+j} = a_{1+j} \oplus a_j$, j=0,1,2,3. Placing of three reading out elements RE_1, RE_2 and RE_3 along a code path of a scale is set according to (11.4) polynomial $r(x) = 1 + x + x^2$.

Table 11.2 Sequence of code combinations three-digit PRCS

Positions PRCS	RE_1	RE_2	RE_3	The decimal code
0	0	0	1	1
1	0	1	0	2
2	1	0	1	5
3	0	1	1	3
4	1	1	1	7
5	1	1	0	6
6	1	0	0	4

Consistently fixing RE a three-digit code combination at scale moving on one quantum against a direction of movement of an hour hand, we receive seven various three-digit code combinations. These code combinations, corresponding to seven various angular positions PRCS are resulted in Table 11.2.

11.1.2 Pseudo-Random Code Scales for Converters of Linear Moving

The method used at construction of circular PRCS, can be used with some additions by working out scales for converters of linear moving [5].

Let us state a method of construction of linear PRCS with resolution $\delta = D/(2^n - 1)$, where D - length of coded moving, and n - word length of a scale. For the set resolution it is necessary to provide possibility of reception with n RE at full moving of a scale of various $M = 2^n - 1$ n-digit code combinations.

It is provided by the solution of a problem of placing on PRCS RE which is reduced to a finding of suitable linearly independent set from n cyclic shifts of M-sequence.

In difference from circular, linear PRCS is opened. Therefore, for maintenance of the set resolution of a scale $\delta = D/M$, it is necessary to receive corresponding sequence of symbols $\boldsymbol{l}$, suitable for synthesis of an information path of linear PRCS. A problem of generation of sequence $\boldsymbol{l}$ we will solve in a general form with use of a recursive parity (11.2) and having assumed, that placing RE on PRCS is correct and is set by a polynomial (11.4). For definiteness initial values of symbols of sequence $\boldsymbol{l}$, we will choose the following $l_0=l_1=\ldots=l_{n-2}=0$, $l_{n-1}=1$.

The sequence $\boldsymbol{l}$ completely includes sequence $\boldsymbol{a}$, and also some additional symbols, which number depends on placing on PRCS RE.

Let us define a difference between numbers of cyclic shifts of the M-sequence corresponding to placing on a scale two next RE, as $d_i = j_m - j_{m-1}$, where $i = 1,2\ldots, n-1$, $m = 2,3\ldots, n$.

Then the number of applications of a recursive parity (11.2), at the set entry conditions, necessary for sequence generation $\boldsymbol{l}$ can be received under the following formula

$$t = 2^n - (n+1) + \sum_{i=1}^{n-1} d_i \tag{11.7}$$

Taking into account that

$$\sum_{i=1}^{n-1} d_i = d_1 + \ldots d_i + \ldots d_{n-1} = (j_2 - j_1) + \ldots + (j_m - j_{m-1}) + \ldots + (j_n - j_{n-1}) = j_n,$$

the parity (11.7) in a final form becomes

$$t = 2^n - (n+1) + j_n. \tag{11.8}$$

The general number of symbols of sequence $\boldsymbol{l}$ with the account n set initial values can be found from a parity

$$T = 2^n + j_n - 1. \tag{11.9}$$

Let us formulate a method of construction of n-digit linear PRCS.

1. Depending on demanded word length n linear PRCS from Table 11.1, the polynomial $h(x)$ degrees n gets out.
2. Taking into account requirements to placing on a scale of reading out elements according to (11.4), the placing polynomial $r(x)$ is formed.
3. Using a recursive parity (11.2), taking into account (11.8) and (11.9), the sequence $\boldsymbol{l}$ is generated.
4. Elementary sites (quanta) linear PRCS are carried out according to sequence symbols $\boldsymbol{l}$ where to symbols of 1 sequence correspond active, and to symbols 0 - passive sites of an information path. For definiteness sequence symbols $\boldsymbol{l}$ are displayed on an information path of a scale from left to right in sequence $l_0 l_1 \ldots l_{T-1}$.

Let us explain the proposed method of construction linear PRCS on an example of a four-digit scale which is resulted on Fig. 11.2.

The information path of a scale is executed according to sequence symbols $\boldsymbol{l}=l_0 l_1 \ldots l_{23}$=000100110101111000100110 lengths for $T = 2^n + j_n - 1 = 2^4 + 9 - 1 = 24$ which construction the primitive not resulted polynomial is used, $h(x) = x^4 + x + 1$, and symbols l_{4+j} sequences $\boldsymbol{l}$ at initial values $l_0=l_1=l_2=l$, $l_3=1$ satisfy to a recursive parity, $l_{4+j} = l_{1+j} \oplus l_j$ j=0,1..., (t-1), where $t = 2^n - (n+1) + j_n = 2^4 - (4+1) + 9 = 20$. In the resulted example placing of four RE along a path linear PRCS is defined by a polynomial $r(x) = 1 + x^2 + x^5 + x^9$.

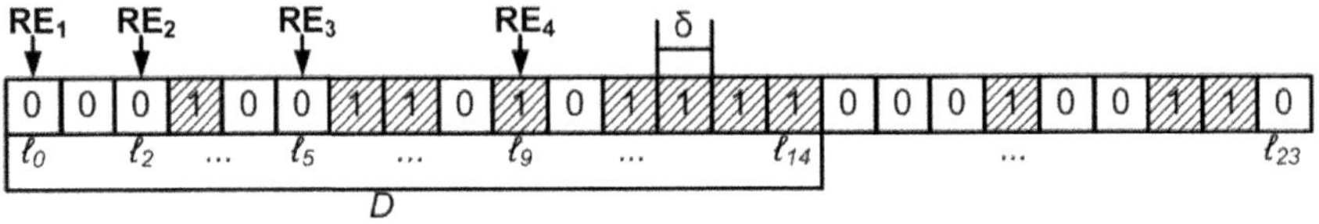

Fig. 11.2 Four-digit linear PRCS

Table 11.3 Sequence of code combinations four-digit linear PRCS

Positions PRCS	RE_1	RE_2	RE_3	RE_4	The decimal code	Positions PRCS	RE_1	RE_2	RE_3	RE_4	The decimal code
0	0	0	0	1	1	8	0	0	1	0	2
1	0	1	1	0	6	9	1	1	1	1	15
2	0	0	1	1	3	10	0	1	0	0	4
3	1	0	0	1	9	11	1	1	0	0	12
4	0	1	1	1	7	12	1	1	0	1	13
5	0	1	0	1	5	13	1	0	1	1	11
6	1	0	1	0	10	14	1	0	0	0	8
7	1	1	1	0	14	-	-	-	-	-	-

Fixing reading out elements RE_1, RE_2, RE_3 and RE_4, consistently code combination at moving linear PRCS on one elementary site (it is from right to left)

received fifteen various four-digit code combinations. These code combinations, corresponding to fifteen various positions PRCS are resulted in Table 11.3.

11.2 Composite Code Scales

11.2.1 Composite Code Scales for Converters of Angular Moving

Resolution angular and linear PRCS on the basis of M-sequences are defined by size of period *M* and are accordingly equal to $\delta = 360^0 / M$ and $\delta = D / M$. For synthesis CS, allowing building on the basis converters of moving with wider spectrum of resolution, we will enter concept of composite binary sequence of *p* order (Cp-sequence) which we will use for reception of drawing of a code path of a scale. Scales on the basis of Cp-sequences we name composite code scales (CCS) [6].

For Cp-sequence synthesis we define polynomial *H (x)* degrees *N* with factors of field Galua GF (2), as kind expression

$$H(x) = \prod_{k=1}^{p} h_k(x) . \qquad (11.10)$$

Here h_k *(x)* are defined according to (11.1) and allow to receive M-sequences of length, $M_k = 2^{n_k} - 1$ and

$$N = \sum_{k=1}^{p} n_k , \qquad (n_a \neq n_b) \cdot \qquad (11.11)$$

Let us notice, that at *p*=1 the composite sequence turns to classical M-sequence.

Symbols of Cp-sequence ***A*** satisfy to a recursive parity

$$A_{N+j} = \underset{i=0}{\overset{N-1}{\Xi}} A_{i+j} H_i \text{ , j=0,1,2...,} \qquad (11.12)$$

where the sign Ξ denotes summation on the module two, and indexes at Cp-sequence symbols undertake on module *R*. Initial values of symbols of Cp-sequence $A_0A_1 ... A_{N-1}$ get out taking into account that GGD $[t_j (x), H (x)] = 1$, where GGD is the greatest general divider, and

$$t_j(x) = \sum_{i=0}^{N-1} A_{i+j} x^i \text{ , j=0,1..., R-1.} \qquad (11.13)$$

For the majority of practical applications (11.12) it is enough to take $t_j (x) = 1$, i.e. $A_0 = A_1 = ... A_{N-2} = 0, A_{N-1} = 1$.

Period *R* of Cp-sequence depends on degrees of polynomials h_k *(x)* and from a polynomial of initial values of symbols of Cp-sequence t_j *(x)*. If all n_k is represented by mutually simple numbers, and GGD $[t_j (x), H (x)] = 1$,

$$R = \prod_{k=1}^{p} M_k \qquad (11.14)$$

Cp-sequences, as well as M-sequences, belong to the class of cyclic codes and can be set by a generating polynomial of a kind

$$G(x) = (x^R + 1) / H(x), \tag{11.15}$$

where $H(x)$ and R are calculated according to expressions (11.10) and (11.14). Therefore, for each Cp-sequence of length R exists exactly R various cyclic shifts which can be received by multiplication of generating polynomial $G(x)$ on x^j, where j=0,1..., R-1.

The analysis of (11.10) and (11.11) allows to draw a conclusion that the number of polynomials of N-th degree $H(x)$ depends both on number and degrees of primitive not resulted polynomials $h(x)$, participating in construction $H(x)$. From (11.12) and (11.14) it is visible also, that periods R of Cp-sequences received with use various $H(x)$ of identical degree, are various.

In Table 11.4 values n_k degrees of polynomials $h(x)$ and periods M_k constructed on their basis of M-sequences, and also possible variants of reception of degrees N of polynomials $H(x)$ and values R of the periods of Cp-sequences which can be used for construction of composite code scales to word length 16 inclusive are resulted.

Table 11.4 Degree and the periods of Cp-sequences

n_k, N	M_k	$N = \sum_{k=1}^{p} n_k$	$R = \prod_{k=1}^{p} M_k$
2	3	-	-
3	7	-	-
4	15	-	-
5	31	(2+3)	21
6	63	-	-
7	127	(2+5), (3+4)	93, 105
8	255	(3+5)	217
9	511	(2+7), (4+5)	381, 465
10	1023	(2+3+5), (3+7)	651, 889
11	2047	(2+9), (3+8), (4+7), (5+6)	1533, 1785,1905, 1953
12	4095	(2+3+7), (3+4+5), (5+7)	2667, 3255, 3937
13	8191	(2+11), (3+10), (4+9), (5+8), (6+7)	6141, 7161, 7665, 7905, 8001
14	16383	(2+5+7), (3+4+7), (3+11), (5+9)	11811, 13335, 14329, 15841
15	32767	(2+13), (3+5+7), (4+11), (5+8)	24573, 27559, 30705, 32385
16	65535	(2+3+11), (2+5+9), (5+11), (3+5+8), (4+5+7), (7+9), (3+13)	42987, 47523, 63457, 55335, 59055, 64897, 57337

Let's formulate a method of construction N-digit circular CCS.

1. Depending on demanded word length of N and resolution R of a scale under tables 11.1 and 11.4 according to (11.10) polynomial $H(x)$ degrees N is formed.
2. Using a recursive parity (11.12) sequence **A** is generated.

3. Elementary sites (quanta) of a scale are made according to symbols of Cp-sequence **A** where to symbols of 1 sequence correspond active, and to symbols 0 – passive sites of an information code path. For definiteness Cp-sequence symbols are displayed on information CP in a direction of movement of an clock hand in sequence $A_0A_1 \ldots A_{R-1.}$
4. Placing on a scale of reading out elements is carried out. As CCS are under construction according to Cp-sequence symbols, it is possible to set by cyclic shifts a placing order on scale N of reading out elements. Differently, m RE (m=1,2, ..., N) j_m-th cyclic shift of Cp-sequence $x^{j_m} g_k(x)$ is put in conformity $x^{j_m} g_k(x)$.

Then the polynomial defining an order of placing N of reading out elements on CCS, looks like

$$r(x) = \sum_{m=1}^{N} x^{j_m} , \tag{11.16}$$

where $j_m \in 0,1..., R\text{-}1$. Having put j_1=0, according to (2.16) we will have the second, the third..., N-th RE displaced (for definiteness in a direction of movement of an clock hand) concerning the first RE on j_2, j_3..., j_N elementary sites of an information path of a scale accordingly.

The problem of placing RE on CCS will be considered and solved in (11.3). In a basis of its solution the finding of linearly independent set of cyclic shifts of Cp-sequence is necessary.

Let us show, that circular CCS allow to build on the basis converters the moving using a method of parallel reading. For this purpose we will formulate the following statements.

Statement 11.3. Composite code scales allow to receive exactly R various N-digit code combinations corresponding to sequence from R of quanta of moving.

Statement 11.4. Resolution circular CCS is defined by a parity

$$\delta = 360^0 / R . \tag{11.17}$$

Proofs of these statements are executed similarly to proofs of statements 11.1 and 11.2.

Let us show a method of construction circular CCS as an example, for simplicity having limited to five categories of transformation. From Tables 11.1 and 11.4 it is visible that five-digit CCS can be received only in one way with use of two polynomials *h (x)*, according to the second and third degrees used for generation of symbols of composite sequence of the second order with period R=21.

The resulted five-digit circular CCS is shown in Fig. 11.3.

At scale moving cyclically on one elementary site, for example, against a direction of movement of an clock hand, from exits of reading out elements RE_1, RE_2, RE_3, RE_4 and RE_5 five-digit code combinations are formed. These code combinations corresponding to twenty one various angular positions CCS are resulted in Table 11.5.

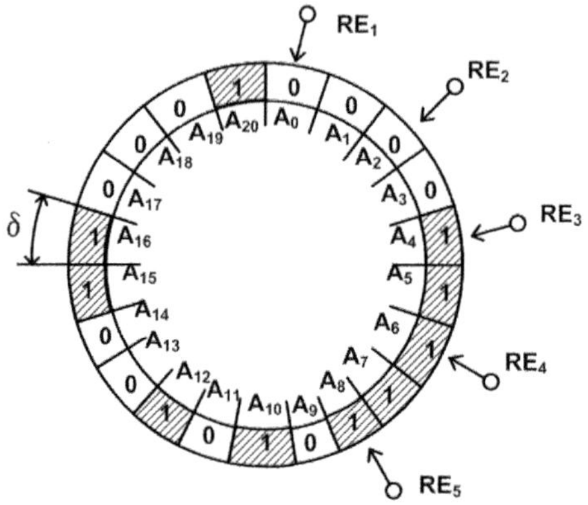

Fig. 11.3 Five-digit CCS

Table 11.5 Sequence of code combinations five- digit CCS

Positions CCS	RE_1	RE_2	RE_3	RE_4	RE_5	The decimal code	Positions CCS	RE_1	RE_2	RE_3	RE_4	RE_5	The decimal code
0	0	0	1	1	1	7	11	0	0	1	0	0	4
1	0	0	1	1	0	6	12	1	0	1	0	1	21
2	0	1	1	1	1	15	13	0	1	0	0	0	8
3	0	1	1	0	0	12	14	0	1	0	1	0	10
4	1	1	1	1	1	31	15	1	0	0	0	0	16
5	1	1	0	0	0	24	16	1	0	1	0	0	20
6	1	1	1	1	0	30	17	0	0	0	0	1	1
7	1	0	0	0	1	17	18	0	1	0	0	1	9
8	1	1	1	0	1	29	19	0	0	0	1	1	3
9	0	0	0	1	0	2	20	1	0	0	1	1	19
10	1	1	0	1	0	26	-	-	-	-	-	-	-

The information path of a scale is executed according to symbols K_2-sequence **A**=A_0 A_1... A_{20} = 000011111010100110001 lengths R=21 for which construction the polynom is used, $H(x) = h_1(x)h_2(x) = (x^2 + x + 1)(x^3 + x + 1) = x^5 + x^4 + 1$ and symbols A_{5+j} K_2- sequence at initial values A_0=A_1=A_2=A_3=0, A_4=1 satisfy to a recursive parity, $A_{5+j} = A_{4+j} \oplus A_j$ (j=0,1..., 15).

In an example placing of five RE along the path, CCS is defined by a polynomial $r(x) = 1 + x^2 + x^4 + x^6 + x^8$.

11.2.2 Composite Code Scales for Converters of Linear Moving

Let us consider construction of N-digit linear CCS with resolution $\delta = D / R$, where D is the length of coded moving, and R is defined according to expression (11.14). For the set resolution it is necessary to provide possibility of reception with N RE at full linear moving of scale R of various N-digit code combinations.

It is provided by the solution of a problem of placing on CCS RE which is reduced to a finding of suitable linearly independent set from N cyclic shifts of Cp-sequence.

For achievement of resolution of a scale δ, it is necessary to receive corresponding sequence of symbols $\boldsymbol{L}$, suitable for synthesis of an information path linear CCS. For reception of sequence $\boldsymbol{L}$ we will take advantage of a recurrent parity (11.12) and we will assume that placing RE on CCS is correct and is set by a polynomial (11.16). For definiteness initial values of symbols of sequence $\boldsymbol{L}$ we will choose following $L_0=L_1 =... =L_{n-2}=0, L_{n-1}=1$.

Sequence $\boldsymbol{L}$ completely includes sequence $\boldsymbol{A}$, and also some additional symbols, whose number depends on placing on linear CCS RE.

By analogy to a method of construction linear PRCS, considered in 11.2, we will define a difference between numbers of cyclic shifts of the Cp-sequence corresponding to placing on a scale two next RE, as $d_i = j_m\text{-}j_{m-1}$, where $i = 1,2..., N\text{-}1$, $m = 2,3..., N$.

Then the number of applications of a recursive parity (11.12), at the set of initial values of the symbols, necessary for generation of sequence $\boldsymbol{L}$, can be received under the following formula

$$t_k = R - N + \sum_{i=1}^{N-1} d_i \qquad (11.18)$$

Taking into account that

$$\sum_{i=1}^{N-1} d_i = d_1 + \ldots d_i + \ldots d_{N-1} = (j_2 - j_1) + \ldots + (j_m - j_{m-1}) + \ldots + (j_N - j_{N-1}) = j_N \cdot$$

The parity (11.18) in a final form becomes

$$t_k = R - N + j_N . \qquad (11.19)$$

The general number of symbols of sequence $\boldsymbol{L}$ taking into account N set of initial values can be found from a parity

$$T_K = R + j_N . \qquad (11.20)$$

Let us formulate a method of construction of N-digit linear CCS.

1. Depending on demanded word length N and resolution R of a scale from tables 11.1 and 11.4 according to (11.10), polynomial $H(x)$ degrees N is formed.
2. Taking into account requirements to placing on a scale of reading out elements according to (11.16), the placing polynomial $r(x)$ is formed.
3. Using a recursive parity (11.12), taking into account (11.19) and (11.20), sequence $\boldsymbol{L}$ is generated.
4. Elementary sites (quanta) of a scale are executed according to symbols of sequence $\boldsymbol{L}$ where to symbols of 1 sequence correspond active, and to symbols 0 passive sites of an information path.

For definiteness symbols of sequence $\boldsymbol{L}$ are displayed on an information code path from left to right in sequence $L_0\, L_1 \ldots L_{tk-1}$.

Let us explain the method of construction stated above linear CCS on an example of a five- digit scale which is resulted on Fig. 11.4.

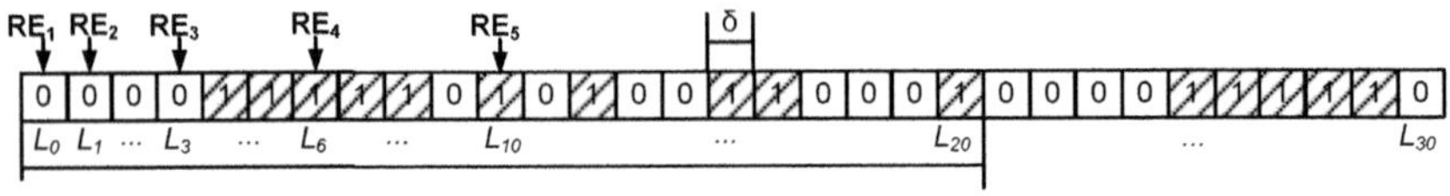

Fig. 11.4 Five- digit linear CCS

The information path of a scale is executed according to symbols of sequence $\boldsymbol{L} = L_0\, L_1 \ldots L_{30} =$ 0000111110101001100010000111110 lengths T_k=31 for which construction the following polynomial should be used, $H(x) = h_1(x)h_2(x) = (x^2+x+1)(x^3+x+1) = x^5+x^4+1$ and symbols L_{5+j} of sequence $\boldsymbol{L}$ at initial values $L_0=L_1=L_2=L_3=0$, $L_4=1$ satisfy to a recursive parity, $L_{5+j} = L_{4+j} \oplus L_j$ (j=0,1..., 25).

In an example placing of five RE along a code path is defined by a polynomial $r(x) = 1+x+x^3+x^6+x^{10}$.

At scale moving cyclically on one elementary site, for example from right to left, from exits of reading out elements RE$_1$, RE$_2$, RE$_3$, RE$_4$ and RE$_5$ five-digit code combinations are formed. These code combinations corresponding to twenty one various position linear CCS are resulted in Table 11.6.

Table 11.6 Sequence of code combinations five-digit linear CCS

Positions CCS	RE$_1$	RE$_2$	RE$_3$	RE$_4$	RE$_5$	The decimal code	Positions CCS	RE$_1$	RE$_2$	RE$_3$	RE$_4$	RE$_5$	The decimal code
0	0	0	0	1	1	3	11	0	1	0	0	0	8
1	0	0	1	1	0	6	12	1	0	1	0	0	20
2	0	0	1	1	1	7	13	0	0	1	0	0	4
3	0	1	1	0	0	12	14	0	1	0	1	0	10
4	1	1	1	1	0	30	15	1	1	0	0	1	25
5	1	1	1	0	1	29	16	1	0	0	0	1	17
6	1	1	0	1	1	27	17	0	0	1	0	1	5
7	1	1	1	0	0	28	18	0	0	0	0	1	1
8	1	0	0	0	0	16	19	0	1	0	1	1	11
9	0	1	1	1	0	14	20	1	0	0	1	0	18
10	1	0	0	1	1	19	-	-	-	-	-	-	-

11.3 Placing of Reading Elements on a Recursive Code Scale

Not any variant of placing RE on RCS allows to receive at a full turn of a scale distinguishable code combinations, i.e. the recursive sequences corresponding to signals, removed with RE at moving RCS, appear linearly dependent. Hence, the solution of the problem of placing RE on RCS is reduced to a finding of linearly independent set from cyclic shifts RS, and capacity of set is equal to number of RE.

11.3.1 Algorithm of Placing of Reading Elements on a Recursive Code Scale

The algorithm is based on use of property RS consisting that any cyclic shift RS is unequivocally defined by its initial block from among the symbols, equal to word length RCS. As M-sequences are a special case of Cp-sequences (for $p = 1$), we will consider algorithm in designations accepted for composite sequences. As pseudo-random and composite sequences are linear in relation to the operator of summation on the module two, the algorithm of placing N RE on RCS is reduced to finding of a suitable linearly independent set from N cyclic shifts RS and includes following steps [7].

1. The location of reading elements on RCS pursuant to expression (11.16) is executed.
2. Each cyclic shift RS $x^{If}G\ (x)$, $f = 1, 2, \ldots, N$, initial block from N symbols $\boldsymbol{B}_{If} = A_{If}\,A_{1+If}\ldots A_{N-1+If}$, where a sum of indexes for sequence symbols takes on the modulo R.
3. The formation of the square matrix **B** is executed, where lines are initial blocks $\boldsymbol{B}_{If}$, i.e

$$B = \begin{Vmatrix} B_{I_1} \\ \ldots \\ B_{I_f} \\ \ldots \\ B_{I_N} \end{Vmatrix} = \begin{Vmatrix} A_{I_1} A_{1+I_1} \ldots A_{m-1+I_1} \\ \ldots \\ A_{I_f} A_{1+I_f} \ldots A_{m-1+I_f} \\ \ldots \\ A_{I_N} A_{1+I_N} \ldots A_{m-1+I_N} \end{Vmatrix}. \quad (11.21)$$

4. The determinant of the matrix (11.21) is calculated.
5. If the determinant of the matrix is not to equal zero, the variant of the RE location accepted pursuant to (11.16) is correct. If determinant is equal to zero, it is necessary to execute a choice of other variant.

Let us explain the algorithm stated above on an example of four-digit linear PRCS which is resulted in Fig. 11.2.

It is necessary to place on a scale four RE so that at full moving of a scale to receive fifteen various four-digit code combinations.

Let placing of four RE along a path of linear PRCS is set by a polynomial $r(x) = 1 + x^2 + x^8 + x^{12}$, where $j_1=0$, $j_2=2$, $j_3=8$, $j_4=12$.

To each cyclic shift $x^{j_1}, x^{j_2}, x^{j_3}, x^{j_4}$ of M-sequence we will put in conformity the initial block from four symbols (fig. 11.2 see), i.e.

$$x^{j_1} \Rightarrow 1_{j_1} = l_0 l_1 l_2 l_3 = 0001,\ x^{j_2} \Rightarrow 1_{j_2} = l_2 l_3 l_4 l_5 = 0100,$$
$$x^{j_3} \Rightarrow 1_{j_8} = l_8 l_9 l_{10} l_{11} = 0101, x^{j_4} \Rightarrow 1_{j_{12}} = l_{12} l_{13} l_{14} l_{15} = 1110.$$

From the found initial blocks $1_{j_1}, 1_{j_2}, 1_{j_8}, 1_{j_{12}}$ we will generate a matrix

$$B = \begin{Vmatrix} 1_{j_1} \\ 1_{j_2} \\ 1_{j_8} \\ 1_{j_{12}} \end{Vmatrix} = \begin{Vmatrix} 0001 \\ 0100 \\ 0101 \\ 1110 \end{Vmatrix}.$$

Analyzing the received matrix, we can see, that its third line turns out as a result of summation on the module of the two for first and second lines. Hence, by means of elementary operations over lines of a matrix in it two identical lines or a line equal to zero can be received.

As the determinant of such matrix is equal to zero between its lines, there is a linear dependence which testifies to presence of linear communication between corresponding cyclic shifts of M-sequence. It means, that the considered placing of four RE on PRCS with the number of quanta equal to 15, does not allow to receive at full moving of a scale of 15 various four-digit code combinations.

In Table 11.7 the sequence of code combinations four-digit PRCS (Fig. 11.2) is resulted at a variant of placing RE along a scale according to a polynomial $r(x) = 1 + x^2 + x^8 + x^{12}$.

Table 11.7 Sequence of code combinations four-digit linear PRCS

Positions PRCS	RE_1	RE_2	RE_3	RE_4	The decimal code	Positions PRCS	RE_1	RE_2	RE_3	RE_4	The decimal code
0	0	0	0	1	1	8	0	0	0	0	0
1	0	1	1	1	7	9	1	1	0	1	13
2	0	0	0	1	1	10	0	1	1	1	7
3	1	0	1	0	10	11	1	1	0	0	12
4	0	1	1	0	6	12	1	1	0	1	13
5	0	1	1	0	6	13	1	0	1	0	10
6	1	0	1	1	11	14	1	0	1	1	11
7	1	1	0	0	12	-	-	-	-	-	-

The table analysis shows, that to some various positions of the coded element there correspond the same code combinations, for example, zero and to the second, to the first and the tenth, the third and the thirteenth, the fourth and the fifth, the sixth and the fourteenth, the seventh and the eleventh, the ninth and the twelfth.

As the problem of placing RE on PRCS has appeared not solved having excluded from consideration those variants of placing which are equivalent previous, it is necessary to choose a new variant of placing.

Let now placing of four RE along a path linear PRCS is set by a polynomial $r(x) = 1 + x^2 + x^5 + x^9$, where j_1=0, j_2=2, j_3=5, j_4=9.

To each cyclic shift $x^{j_1}, x^{j_2}, x^{j_3}, x^{j_4}$ of M-sequence we will put in conformity the initial block from four symbols (Fig. 11.2), i.e.

$$x^{j_1} \Rightarrow 1_{j_1} = l_0 l_1 l_2 l_3 = 0001,\ x^{j_2} \Rightarrow 1_{j_2} = l_2 l_3 l_4 l_5 = 0100,$$

$$x^{j_3} \Rightarrow 1_{j_5} = l_5 l_6 l_7 l_8 = 0110,\ x^{j_4} \Rightarrow 1_{j_9} = l_9 l_{10} l_{11} l_{12} = 1011.$$

From the found initial blocks $1_{j_1}, 1_{j_2}, 1_{j_5}, 1_{j_9}$ we will generate a matrix

$$\mathrm{B} = \begin{Vmatrix} 1_{j_1} \\ 1_{j_2} \\ 1_{j_5} \\ 1_{j_9} \end{Vmatrix} = \begin{Vmatrix} 0001 \\ 0100 \\ 0110 \\ 1011 \end{Vmatrix}.$$

Having done elementary operations over lines of matrix B, we will receive an individual matrix. As the determinant of such matrix is equal to digit between its lines linear dependence is absent that testifies also to absence of linear communication between corresponding cyclic shifts of M-sequence. It means, that the considered placing of four RE on PRCS with the number of quanta equal 15 allows to receive at full moving of a scale of 15 various four-digit code combinations.

In table 11.3 the sequence of four-digit code combinations for 15 various positions PRCS (fig. 11.2) is resulted, at a variant of placing RE along a scale defined according to a polynomial $r(x) = 1 + x^2 + x^5 + x^9$.

The table analysis shows, that various positions of a scale correspond to various code combinations.

11.3.2 Reading Elements Location on the Pseudo-Random Code Scale with a Constant Step

At the solution of a problem of placing we will apply algebra of final fields GF (2^n) which are generated by means of not resulted polynomials of degree n. At use of primitive not resulted polynomials $h(x)$ simple field GF (2) can be expanded to field GF (2^n) at the expense of root α of a polynomial $h(x)$.

We show that there is mutual conformity between initial blocks $\mathrm{a}_j = (a_j a_{1+j} \ldots a_{n-1+j})$ of M-sequence cyclic shifts and elements α^j, j=0,1..., M-1 of GF (2^n) [8].

We can write symbols of an initial block $\mathbf{a}_j$ as

$$\begin{aligned} a_j &= \oplus \sum_{i=0}^{n-1} a_{j+i-n} h_i \\ &\ldots\ldots\ldots\ldots\ldots\ldots \\ a_{n+j-1} &= \oplus \sum_{i=0}^{n-1} a_{j+i-1} h_i \end{aligned} \qquad (11.22)$$

It can be presented in the matrix form

$$\mathbf{a}_j = \begin{Vmatrix} a_{j-1} \dots a_{n+j-2} \\ a_{j-2} \dots a_{n+j-3} \\ \dots\dots\dots\dots \\ a_{j-n} \dots a_{j-1} \end{Vmatrix} \times \begin{Vmatrix} h_{n-1} \\ h_{n-2} \\ \dots\dots \\ h_0 \end{Vmatrix} = T \times h. \tag{11.23}$$

All possible initial blocks of M-sequence need to be interpreted as the set of the GF (2^n) elements of, i.e.

$$\mathrm{a}_j \Leftrightarrow \{\alpha^j\} = GF(2^n) \setminus \{0\}. \tag{11.24}$$

Elements GF (2^n) are all binary sets which can be considered as the linear space of the dimension n over GF (2). As it is mentioned above, in order to generate (2^n) elements it is convenient to use any primitive root a giving representation of non zero elements of GF (2^n) in form of cyclic multiplicity group on degrees of the elementα. Thus α^0=00... 01, α^1=00... 010...., α^{n-1}=10... 00,

$$\alpha^{n+j} = \oplus \sum_{i=0}^{n-1} \alpha^{j+i} h_i,\ j = 0,1,\dots,M-n. \tag{11.25}$$

As a root $h\ (x)$ we can use a binary set α^1=00... 010 as far as such an element is always primitive.

We define linear isomorphous transformation of elements of GF (2^n) by equation

$$\mathrm{a}_j = T\,\alpha^j,\ j = 0,1,\dots,M-1. \tag{11.26}$$

For the location of m RE on PRCS with a constant step d, it is necessary to ensure linear independence between the same cyclic shifts of M-sequence and elements of GF (2^n).

At the location RE with a constant step the equation (11.4) will have the form

$$r(x) = \sum_{j=0}^{n-1} x^{jd}, \tag{11.27}$$

where $d = 1,2\dots,k$; $k =]\ L/n\ [$.

We shall research the subset of GF (2^n) elements α^0, α^d..., α^{jd}..., $\alpha^{(n-1)\,d}$. As it is known from the theory of algebraic fields [9], all linear dependencies in any cyclic group with $\beta=\alpha^d$ are determined according to minimal polynomial $M_{\beta(x)}$:

$$M_{\beta(x)} = \prod_{i=0}^{l-1} (x + \beta^{2^i}), \tag{11.28}$$

where l is period of a cyclic group containing element β.

From the work [9], it is followed that the RE location on PRCS is correct if, and only if

$$\deg\left[\mathrm{M}_{\beta(x)}\right] = n. \tag{11.29}$$

Thus it is enough to evaluate only the degree of polynomial M $_{\beta(x)}$.

Using Ferma theorem, we can set:

$$\deg\left[\mathrm{M}_{\beta(x)}\right] = \min\{l : l > 0, d\,2^l = d, \mathrm{mod}\,(M = 2^n - 1)\} = l. \tag{11.30}$$

It is known [9] that $l = W^q$ (2), where W^q (2) is multiplicity order 2 to module $q = M$ / GCD [M, d].

For practical using (11.28 - 11.30) it is required to evaluate W^q (2). The methods of the calculation of the multiplicity order are presented in [9]. Following properties are used to simplify the calculations.

1. If $q = \prod q_i$, where $q_i = p_i^{\,l_i}$ are any degrees of prime numbers p_i, then

$$W^q(2) = \mathrm{LCM}\{W^{q_i}(2)\},$$

   where LCM - the least common multiply of numbers W^{q_i} (2).
2. The multiplicity orders for numbers of a form 2^m-1 and 2^m+1 are equal respectively to m and $2m$.
3. The multiplicity order for any prime number q such that GCD [2, q] = 1 is a divider of number q - 1.

We show the application of an equation (11.29) to check the correctness of the constant step d=3 location RE on four-digit PRCS.

Initial data for the account are primitive polynomial $h\,(x) = x^4 + x + 1$, period of the M-sequence $M = 2^4 - 1 = 15$ and polynomial location RE $r\,(x) = x^0 + x^d + x^{2d} + x^{3d} = 1 + x^3 + x^6 + x^9$.

It is necessary to evaluate the size of the multiplicity order W^q (2), where $q = M/GCD[M, d] = 15 / GCD\,[15,3] = 15/3 = 5$.

As number 5 has form $2^m + 1$ where m=2 then W^q (2) $=2^m$=4, that corresponds to correctness of the location on PRCS from 15 quantum of four RE with the constant step d=3.

In Fig. 11.5 of circular PRCS is shown. Along a scale four RE are located with constant step d=3.

The information track of the scale is executed according to symbols of M-sequence $\boldsymbol{a} = a_0 a_1 \ldots a_{14}$ = 000100110101111 of the period M=15 for construction of which the primitive polynomial $h\,(x) = x^4 + x + 1$ is used and the symbols a_{4+j} of the M-sequence at initial values $a_0 = a_1 = a_2 = 0$, $a_3 = 1$ satisfy to the recursive equation $a_{4+j} = a_{1+j} \oplus a_j$, j=0,1..., 10.

Consistently fixing RE a four-digit code combination at scale moving on one quantum against a direction of movement of an hour hand, we receive 15 various four-digit code combinations. These code combinations corresponding to 15 various angular positions PRCS are resulted in Table 11.8.

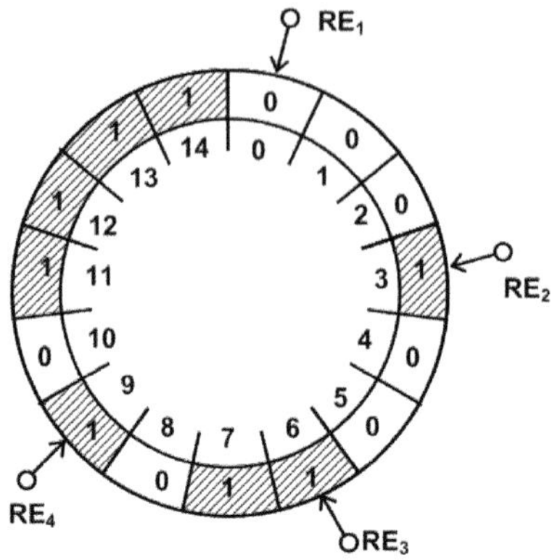

Fig. 11. 5 Circular four-digit PRCS

Table 11.8 Sequence of code combinations four-digit PRCS

Positions PRCS	RE_1	RE_2	RE_3	RE_4	The decimal code	Positions PRCS	RE_1	RE_2	RE_3	RE_4	The decimal code
0	0	1	1	1	7	8	0	1	1	0	6
1	0	0	1	0	2	9	1	1	0	1	13
2	0	0	0	1	1	10	0	1	0	0	4
3	1	1	1	1	15	11	1	1	0	0	12
4	0	1	0	1	5	12	1	0	1	1	11
5	0	0	1	1	3	13	1	0	0	1	9
6	1	1	1	0	14	14	1	0	0	0	8
7	1	0	1	0	10	-	-	-	-	-	-

11.3.3 Reading Elements Locations on the Composite Code Scale with a Constant Step

In common case set of initial blocks $\mathbf{A}_j=A_jA_{1+j}... A_{N-1+j}$ j=0,1..., R-1 of all cyclic shifts for any Cp-sequences isomorphousely to a subset of elements of an extended field Galua GF (2^N) where N is calculated from (11.11) [10].

It is possible to spread all adduced in 11.3.2 to Cp-sequences. Thus equation (11.26) can be written in the following form

$$\mathrm{A}_{\mathrm{j}} = T_k\, S^{j},\ j = 0,1,...,R-1, \tag{11.31}$$

where

$$T_k = \begin{Vmatrix} A_{j-1} \cdots A_{N+j-2} \\ A_{j-2} \cdots A_{N+j-3} \\ \dots\dots\dots\dots\dots \\ A_{j-N} \cdots A_{j-1} \end{Vmatrix} \tag{11.32}$$

and $S^j = \varphi(\alpha^1..., \alpha^k..., \alpha^p)^j$ (φ is the function of linear transformation) α^k-primitive roots of the polynomial $h_k(x)$ are calculated with using of the Chine's Theorem.

Consider location of N RE on CCS with a constant step d. For it we transform the equation (11.26) to following form

$$r(x) = \sum_{j=0}^{N-1} x^{jd}, \tag{11.33}$$

where $d = 1,2..., f$, $f =]\, R / N\, [$. We formulate the following statement. For binary Cp-sequences with period R (R is determined according to (11.14)) set of initial blocks of symbols $\mathbf{A}_{md}$ sequences corresponding cyclic shifts are linearly independent if and only if when

$$W^{q_k}(2) = \deg[h_k(x)], k = 1, 2, ..., \tag{11.34}$$

where $W^{qk}(2)$ - multiplicity order of number 2 on module $q_k = M_k / \mathrm{GCD}\,[M_k, d]$.

Now consider the set of polynomials form

$$R_{md}(x) = x^{md} \bmod H(x), m = 0, 1, ..., N-1. \tag{11.35}$$

It is possible to show that $R_{md}(x)$ represent the polynomial record of a subset S^{nd} of set of elements S^j, and the conformity between S^{md} and A_{md} is defined by equation (11.31).

For any m it is possible to record

$$x^{md} = Q(x)H(x) + R_{md}(x), \tag{11.36}$$

i.e.

$$R_{md}(x) = x^{md} + Q(x)H(x), m = 0, 1, ..., N-1. \tag{11.37}$$

Applying to (11.27) the reduction to module polynomial $h_k(x)$ where $k = 1,2...$, we receive

$$(R_{md})_k(x) = x^{md} \bmod h_k(x) \tag{11.38}$$

for all $m = 0,1..., N\text{-}1$.

It is possible also to show, that $(R_{md})_k(x)$ are records in polynomial form of a subset $(\alpha^{md})_k$ of set of elements α^j, j=0,1..., M_k-1 and the relation between $(\alpha^{md})_k$ and $\mathbf{a}_{md}$ is defined by equation (11.26).

Thus, to check correctness of the location of N RE on CCS with a constant step d, it is necessary to evaluate the linear independence of set of initial blocks $\mathbf{a}_j$ for each k of the M-sequences separately. Conclusion about the correctness of the location (11.33) is done after a check of the condition (11.34).

Now we show application of the equation (11.34) for check correcting of location with the constant step d=4 RE on five-digit CCS.

Initial data are period $M_1 = 2^2\text{-}1 = 3$, the M-sequence constructed on the basis of the primitive polynomial $h_1(x) = (x^2+x+1)$, period $M_2 = 2^3\text{-}1 = 7$ the M-sequences, constructed on the basis of primitive polynomial $h_2(x) = (x^3+x+1)$, the polynomial of five RE location $r(x) = x^0+x^d+x^{2d}+x^{3d}+x^{4d} = 1+x^4+x^8+x^{12}+x^{16}$.

It is necessary to evaluate of multiplicity orders $W^{q_1}(2)$ and $W^{q_2}(2)$, where $q_1=M_1/\mathrm{GCD}[M_1,d]=3/\mathrm{GCD}[3,4]=3/1=3$, $q_2=M_2/\mathrm{GCD}\ [M_2,d]=7/\mathrm{GCD}\ [7,4]=7/1=7$. We have $W^{q_1}(2)=2$, $W^{q_2}(2)=3$ that testifies about the correcting of the location on CCS from 21 elementary quantum five RE with the constant step d=4.

In Fig. 11.6 of circular CCS is shown. Along the scale five RE are located with the constant step d=4.

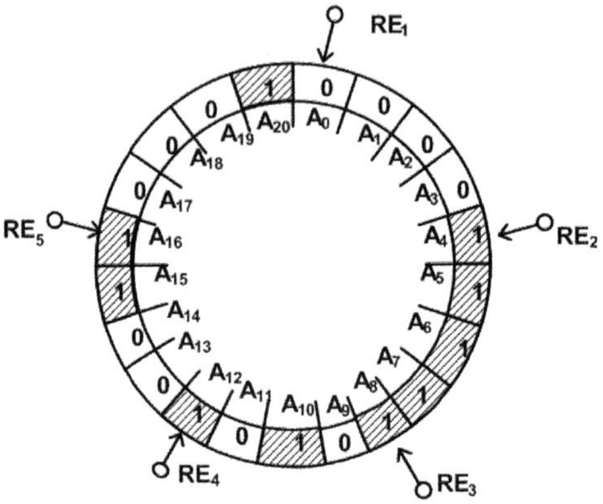

Fig. 11.6 Circular five-digit CCS

Table 11.9 Sequence of code combinations five-digit CCS

Positions CCS	RE_1	RE_2	RE_3	RE_4	RE_5	The decimal code	Positions CCS	RE_1	RE_2	RE_3	RE_4	RE_5	The decimal code
0	0	0	1	1	1	7	11	0	0	1	0	0	4
1	0	0	1	1	0	6	12	1	0	1	0	1	21
2	0	1	1	1	1	15	13	0	1	0	0	0	8
3	0	1	1	0	0	12	14	0	1	0	1	0	10
4	1	1	1	1	1	31	15	1	0	0	0	0	16
5	1	1	0	0	0	24	16	1	0	1	0	0	20
6	1	1	1	1	0	30	17	0	0	0	0	1	1
7	1	0	0	0	1	17	18	0	1	0	0	1	9
8	1	1	1	0	1	29	19	0	0	0	1	1	3
9	0	0	0	1	0	2	20	1	0	0	1	1	19
10	1	1	0	1	0	26	-	-	-	-	-	-	-

The code track of the scale is formed according to symbols C_2-sequences $\mathbf{A} = A_0\ A_1 \ldots A_{20}$ =000011111010100110001 with the period R=21 for construction of which the polynomial $H(x) = h_1(x)\ h_2(x) = (x^2+x+1)\ (x^3+x+1) = x^5+x^4+1$ is used and the symbols A_{5+j} C_2-sequences at initial significance $A_0=A_1=A_2=A_3=0$, $A_4=1$ satisfy to the recursive equation $A_{5+j}=A_{4+j}\oplus A_j$, $j=0,1\ldots, 15$.

Consistently fixing RE a five-digit code combination at scale moving on one quantum against a direction of movement of an clock hand, we receive 21 various five-digit code combinations. These code combinations corresponding to 21 various angular positions CCS are resulted in Table 11.9.

11.4 Correcting Possibilities of Recursive Code Scales

Increase of converters reliability parameters can be reach by using CS with a possibility of formation of correcting codes. Known methods do not permit to solve this task without increase of converters dimensions as the correcting digits of conventional scales can be realized at the expense of additional control tracks using and the introduction of redundancy on the number RE. Recursive CS permit to form codes correcting and (or) discovering error of reading only at the expense of the redundancy introduction on number RE without using of additional control tracks.

So, that the code had correcting possibilities, it alongside with information tracks should contain the certain number of correcting symbols. Values of such symbols are determined of modulo-two addition of some fixed information symbols. The number of correcting symbols in a code is determined by the number of information symbols and the given number found out and (or) corrected errors. The methods of correcting sequences construction are known from the literature on the coding theory.

In the basis of correcting sequences construction using in RCS property of «shift and addition» of pseudo-random and composite sequences is put. Using this property, we formulated the technique of the number determination and location for additional correcting RE along an information track which consists in the following [11].

1. Pursuant to technical requirements, the code choice is executed which should be formed by information and correction RE. The code can execute discovering and correction of determined number of errors.
2. The correcting sequence which represents the binary word consisting from q symbols is constructed. The number of correcting symbols q in a correction code is determined by the number of information symbols N and the given number discovering and (or) correcting errors. Numbers N and q correspond to numbers of information and correction RE.
3. Pursuant to the correcting sequence from (11.6), sums $r^l(x)$, where l=1,2, …,q, are formed including cyclic shifts RS. Shifts appropriate these information RE which are used for the formation of information symbols entering in l-th correcting rule.
4. Each sum $r^l(x)$ is divided on multinomial form (11.10) on the part of junior degrees on modulo-two up to receptions the rest in a form of the one-member x^{S_l}. Degree S_l if it exceeds size R-1 undertakes on a module R.
5. The correction of l-th RE location along an information track RCS displaces about first information RE on number S_l elementary sites of a scale δ.

In Fig. 11.7 the circular four-digit PRCS with four information RE and three correction RE_c is shown.

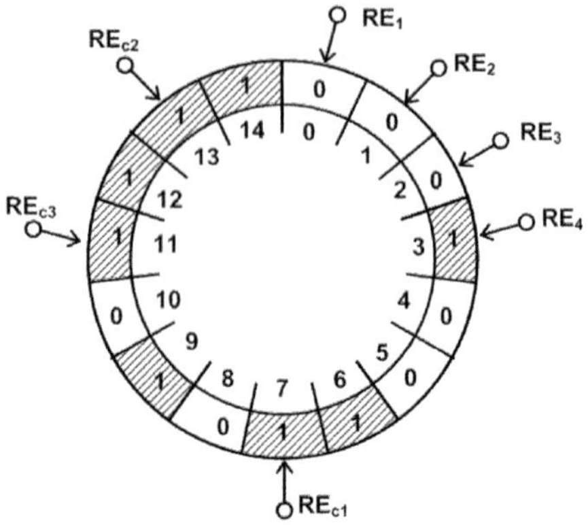

Fig. 11.7 Four-digit PRCS with four information RE and three correction RE_c

The information track of the scale is formed according to symbols of M-sequence $\boldsymbol{a} = a_0a_1 \ldots a_{14}$ = 000100110101111 of the period M=15 for construction of which the primitive polynomial $h(x) = x^4+x+1$ is used and the symbols a_{4+j} of the M-sequence at initial values $a_0=a_1=a_2=0$, $a_3=1$ satisfy to the recursive equation $a_{4+j}=a_{1+j} \oplus a_j$ (j=0,1..., 10).

The location order of four information RE along the track PRCS is determined by multinomial $r(x) = 1 + x + x^2 + x^3$, here the second, third and fourth RE are biased concerning first RE ($j_1 = 0$) on $j_2 = 1$, $j_3 = 2$ and $j_4 = 3$ elementary sites δ=360°/15 of a scale respectively. The correction RE number and their location along CT is determined pursuant to technique adduced above. The correcting sequence is received with the help of Hamming equations for the code correcting the single error. First, second and third correction RE are locate along the information track PRCS in a direction of clock hand on S_1=7, S_2=13 and S_3=11 of elementary sites.

Consistently fixing RE a seven-digit code combination at scale moving on one quantum against a direction of movement of an clock hand, we receive 15 various seven-digit code combinations. These code combinations corresponding to 15 various angular positions PRCS are resulted in Table 11.10.

Table 11.10 Sequence of seven-digit code combinations four-digit PRCS

Positions PRCS	RE_{c1}	RE_{c2}	RE_1	RE_{c3}	RE_2	RE_3	RE_4	Positions PRCS	RE_{c1}	RE_{c2}	RE_1	RE_{c3}	RE_2	RE_3	RE_4
0	1	1	0	1	0	0	1	8	0	1	0	0	1	0	1
1	0	1	0	1	0	1	0	9	0	1	1	0	0	1	1
2	1	0	0	1	1	0	0	10	0	0	0	1	1	1	1
3	0	0	1	1	0	0	1	11	1	1	1	1	1	1	1
4	1	0	0	0	0	1	1	12	0	0	1	0	1	1	0
5	1	1	0	0	1	1	0	13	0	1	1	1	1	0	0
6	1	0	1	0	1	0	1	14	1	1	1	0	0	0	0
7	1	0	1	1	0	1	0	-	-	-	-	-	-	-	-

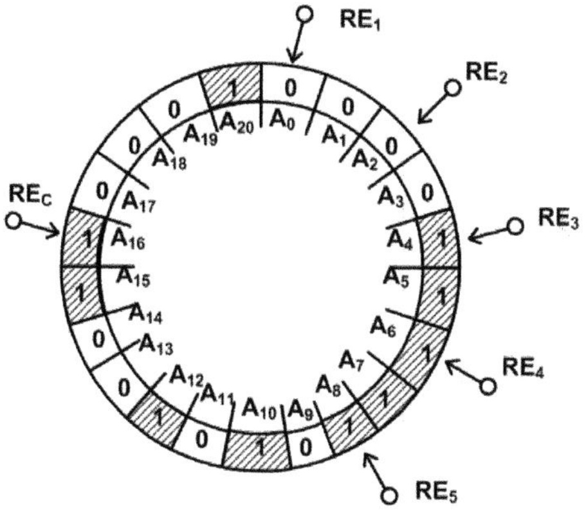

Fig. 11.8 Five-digit RCS with five information RE and one control RE_c

There are the first, second and fourth positions of the code corresponding to correcting symbols, and the third, fifth, sixth and seventh positions to information symbols. The analysis of code combinations shows that between the correcting and information positions execute Hamming equations for a code correcting single error.

Application of other correcting sequences using rule of check on parity with the introduction of appropriate redundancy on number RE will allow forming codes with required correction properties.

In Fig. 11.8 the circular five-digit RCS with five information RE and one control RE_c is shown.

The code track of the scale is formed according to symbols C_2-sequences $\boldsymbol{A} = A_0\, A_1 \ldots A_{20}$ =000011111010100110001 with the period R=21 for construction of which the polynomial $H(x) = h_1(x)\, h_2(x) = (x^2+x+1)\,(x^3+x+1) = x^5+x^4+1$ is used and the symbols A_{5+j} C_2-sequences at initial significance $A_0=A_1=A_2=A_3=0$, $A_4=1$ satisfy to the recursive equation $A_{5+j}=A_{4+j}\oplus A_j$, j=0,1...,15. The location order of five information RE along the track CCS is determined by multinomial $r(x) = 1 + x^2 + x^4 + x^6 + x^8$.

Table 11.11 Sequence of code combinations five-digit CCS with the control on parity

Positions CCS	RE_1	RE_2	RE_3	RE_4	RE_5	RE_c	Positions CCS	RE_1	RE_2	RE_3	RE_4	RE_5	RE_c
0	0	0	1	1	1	1	11	0	0	1	0	0	1
1	0	0	1	1	0	0	12	1	0	1	0	1	1
2	0	1	1	1	1	0	13	0	1	0	0	0	1
3	0	1	1	0	0	0	14	0	1	0	1	0	0
4	1	1	1	1	1	1	15	1	0	0	0	0	1
5	1	1	0	0	0	0	16	1	0	1	0	0	0
6	1	1	1	1	0	0	17	0	0	0	0	1	1
7	1	0	0	0	1	0	18	0	1	0	0	1	0
8	1	1	1	0	1	0	19	0	0	0	1	1	0
9	0	0	0	1	0	1	20	1	0	0	1	1	1
10	1	1	0	1	0	1	-	-	-	-	-	-	-

At the expense of the introduction one additional RE_c control of the information on parity is executed. The control element is biased about first information RE on 16 sites of the scale δ.

At scale moving cyclically on one elementary site, for example, against a direction of movement of the clock hand, from exits of reading out elements RE_1, RE_2, RE_3, RE_4, RE_5 and RE_c six-digit code combinations are formed. These code combinations containing even number of digits and corresponding to twenty one various angular position CCS are resulted in Table 11.11.

11.5 Conclusions

In Fig. 11.9 classification of RCS is resulted, binary sequences are put in a basis of construction of which code paths, and symbols of sequences turn out by a rule recursion. Classification concerns to RCS converters of angular and linear moving.

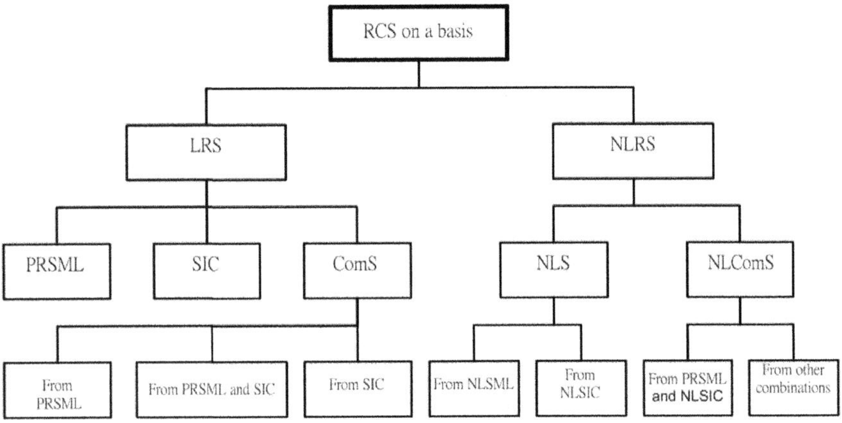

Fig. 11.9 Classification of recursive code scales

As the first classification sign property of linearity of recursive sequences in relation to the operator of summation on the module two is used. To this sign RCS share on two groups: CS on the basis of linear RS and CS on the basis of nonlinear RS.

In turn, CS on the basis of linear RS are subdivided into three groups: CS on the basis of pseudo-random sequences of the maximum length (PRSML), CS on the basis of sequences of incomplete cycles (SIC), and CS on the basis of composite sequences (ComS).

Principles of construction of CS on the basis of PRSML (M-sequences), are considered in 11.1. At reception of the recursive parity used for generation of symbols of M-sequence, primitive polynomials undertake only. Code scales on the basis of PRSML have received name PRCS.

Principles of construction of CS on the basis of SIC in the given chapter are not considered. Sequences of incomplete cycles are under construction on the basis of prime (over field GF (2)) polynomials. The polynomial of the fourth degree $h(x)=x^4+x^3+x^2+x+1$ can be an example of such prime polynomial. The recursive parity corresponding to this polynomial, looks like $z_{4+j}=z_{3+j}\oplus z_{2+j}\oplus z_{1+j}\oplus z_j$, $j=0,1...$. At nonzero entry conditions the given recursive parity allows to generate linear recursive sequence of length five (00011). If this RS is used for reception of drawing CT of a scale at placing on it RE, for example with step to one quantum, it is possible to receive five various four-digit code combinations from a scale. It is obvious, that such CS will have considerably smaller, in comparison with PRCS, resolution.

Code scales on the basis of composite sequences can be divided into three groups: CS on the basis of composite sequences, where ComS turn out from various combinations of M-sequences; CS on the basis of composite sequences, where ComS turn out from combinations of M-sequences and SIC; CS on the basis of composite sequences, where ComS turn out from various combinations SIC.

Principles of construction CS on the basis of ComS, where sequences turn out from various combinations of M-sequences, are considered in subsections 11.2. Such code scales are named CCS.

Principles of construction CS of the second and third groups on the basis of ComS in the given chapter are not considered. It is easy to see, that the least resolution possess CS on the basis of RS the third group, the greatest - CCS, and CS the second group occupy on this parameter intermediate position. Code scales on the basis of nonlinear RS are subdivided on two groups: CS on the basis of nonlinear sequences (NS) and CS on the basis of nonlinear ComS (NComS).

In turn, CS on the basis of NS are also shared on two groups: CS on the basis of NS the maximum length (NSML), and CS on the basis of NS incomplete cycles (NSIC).

At reception of the recursive parity necessary for generation of symbols of nonlinear sequence of the maximum length, primitive polynomials with introduction in a recursive parity composed, providing nonlinearity of synthesized sequence undertake.

Principles of construction CS on the basis of NSIC in the given chapter are not considered. Reception NSIC is in detail considered in work [12]. Sequences of incomplete cycles, also as well as NSML allow their resolution obviously less, than resolution CS on the basis of NSML will build on basis CS, however.

Code scales on the basis of NComS can be divided on two groups: CS on the basis of NComS where NComS turn out from combinations of M-sequences both NSML, and CS on the basis of NComS, where NComS turn out from various combinations of linear and nonlinear sequences both maximum and incomplete cycles.

Principles of construction CS on the basis of NComS in the given chapter are not considered. It is easy to see, that the best resolution will possess CS on the basis of NComS where NComS turn out from combinations of M-sequences and NSML, and CS on the basis of NComS where NComS turn out from other combinations of linear and nonlinear sequences, possess, in comparison with the first, smaller resolution.

References

[1] MacWilliams, F., Sloane, N.: Pseudo-Random Sequences and Tables. Proc. IEEE 64(12) (1976)
[2] Mutter, V.: Principles of noise-immune telecommunications. Energoatomizdat, Leningrad (1990) (in Russian)
[3] Ojiganov, A.A.: Pseudorandom encoded scales. Instrument engineering 2, 40–43 (1987) (in Russian)
[4] Azov, A.K., Ojiganov, A.A.: Principles of Construction of One-Track Code Scales on the Basis of Recursive Sequences. In: Proc. of the 9th International Conference on Systems for Automation of Engineering and Research (SAER 1995) and DECUS National Users Group Seminar 1995, Sofia (1995)
[5] Ojiganov, A.A.: Pseudorandom encoded scales, Instrument engineering (11-12) (1995) (in Russian)
[6] Ojiganov, A.A., Tarasjuk, M.V.: Compositional encoded scales. Instrument engineering 5-6, 26–29 (1994) (in Russian)
[7] Ojiganov, A.A.: Algorithm for read-out element placement on pseudorandom encoded scales. Instrument engineering 2, 22–27 (1994) (in Russian)
[8] Ojiganov, A.A., Tarasjuk, M.V.: Distribution of read-out elements for the scale with constant step. Instrument engineering, 11-12 (1994) (in Russian)
[9] Berlekamp, E.R.: Algebraic coding theory. McGraw-Hill, New York (1968)
[10] Azov, A.K., Ojiganov, A.A., Tarasyik, M.V.: Reading Digit Constant Step Location in One-Track Recursive Code Scales. In: Proc. of the 10th International Conference on Systems for Automation of Engineering and Research (SAER 1996) and DECUS National Users Group Seminar 1996, Sofia (1996)
[11] Ojiganov, A.A.: Correcting possibilities of pseudorandom encoded scales. Instrument engineering 7, 26–30 (1988) (in Russian)
[12] Agulnik, A.R., Musaeljan, S.S.: Construction of nonlinear binary sequences. Radioelectronics 4, 19–28 (1983) (in Russian)

12 Infrastructure Intellectual Property for SoC Simulation and Diagnosis Service

Vladimir Hahanov

Kharkov National University of Radioelectronics,
Apt. 321, Lenin Avenue, 14, Kharkov 61166, Ukraine
e-mail: hahanov@kture.kharkov.ua

Abstract. The models and methods for creating Infrastructure Intellectual Property (I-IP) service for the functionalities System on Chip (SoC), which has a minimum set of the real time Built-In Self Test (BIST) tools, are proposed in this chapter. The means I-IP provide an opportunity to services: fault modeling and simulation for the functional primitives to evaluate the test quality and to build Fault Detection Table (FDT); diagnosis of a given defects search depth in the SoC; repairing embedded memory functionality, by using spare row and column components. High performance deductive-parallel fault analysis method for building FDT and tests quality assessment is offered. Algebra logical methods of fault diagnosis and embedded memory repair by synthesis Disjunctive Normal Form (DNF) completing all decisions for diagnosis SoC functionalities in the real time are represented.

12.1 Infrastructure IP

Computational and hardware complexity of modern System-on-Chip is characterized by millions of equivalent gates and requires the creation and implementation of new technologies high-level design declared in Electronic Design Automation market like Electronic System Level (ESL) Design, Transaction Level Modeling (TLM) [6,11,12] and Infrastructure Intellectual Property (I-IP) [5,9-12,39, 49,50,53]. This means that the searching high-performance methods and testing approaches [1,6,11,12,19,29,34,35] leads all researchers for the needs to increase the models level abstraction for the created custom functionalities – are said to be Functional Intellectual Property (F-IP), embedded in the silicon. EDA software world market is already provides the convenient tools to automate processes of SoC simulation and verification [8,13-17,20,24-26,28,30-33,36,42-44] for system-level designs, starting with HDL Compiler languages (C++, SystemC, SystemVerilog, UML, SDL) [1,6,11,34] and ending with graphical environments (Simulink, LabView, Xilinx EDK). These tools allows to create SoC projects by using ESL mapping with TLM interface establishing based on the existing library components [11]. The EDA-market attractiveness of the SoC implementation in Field

M. Adamski et al. (Eds.): Design of Digital Systems and Devices, LNEE 79, pp. 289–330.
springerlink.com

Programmable Gate Array (FPGA) is determined: the application of relatively low-cost chips instead of the universal processors, low power consumption, small dimensions, high quality and reliable core functions. These properties are possible because I-IP infrastructure embedding in SoC [49], which is very relevant in the spread age of mobile computing devices. The leveraging I-IP represents the possibility for higher yield and reliability, Time-to-Volume (TTV) Acceleration, but it may require external support, automated tools and special equipment. Yield optimization loops leveraged at different product realization steps during design, fabrication, test and in-field. Collaborative environment is necessary to achieve Yield, Quality and TTV goals.

The problem of fault simulation, diagnosis [2-4,7,22,27,40,41,47,48] of digital system components, and memory repair [18,21,23,37,38,45,46,51,52], linked to the trend for permanent reduction SoC silicon square pointed for the original and standardized logic with a simultaneous increase of the embedded memory. As shown in Fig. 12.1, an increase in the share of memory on silicon leads for its full domination to store data and programs, which by 2014 will reach 94% [9,10,49,50,53]. This will provide not only high performance with functionality operation, but also the flexibility inherent in the product design error correction. A feature of the memory element is the fact that in the process of their construction and operation separate cells under the influence of defects can go out of normal functionality. This fact does not necessarily lead to a matrix memory critical condition, when the restoration is impossible. Therefore, such technical memory state is considered when the total number of defective cells does not exceed the capacity of the rows and columns spare intended for the repairs.

The purpose of the study – the technology development of built-in-service functionality for digital system on chip designed for modeling, fault simulation, diagnosis and repair of SoC components, including embedded memory matrix, in real time.

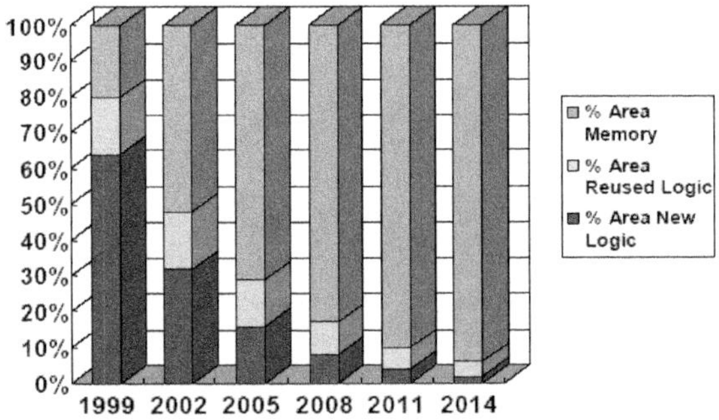

Fig. 12.1 Share memory on SoC

Objectives: 1) The state of the I-IP market technology [5,9-12,39,49,50,53]; 2) Deductive-parallel fault simulation [14-17,20,28,30-33,36,42–44]; 3) Algebra Logical (AL) method of the embedded service based on the matrix coverage; 4) AL-method application for the diagnosis of SoC components; 5) Adapting AL-method for the memory repair, 6) Practical results of the investigations.

Modern technologies for the design of digital systems on chip offer, along with the functional blocks of F-IP, the development of service modules I-IP oriented on the integrated solution of the problem of improving the project quality and Yield increasing in the manufacturing process, which is defined by implementation to silicon the following services [9-12,39,49,50,53]:

1. Monitoring the internal and output lines in the operation, verification and testing of functional blocks on the basis of IEEE 1500 boundary scan standard [12];
2. Testing the functional modules by applying different test generators, targeting fault detection or behavior checking;
3. Diagnosis failures and defects by analyzing the information received from testing phase and by using the special embedded methods for troubleshooting based on the IEEE 1500 standard [12];
4. Repair of functional modules and memory after fixing a negative test result and determine the location and type of defect in the executing phase of the diagnosis;
5. Built-in-measurement of parameters and characteristics of the SoC operation, allowing the temporal and volt-ampere measurements;
6. The reliability and fault tolerance of SoC in the operation, which are achieved by using diversification of functional blocks, duplication and recovery SoC efficiency in a real time.

The truncated I-IP-structure represented on Fig. 12.2 [49-50,53] is oriented to the execution of the following tasks: 1. Testing functionalities based on the generated test patterns by using Automated Test Pattern Generators (ATPG), and on the analysis of output responses. 2. Modeling and Fault Simulation [13-17] to provide diagnosis goals and repair SoC modules on the basis of the Fault Detection Table (FDT). 3. Diagnosis defects with a prior given fault localization depth by using boundary scan register as the troubleshooter from standard IEEE 1500. 4. Built-in-repair matrix memory through the use of spare components (columns and rows) [18,21,37,38,51,52]. The first two items are considered more conceptually, and the latter two formally constitute the essence of the proposed study.

The test synthesis module (Fig. 12.2) for check of functionality and single faults consists of a set of input patterns generators, which provide creation of the following tests [1,6,11,12,19,29,34,35]: 1. PRTG is pseudo-random test generator of input stimulus with even distribution law of zero and one signals at input variables; 2. SATG is a test generator of hexadecimal codes on basis of the signature analysis; 3. SPTG is an algorithmic generator of test vectors, activating logical single paths targeted for detection specified single fault; 4. ADTG is a test generator designed to testing Arithmetic-Logical Unit (ALU) 5. BSTG is a test generator of the bus structures for the reception and transmission of data; 6. METG is test generator aimed at verifying the memory matrix; 7. DFTG is test synthesizer for automata represented in the form of design flow. 8. RCTG is ad-hoc test generator for sequential circuits like registers, counters structures and flip-flops.

ATPG module has the features to analyze the structural-functional model of SoC functionality to be testing, and assign a subset of these test generators, which provide desired quality coverage (F^c) of faults and functional modes (P^c):

$$F^c\left(\bigcup_{i=1}^{n_{min}} T_i\right) \geq F^c_{min};\ P^c\left(\bigcup_{i=1}^{n_{min}} T_i\right) \geq P^c_{min}, \tag{12.1}$$

$$T = \{T_1^{PR}, T_2^{SA}, T_3^{SP}, T_4^{AD}, T_5^{BS}, T_6^{ME}, T_7^{DF}, T_8^{RC}\}.$$

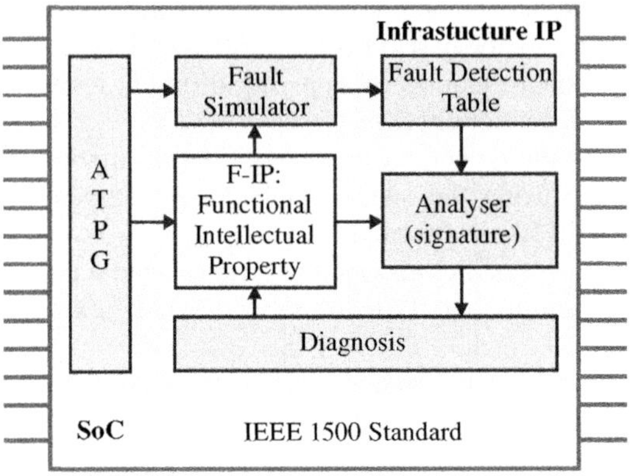

Fig. 12.2 Infrastructure IP for the SoC

Generalized structure of testbench synthesis [6] presented in Fig. 12.3, also includes HDL code generator, which is designed for testing and verification SoC functionalities at the stage of project development.

The number of test generators for SoC design phase can be significantly higher than subset, which further implements into the silicon. Therefore, in the simulation and verification process it is analyzed coverage properties of each test generator in order to find their total configuration minimum that meets the expression (12.1). It is important to note that in the next 5 years the tests synthesis ideology for digital systems on chip will borrow the best tradition of ESL- and TLM-design [11].

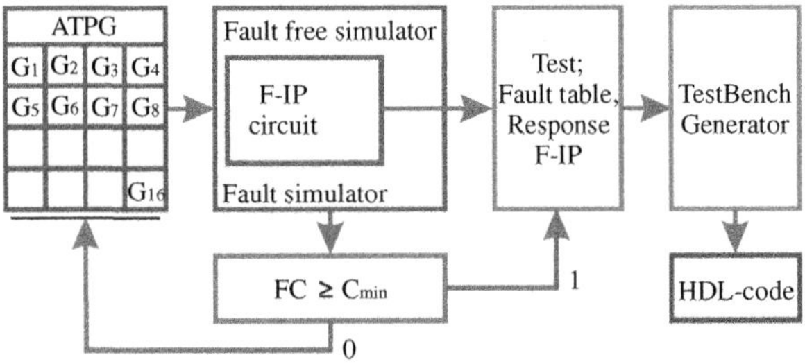

Fig. 12.3 Testbench synthesis process structure for the F-IP

This means: 1) Using the library tests (Testbench) from the world leading companies for the testing and verification of standardized functionalities identified as F-IP. 2) The use of standard solutions of service I-IP testing components for embedded systems on chip. 3) Create your own testbench library for the newly developed functionalities. 4) The providing new technology of a test synthesis for digital systems based on discrete mapping [11] (Figure 12.4) for the covering defects and functionalities original specifications by using the minimum Testbench set, from the library of tests. 5) Use of built-in-testing and diagnosis means such as IEEE 1500 SoC boundary scan standard and six components of I-IP technological procedures to enhance the tests synthesis and diagnosis procedure quality in the real time.

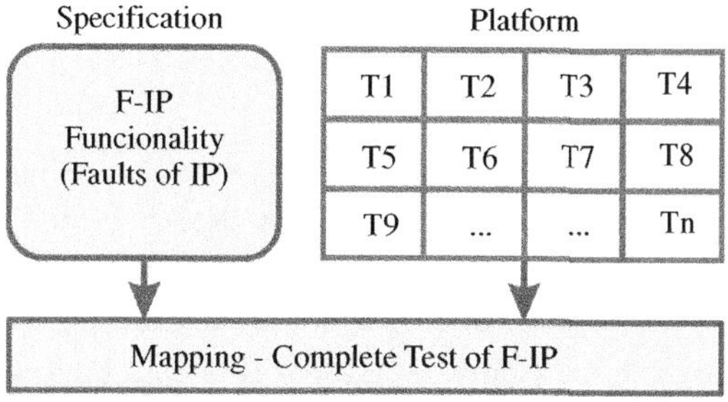

Fig. 12.4 Test synthesis mapping model for the F-IP

12.2 The Theoretical Foundations of Deductive Fault Analysis

A deductive-parallel fault simulation method [13-17,20,24-26,28,30-33,36,42-44] focused on digital projects of the large dimension with gate, register and system levels for the purpose of obtaining fault detection table, fault coverage for test quality assessment of given defects class, is offered. The Unit under Test is represented in form of structures, tables, Boolean equations, cubic coverage which is implementing as a complex digital system in the silicon. The proposed method combines fault simulation advantages of deductive definition of the fault list, effective from the point of mathematics view, targeting the high-speed parallel procedures digital devices of the gate, register and system levels of SoC descriptions.

The goal is to create the high performance deductive-parallel fault simulation method targeting to assess the quality (stack-at-faults fault coverage) of synthesized test of digital systems implemented in silicon containing millions of gates.

The background of deductive-parallel fault simulation are the methods of enhancing performance for fault analysis [28], deductive model fault propagation, a parallel method of fault lists processing trough the functional elements back traced algorithm of the primitives evaluation in the simulating digital devices faults [1].

Deduction is reasoning in the mathematical evidence system coming from general to specific. In terms of application to the fault analysis such algebra-logical means finding formal patterns, which can, once received complex models, use repeatedly for the fault simulation processing of digital systems. In doing so, each defect is to be initially described by using truth tables, Boolean equations, and the flow chart. In fact, deductive model of the functionality fault analysis allows simulating arbitrary digital circuit, with one iteration (several – for sequential circuits with global feedbacks), all the faults detecting by a test-vector. Mathematical model $T \oplus C = L$ of digital systems fault deductive analysis can be represented matrix equation [19,23-25, 27-29]:

$$(T_1, T_2, ..., T_t, ..., T_n) \oplus \begin{bmatrix} C_{11}, C_{12}, ..., C_{1i}, ..., C_{1n} \\ .... \\ C_{t1}, C_{t2}, ..., C_{ti}, ..., C_{tn} \\ .... \\ C_{k1}, C_{k2}, ..., C_{ki}, ..., C_{kn} \end{bmatrix} = \begin{bmatrix} L_{11}, L_{12}, ..., L_{1i}, ..., L_{1n} \\ .... \\ L_{t1}, L_{t2}, ..., L_{ti}, ..., L_{tn} \\ .... \\ L_{k1}, L_{k2}, ..., L_{ki}, ..., L_{kn} \end{bmatrix} \qquad (12.2)$$

Where C is fault-free behavioral cubic coverage of device model with n lines; $T = (T_1, T_2, ..., T_t, ..., T_n)$ is test-vector to faults detection, distorting the functionality C in operation, redefined in the fault-free simulation procedure on the set of input, output and internal lines. Coordinates of faulty matrix is determined by the execution of the logical XOR operation $L_{ti} = T_t \oplus C_{ti}$. Matrix $L = |L_{ti}|$ is a deductive function (DF) of fault simulation at the test-vector T corresponding good behavior element described with C-coverage, which allows calculating the input fault list propagated to the element outputs [1].

In general, the digital device in operation presented with a truth table, and use of deductive formula (12.2) allows generating fault detection table for a certain test- vector T, which (FDT) can be written analytical deductive fault simulation formula. Examples of such deductive functions obtaining presented below in the form of (test-vector, truth table, and fault detection table):

X_1	X_2	Y_1
0	1	0

$\oplus$

X_1	X_2	Y_1
0	0	0
0	1	0
1	0	1
1	1	1

$=$

X_1	X_2	L_1
0	1	0
0	0	0
1	1	1
1	0	1

$\rightarrow L_1 = X_1X_2 \vee X_1\overline{X}_2;$

X_1	X_2	Y_2
1	1	1

$\oplus$

X_1	X_2	Y_2
0	0	0
0	1	0
1	0	0
1	1	1

$=$

X_1	X_2	L_2
1	1	1
1	0	1
0	1	1
0	0	0

$\rightarrow L_2 = X_1X_2 \vee X_1\overline{X}_2 \vee \overline{X}_1X_2.$

Deductive functions are written in the disjunctive normal form according to the outputs unit values of fault detection table of the primitive element. Formula (12.2) makes it possible to perform a fault analysis of arbitrarily complex digital device, represented with the gate, register, and system levels of descriptions to be presented in a truth table form as cubic coverage. If the model is represented as a product of logical elements structures or structure of larger components, the deductive analysis of each primitive of digital device is fulfilling according with the following expression:

$$L_{ti} = T_t \oplus F_i = f_{ti}[(X_{i1} \oplus T_{t1}), (X_{i2} \oplus T_{t2}), ..., (X_{ij} \oplus T_{tj}), ..., (X_{in_i} \oplus T_{tn_i})] \oplus T_{ti}. \qquad (12.3)$$

The last expression is isomorphic formula (12.2). Thus, the formula (12.2) and (12.3) cover all digital systems, which are presented as a high level description (system, register) and the lowest (gate) level.

12.3 Deductive Components Synthesis for SoC Functions

Gate level of circuit description is characterized by logical elements, functioning of them is specified by the truth tables, cubic coverage or logical equations. In this case consideration of synthesis procedures on basis of analytical form using is technological. At that two-input logical element is transformed to four-input one, where two additional inputs (a,b) are register, and they intended for fault lists transferring. The Boolean inputs (x,y) are control for carrying out operations at external fault lists. Lets there is the logical element And, its deductive function is specified by the Karnaugh map [35]:

$L = f(x,y,a,b) =$ (12.4)

(x, y)\(a, b)	00	01	11	10
00	0	0	1	0
01	0	0	0	1
11	0	1	1	1
10	0	1	0	0

Minimization of the primitive, specified by (12.4), results in two variants of the deductive function with different complexity (quantity of variables and terms) by Quine (19, 15 and 17):

$$
\begin{aligned}
&1)\, L = f(x,y,a,b) = (\bar{x}\,\bar{y} \wedge ab) \vee (y \wedge a\bar{b}) \vee (x \wedge \bar{a}b) \vee (xy \wedge b) \vee (xy \wedge a) = \\
&\quad = (\bar{x}\,\bar{y} \wedge ab) \vee (x \wedge \bar{a}b) \vee (y \wedge a\bar{b}) \vee [(xy \wedge (a \vee b)]; \\
&2)\, L = f(x,y,a,b) = (\bar{x}\,\bar{y} \wedge ab) \vee (y \wedge a\bar{b}) \vee (x \wedge \bar{a}b) \vee (xy \wedge a) = \\
&\quad = (\bar{x}\,\bar{y} \wedge ab) \vee (x \wedge \bar{a}b) \vee [(ya \wedge (x \vee \bar{b})]; \\
&3)\, L = f(x,y,a,b) = (\bar{x}\,\bar{y} \wedge ab) \vee (y \wedge a\bar{b}) \vee (x \wedge \bar{a}b) \vee (xy \wedge b) \vee (xy \wedge a) = \\
&\quad = (\bar{x}\,\bar{y} \wedge ab) \vee [(ya \wedge (x \vee \bar{b})] \vee [(xb \wedge (y \vee \bar{a})].
\end{aligned}
\tag{12.5}
$$

Choice of the best one results in the formula (12.6). The analogous transformations applied to the elements Or, Not enable to synthesis of the Boolean equations and circuit structures. The element Or. Synthesis of its deductive function is specified by the following transformations:

$L = f(x,y,a,b) =$

(x, y)\(a, b)	00	01	11	10
00	0	1	1	1
01	0	1	0	0
11	0	0	1	0
10	0	0	0	1

$$
\begin{aligned}
&1)\, L = f(x,y,a,b) = (\bar{x}\,\bar{y} \wedge a) \vee (\bar{y} \wedge a\bar{b}) \vee (\bar{x} \wedge \bar{a}b) \vee (xy \wedge ab) = \\
&\quad = [\bar{y}a \wedge (\bar{x} \vee \bar{b})] \vee (\bar{x} \wedge \bar{a}b) \vee (xy \wedge ab); \\
&2)\, L = f(x,y,a,b) = (\bar{x}\,\bar{y} \wedge a) \vee (\bar{x}\,\bar{y} \wedge b) \vee (\bar{y} \wedge a\bar{b}) \vee (\bar{x} \wedge \bar{a}b) \vee (xy \wedge ab) = \\
&\quad = [\bar{x}\,\bar{y}(a \vee b)] \vee (\bar{y} \wedge a\bar{b}) \vee (\bar{x} \wedge \bar{a}b) \vee (xy \wedge ab); \\
&3)\, L = f(x,y,a,b) = (\bar{x}\,\bar{y} \wedge a) \vee (\bar{y} \wedge a\bar{b}) \vee (\bar{x} \wedge \bar{a}b) \vee (xy \wedge ab).
\end{aligned}
\tag{12.6}
$$

Similarly synthesis of Xor element deductive function is carried out.

Results of hardware realization of minimal deductive functions by Quine of three above elements are implemented to the following circuits (Fig. 12.5).

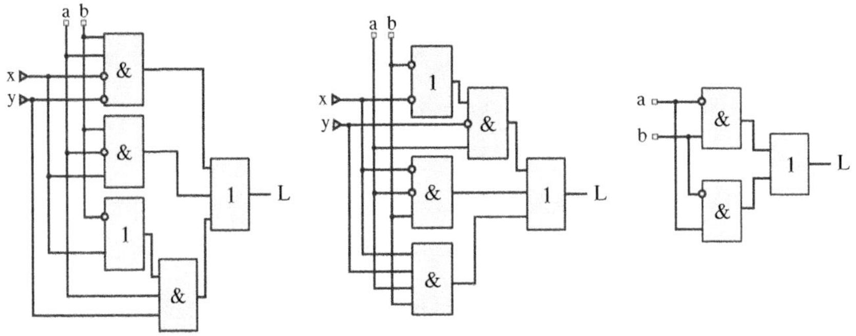

Fig. 12.5 Deductive primitives of logical elements (And, Or, Xor)

Register description level of digital system component differs by the functional complexity that influences on true table or cubic coverage dimension. Such functionalities as flip-flops, latches, counters, multiplexers, registers and bus structures are considered here. Analogous transformations, intended for deductive function synthesis by the flip-flop true table (there are three Boolean inputs and three register ones) $Q = DC \vee \overline{C}DQ(t-1)$ give the following result - (12.9).

$L = f(x,y,a,b) =$ (12.7)

(x, y)\(a, b)	00	01	11	10
00	0	1	0	1
01	0	1	0	1
11	0	1	0	1
10	0	1	0	1

$$L = f(x,y,a,b) = \bar{a}b \vee a\bar{b}. \tag{12.8}$$

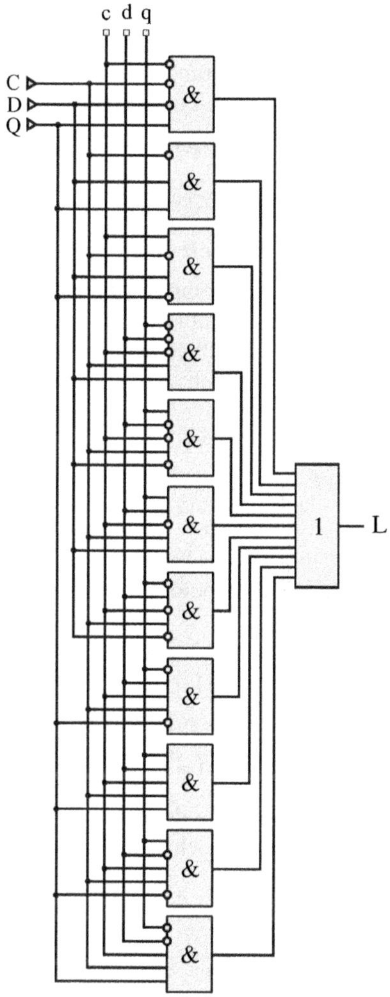

Fig. 12.6 Deductive function of fault analysis for D-flip-flop

$L = f(T, X) =$ (12.9)

(T)\(X)	000	001	011	010	110	111	101	100
000	0	1	1	0	1	1	0	0
001	0	1	1	0	0	0	1	1
011	0	1	1	0	1	1	0	0
010	0	1	1	0	0	0	1	1
110	0	0	1	1	1	0	0	1
111	0	0	1	1	0	1	1	0
101	0	0	1	1	1	0	0	1
100	0	0	1	1	0	1	1	0

The flip-flop hardware realization (the Quine estimation is 62) is represented in Fig. 12.6. The flip-flop deductive function has more then 10-fold hardware redundancy in comparison with fault-free functionality. Though this representation enables to gain in the speed of deductive fault simulation in hundreds times.

$$\begin{aligned} L = f(c,d,q,C,D,Q) = (\bar{c} \wedge \bar{C}\bar{D}Q) \vee (\bar{C}DQ) \vee (c \wedge \bar{C}D\bar{Q}) \vee \\ \vee (\bar{c}\bar{d}\bar{q} \wedge CD) \vee (\bar{c}\bar{d}q \wedge C\bar{D}) \vee (\bar{c}dq \wedge CD) \vee (\bar{c}d\bar{q} \wedge C\bar{D}) \vee \\ \vee (cd\bar{q} \wedge C\bar{Q}) \vee (cdq \wedge CQ) \vee (c\bar{d}q \wedge C\bar{Q}) \vee (c\bar{d}\bar{q} \wedge CQ). \end{aligned} \quad (12.10)$$

Concerning analysis of system level components, in general case the true table (transitions-outputs) is partially or completely defined. It means that the table coordinate definition alphabet contains three symbols (0,1,X) at least. For this case it is necessary modification of the deductive fault analysis procedure that is in the ternary alphabet:

$$L_r = \bigvee_{\forall i(T_r \oplus C_{ir}^z = 1)} [(\bigotimes_{\forall j(T_j \oplus C_{ij}^x = 1)} L_j) \setminus (\bigvee_{\forall j[(T_j \oplus C_{ij}^x = 0) \& (C_{ij}^x \neq X)]} L_j), \quad (12.11)$$

where n is a number of rows (cubes); m is a number of input lines; k is a number of output lines in a device (primitive); L_r is a fault list that is formed for the output r in the form of faults, transferred through a primitive or a digital system from external inputs.

The main operations in the ternary alphabet are:

$$\mathrm{Xor} = [0 \oplus 0 = 0; 0 \oplus 1 = 1; 1 \oplus 0 = 1; 1 \oplus 1 = 0; 0 \oplus X = X; 1 \oplus X = X; X \oplus X = X];$$

$$\mathrm{Or} = [0 \vee 0 = 0; 0 \vee 1 = 1; 1 \vee 0 = 1; 1 \vee 1 = 1; 0 \vee X = X; 1 \vee X = 1; X \vee X = X];$$

$$\mathrm{And} = [0 \wedge 0 = 0; 0 \wedge 1 = 0; 1 \wedge 0 = 0; 1 \wedge 1 = 1; 0 \wedge X = 0; 1 \wedge X = X; X \wedge X = X].$$

Subject to the introduced definitions the deductive function synthesis for system level functionality, specified by the flow chart in Fig. 12.7, is proposed below.

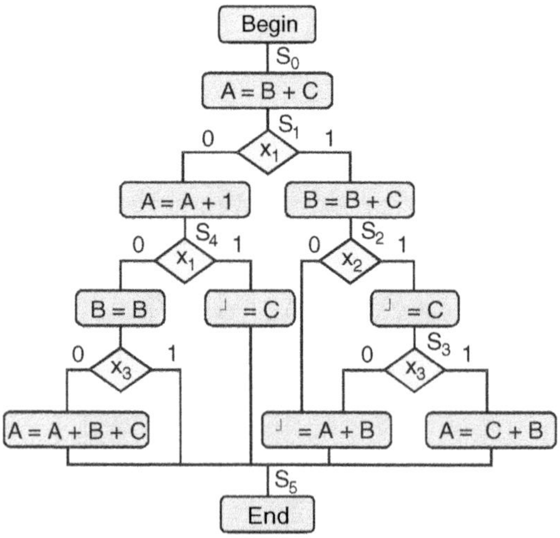

Fig. 12.7 Functionality flow chart

The transition matrix of abstract automata, corresponding to the flow chart in Fig. 12.7, as well as the transitions-outputs table of a structure automata with coded states of input, internal and output variables are represented in the following table:

$C =$

X	S_i	S_{i+1}	Y
–	S_0	S_1	$Y_0: A = B + C$
X_1	S_1	S_2	$Y_1: B = B + C$
$\overline{X_1}$	S_1	S_4	$Y_2: A = \overline{A} + 1$
X_2	S_2	S_3	$Y_3: C = \overline{C}$
$\overline{X_2}$	S_2	S_6	$Y_4: C = A + B$
X_3	S_3	S_6	$Y_5: A = \overline{C} + B$
$\overline{X_3}$	S_3	S_6	$Y_4: C = A + B$
X_1	S_4	S_6	$Y_3: C = \overline{C}$
$\overline{X_1}$	S_4	S_5	$Y_6: B = \overline{B}$
X_3	S_5	S_6	–
$\overline{X_3}$	S_5	S_6	$Y_7: A = A + \overline{B} + C$

$=$

X	S_i	S_{i+1}	Y
XXX	000	001	000
1XX	001	010	001
0XX	001	100	010
X1X	010	011	011
X0X	010	110	100
XX1	011	110	101
XX0	011	110	100
1XX	100	110	001
0XX	100	101	110
XX1	101	110	XXX
XX0	101	110	111

(12.12)

In this case the input variables are vectors, which are concatenated by the variables (XS_i), the output lines are $(S_{i+1}Y)$. To form the deductive matrix that defines a simulation primitive of all faults, corresponding to the structure automata, it is necessary to construct the true table on a set of rows or coverage cubes. This procedure is hard to realize manually. It is not difficult in computer realization. For one input vector the deductive fault analysis matrix that is result of Xor operation under an input pattern and all coordinates of a fault-free behavior matrix is:

$T \oplus C = L \rightarrow (100100\ 110001) \oplus$

$\oplus$

X	S_i	S_{i+1}	Y
XXX	000	001	000
1XX	001	010	001
0XX	001	100	010
X1X	010	011	011
X0X	010	110	100
XX1	011	110	101
XX0	011	110	100
1XX	100	110	001
0XX	100	101	110
XX1	101	110	XXX
XX0	101	110	111

$=$

X	S_i	L_{i+1}	L_Y
XXX	100	111	001
0XX	101	100	000
1XX	101	010	011
X1X	110	101	010
X0X	110	000	101
XX1	111	000	100
XX0	111	000	101
0XX	000	000	000
1XX	000	011	111
XX1	001	000	XXX
XX0	001	000	110

(12.13)

The specified deductive model is a structure of register level that can be realized in FPGA, where the true tables are used for function definition directly. Though circuit realization of the deductive functions $(S_{i+1}Y)$, written as DNF by constituent of unity of corresponding column, is possible.

$$\begin{aligned}
L_{i+1}^1 &= S_i^1\overline{S_i^2 S_i^3} \vee \overline{X_1} S_i^1 \overline{S_i^2} S_i^3 \vee X_2 S_i^1 S_i^2 \overline{S_i^3};\\
L_{i+1}^2 &= S_i^1\overline{S_i^2 S_i^3} \vee X_1 S_i^1 \overline{S_i^2} S_i^3 \vee X_1 \overline{S_i^1 S_i^2 S_i^3};\\
L_{i+1}^3 &= S_i^1\overline{S_i^2 S_i^3} \vee X_2 S_i^1 S_i^2 \overline{S_i^3} \vee X_1 \overline{S_i^1 S_i^2 S_i^3}.
\end{aligned} \tag{12.14}$$

$$\begin{aligned}
L_Y^1 &= \overline{X_2} S_i^1 S_i^2 \overline{S_i^3} \vee X_3 S_i^1 S_i^2 S_i^3 \vee \overline{X_3} S_i^1 S_i^2 S_i^3 \vee X_1 \overline{S_i^1 S_i^2 S_i^3} \vee \overline{X_3}\, \overline{S_i^1 S_i^2 S_i^3};\\
L_Y^2 &= X_1 S_i^1 \overline{S_i^2} S_i^3 \vee X_2 S_i^1 S_i^2 \overline{S_i^3} \vee X_1 \overline{S_i^1 S_i^2 S_i^3} \vee \overline{X_3}\, \overline{S_i^1 S_i^2} S_i^3;\\
L_Y^3 &= S_i^1 \overline{S_i^2 S_i^3} \vee X_1 S_i^1 \overline{S_i^2} S_i^3 \vee \overline{X_2} S_i^1 S_i^2 \overline{S_i^3 \vee \overline{X_3} S_i^1 S_i^2 S_i^3} \vee X_1 \overline{S_i^1 S_i^2 S_i^3}.
\end{aligned} \tag{12.15}$$

Equations (12.14) and (12.15) define the fault lists forming conditions at six outputs on the test pattern (100100 110001). The complex digital circuit comes out even on a single vector (Fig. 12.8); its hardware costs by Quine are 42. The output function that is realized by 84 inputs and 17 logical elements has more complex result in the form of circuit.

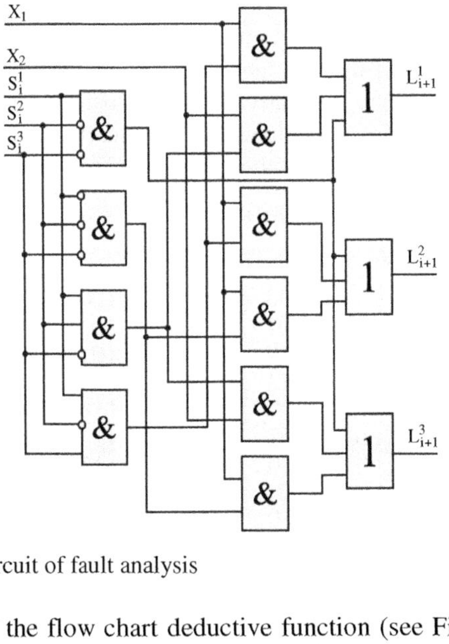

Fig. 12.8 Deductive circuit of fault analysis

So, realization of the flow chart deductive function (see Fig. 12.7) on a single input pattern has computational complexity that is equal to 84 + 42 = 126. If to multiply such combinational circuit on 2^{12} patterns, at worst the hardware costs result in the structure, defined by estimation:

$$Q = Q^t \times 2^{2\times(|X|+|S_i|)} = 126 \times 2^{12} = 516\ 096. \tag{12.16}$$

Naturally that half a million gates are inadmissible quantity for fault simulation, even if the simulation speed greater in hundreds times then software analog. In this case the problem solution is hybrid one – firmware complex of fault simulation that is flexible with respect to test vectors. In this case software oriented deductive analysis model is generated in real time as function of fault-free behavior and the test patterns $L = f(T, C)$. In this case the automaton model of fault analysis expanded in time (X,Z,Y are sets of input, internal and output variables respectively) can be represented as follows:

$$M = \langle L, T, C, X, Z, Y \rangle,$$
$$\begin{cases} L_z^t = f(T_x^t, T_z^{t-1}, C); \\ L_y^t = f(T_x^t, T_z^{t-1}, C). \end{cases} \quad (12.17)$$

So the technology of hardware embedded simulation comes back to the software oriented solutions. Actually in the near future the electronic technology market will go to flexible reusable software solutions. This direction has the following reasons: 1. System-on-a-chip realization becomes more software oriented, because in a 5 years memory will occupy 94% of the chip area. 2. To control of computational processes relating to simulation it is necessary to have microprocessor on a chip that is realized by flexible software technology and embedded into memory or by hardware technology and realized in a chip.

12.4 Structure Models of Simulator Primitives

In general case obtainment of deductive primitives for parallel fault simulation is related to the function synthesis on the exhaustive test. The complexity of deductive primitives depends on functionality representation level. The gate level structures in the form of basis of logical elements And, Or, Not are the simplest.

By means of the main expression (12.3) of deductive function synthesis, which transport faults through a logical element, construction of all basic components (And, Or, Not) [14-17] is carried out - (12.18).

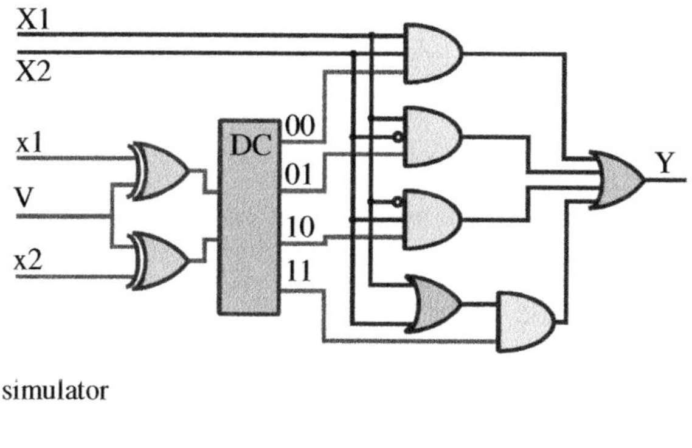

Fig. 12.9 Fault simulator

In the equations $T_t = (T_{t1}, T_{t2}, T_{t3}), (t = \overline{1,4})$ is a test vector that has 3 coordinates, and last one defines an output state of the elements And, Or. For the inverter the test vector has 2 coordinates: $T_t = (T_{t1}, T_{t2}), (t = \overline{1,2})$, last one is output state of an element. The equation for an inverter shows the immateriality of inversion operation at an element output for fault transfer. So there is not this function (Not) on deductive primitives' outputs. Hardware realization of the deductive functions [16-17] for two-input elements (And, Or) on the exhaustive test is represented in Fig. 12.9 by deductive parallel fault analysis circuit.

$$
\begin{aligned}
&L_{And}[T=(00,01,10,11), F=(X_1 \wedge X_2)] = L\{(\bar{x}_1\bar{x}_2 \vee \bar{x}_1 x_2 \vee x_1 \bar{x}_2 \vee x_1 x_2) \wedge \\
&\wedge [(X_1 \oplus T_{t1} \wedge X_2 \oplus T_{t2}) \oplus T_{t3})]\} = (\bar{x}_1\bar{x}_2)\{[(X_1 \oplus 0) \wedge (X_2 \oplus 0)] \oplus 0\} \vee \\
&\vee (\bar{x}_1 x_2)\{[(X_1 \oplus 0) \wedge (X_2 \oplus 1)] \oplus 0\} \vee (x_1 \bar{x}_2)\{[(X_1 \oplus 1) \wedge (X_2 \oplus 0)] \oplus 0\} \vee \\
&\vee (x_1 x_2)\{[(X_1 \oplus 1) \wedge (X_2 \oplus 1)] \oplus 1\} = (\bar{x}_1\bar{x}_2)(X_1 \wedge X_2) \vee (\bar{x}_1 x_2)(X_1 \wedge \bar{X}_2) \vee \\
&\vee (x_1 \bar{x}_2)(\bar{X}_1 \wedge X_2) \vee (x_1 x_2)(X_1 \vee X_2); \\
&L_{Or}[T=(00,01,10,11), F=(X_1 \vee X_2)] = (\bar{x}_1\bar{x}_2)(X_1 \vee X_2) \vee (\bar{x}_1 x_2)(\bar{X}_1 \wedge X_2) \vee \\
&\vee (x_1 \bar{x}_2)(X_1 \wedge \bar{X}_2) \vee (x_1 x_2)(X_1 \wedge X_2); \\
&L_{Not}[T=(0,1), F=\bar{X}_1] = L\{(\bar{x}_1 \vee x_1)[(\overline{X_1 \oplus T_{t1}}) \oplus T_{t2}]\} = \bar{x}_1 \wedge \\
&\wedge [(\overline{X_1 \oplus 0}) \oplus 1] \vee x_1[(\overline{X_1 \oplus 1}) \oplus 0] = \bar{x}_1 \bar{\bar{X}}_1 \vee x_1 \bar{\bar{X}}_1 = \bar{x}_1 X_1 \vee x_1 X_1.
\end{aligned}
\tag{12.18}
$$

There are the Boolean variables (x1,x2) and the register ones (X1,X2), the select input V of fault-free function type: V=0 (And), V=1 (Or), the output register variable Y. The binary input (x1,x2 and V) states form one of four deductive functions for obtainment of the testable fault vector Y. Implementation of the deductive model in HDL-code is represented by listing 1.

Listing 1 VHDL-model of a sequencer

```
library IEEE;
use IEEE.STD_LOGIC_1164.all;
entity Fub1 is
      port( i0, i1  : in STD_LOGIC;
               o00, o01, o10 , o11 : out STD_LOGIC);
end Fub1;

architecture Fub1 of Fub1 is
begin
      o00 <= not i0 and not i1;
      o01 <= not i0 and i1;
      o10 <= i0 and not i1;
      o11 <= i0 and i1;
end Fub1;
library IEEE;
use IEEE.std_logic_1164.all;
entity sequencer is
```

```
    port( V ,  X1_s, X2_s , x1, x2 : in STD_LOGIC;
            Y : out STD_LOGIC);
end sequencer;
architecture sequencer of sequencer is
component Fub1
    port( i0, i1 : in STD_LOGIC;
            o00, o01, o10 , o11 : out STD_LOGIC);
end component;
signal a0, a1, a2, a3, a4 : STD_LOGIC;
signal o00, o01, o10, o11 : STD_LOGIC;
signal x3, x4 : STD_LOGIC;
begin
    U1 : Fub1 port map(i0 => x3, i1 => x4, o00 => o00,
                 o01 => o01, o10 => o10, o11 => o11 );
    a0 <= o00 and X2_s and X1_s;
    a1 <= not(X2_s) and o01 and X1_s;
    a2 <= not(X1_s) and X2_s and o10;
    a3 <= X2_s or X1_s;
    a4 <= o11 and a3;
    Y <= a4 or a2 or a1 or a0;
    x3 <= V xor x1;
    x4 <= x2 xor V;
end sequencer;
```

The simulator operation is demonstrated in the table of parallel simulation of 8-bit input fault vectors to obtain the testable fault vector for the logical elements And, Or on the output Y:

(V, x1, x2) =	000	100	011	111	010	110
X1(RG)	01110001	01110001	10110110	00111011	00101010	10111001
X2(RG)	01111000	01111000	10110101	00110100	10111001	00101010
Y(RG)	01110000	01111001	10110111	00110000	10010001	10010001

Application of the simulator enables to transform the gate model F of fault-free circuit behavior to the deductive one L that is invariant (in terms of universality) to test patterns and does not need to use the model F at simulation. So the simulator as hardware model of DF is oriented on creation of embedded deductive parallel simulation facilities, which raise the analysis speed in 10 – 1000 times in comparison with software realization. But at that the volume ratio of post-simulation fault-free behavior model and fault analysis is 1:16. Hardware fault analysis is directed on functionality enhancement of embedded fault-free behavior simulation facilities (HES™ - Hardware Embedded Simulator) of Aldec company (*www.aldec.com*). Computational complexity of project processing that consists of n gates is equal to $Q = (2n^2\tau)/W$, where τ is the execution time of a register operation (And, Or, Not); W is the register capacity.

The hardware structure, represented in Fig. 12.10 [15], can be used for hardware realization of deductive-parallel simulation on basis of the proposed simulator.

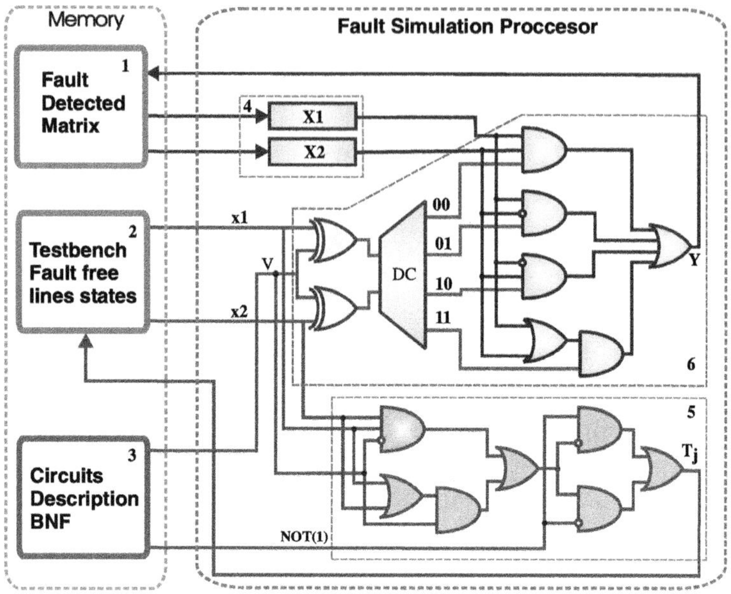

Fig. 12.10 HFS-structure of hardware simulation

The feature of hardware realization consists in combined execution of two operations: single-bit one for functional emulation of the logical elements And, Or, and parallel one for fault detection vectors processing by means of carrying out of conjunction, negation and addition logical operations. The main blocks (processor and memory) functionality is: 1. $M = [M_{ij}]$ is a quadratic fault simulation matrix, where i,j =1,q; q is the total quantity of lines in a processed circuit. 2. Fault-free behavior state vectors, defined in time t-1 and t, are necessary for forming of deductive primitive functions. 3. Memory module is required for storage of circuit definition in the form of logical element structure. 4. Buffer registers of the capacity q are necessary for operand storage and carrying out of parallel register operations at fault vectors, which are read-in form the matrix M. 5. Fault-free simulation block is required for identification the digital state of next processing logical element. 6. Deductive parallel simulator that processes two register variables X1, X2 per a cycle is necessary for definition of a fault vector, which is transferred to the logical element output Y.

The advantages of proposed fault simulation structure are:

1. Essential reduction the quantity of simulated faults, which are defined by the quantity of reconvergent fan-outs that is up to 20 % from the quantity of total lines. 2. Reduction of the memory volume, required for fault matrix storage. 3. Realization simplicity of Hardware Fault Simulator (HFS) that enables to increase the fault simulation speed by order. 4. Use HFS as first stage of the deductive topological method that is based on reconvergent fan-out processing result for high-performance analysis of tree-type structures.

Digital system on a chip simulation algorithm [16] with preliminary splitting of the device model on two structural parts (reconvergent fan-outs and tree subgraphs) is represented in Fig. 12.11.

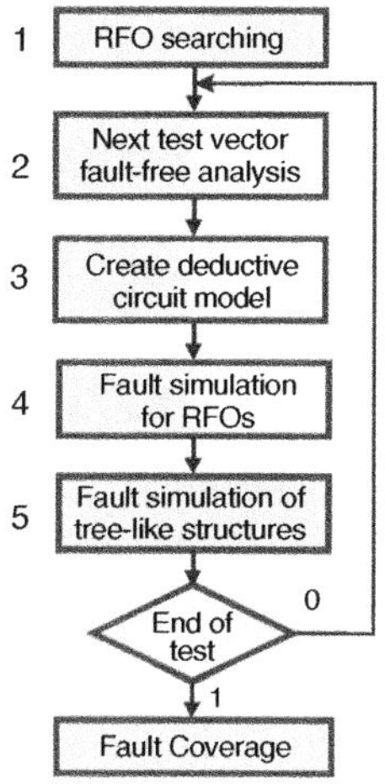

Fig. 12.11 Deductive-parallel simulation model

Resume of the proposed simulation technology with preliminary splitting of a circuit on reconvergent fan-outs and tree subgraphs. Deductive parallel fault analysis on basis of the fault back tracing requires almost linear memory and time costs, dependent on a number of circuit lines. Time performance for reconvergent fan-out processing depend on quantity of them quadratically: $Q=(r^2/W)+n_r+n_p+(n-r-r^0)$. Here (r^2/W) is the fault simulation time of r reconvergent fan-outs, their quantity is determined as $r=0.2\times n$; $n_r=n$ is the reconfiguration time of circuit primitives on an input pattern; $n_p=n$ is the search time of line subgraphs corresponding to the undetectable reconvergent fan-outs; $(n-r-r^0)=n-0.2\times n-0.4\times n=0.4\times n$ is the execution time of solution superposition on a set of circuit lines without reconvergent fan-outs and ancestors for undetectable ones. In consideration of actual parameter values in a function of circuit line quantity the estimated speed of the deductive parallel method can be obtained [1,28]:

$$Q=[(0.2\times n)^2/W]+n+n+(n-0.2\times n-0.4\times n)=[(0.2\times n)^2/W]+2.4\times n).$$

So, gain in speed of proposed method the more then less the percent of reconvergent fan-outs in a digital devise circuit [15,16,28].

For comparison a parallel algorithm has the computational complexity C_p that is defined by the functional dependency on the quantity of nonequivalent faults (b), the length of a computer word (W), quantity of equivalent gates (G):

$C_p = (b^2 / W) \times G^3$. The deductive algorithm has differences in the estimated speed formula: $C_d = b^2 \times Q \times G^2 \big|_{Q=G} = b^2 G^3$, where Q is an average quantity of gates, activated by faults. The speed of deductive parallel method without circuit splitting is defined by expression: $C_{dp} = G^2 + (b^2 / W) \times G^2$. The first summand defines the fault-free simulation time, second one – the fault analysis time of a digital device if their lines are not ranged. The estimated speed of combinational ranged circuit is $C^r_{dp} = G + (b^2 / W) \times G$. The speed of deductive parallel method is greater then the speed of parallel and deductive ones ($C^r_{dp} << \{C_p, C_d\}$) due to separation of fault-free and fault simulation.

Proposed technology of hardware-software deductive parallel fault simulation is oriented on development of deductive primitive models of gate, register and system levels to test digital systems on chips, containing millions gates. The structure model of hardware simulator and simulation device in whole are represented. They are oriented on speedup of the simulation features of high dimensionality digital devices by means of separation of fault-free analysis and determination of testable fault lists on input patterns.

SIGETEST (SImulation, GEneration of TEST) system [16,17] is developed on basis of the technology described above. It is high-speed fault simulation and test generation system, using the models of designed digital systems of interpretatively- compilation type. Some digital structure, defined by the Boolean equations, which are implemented in CPLD, FPGA, ASIC, can be a simulation object. The system processes complex digital projects, which consist of hundred thousand logical gates on post-synthesis stage (gate level description). The system has integrated environment that realizes high level graphical interface. Input of a project is realized in the description form. There are supported operations: AND, OR, NOT, XOR. The bus structures are supported too. A compiler transforms a circuit description to the internal easy-to-simulation data structures. The simulation kernel includes fault-free and fault simulation algorithms: Parallel, Backtraced Quasi Exact, Deductive-Parallel and Backtraced-Deductive-Parallel. A test generator includes a set of test patterns synthesis algorithms (pseudorandom, deterministic, synthesizers). A result of software operation is test-bench in VHDL format. The system displays information about fault-free and fault simulation, the fault coverage quality, simulation statistics. Simulation results are displayed in Fault Coverage Window that is multivalued fault detection table.

There are facilities for test synthesis control and monitoring in SIGETEST. The simulation can be limited in time; a number of simulated test patterns can be specified. The percent of fault coverage by generable patterns can be limited from below. While simulation the system displays information about the simulation progress in percentage of the total quantity of the vectors or specified time interval. SIGETEST system is oriented on integration with modern synthesis and simulation facilities, such as ALDEC Active-HDL, Riviera, SYNOPSYS Design Compiler.

12.5 Algebra-Logical Fault Diagnosis Method

The main attention is paid to the boundary scan technology [12] that is able to make easier solution of all SoC Functional Intellectual Property problems, when it is implemented into a chip. An access controller to internal lines and ports of the boundary scan register uses a cell or a stage of the register. To provide the monitoring the total quantity of such cells should be equal to a number of observed problem lines of a project, which are necessary for exact diagnosis.

The structure of I-IP service modules for fault diagnosis in F-IP functional blocks is represented in Fig. 12.12. Module ($\oplus$) analyses output reactions of a model MUS and a real device DUT on input test vectors, entering from a test generator. Boundary Scan Register is a troubleshooter that is designed for exact diagnosis. Scoreboard performs the function of diagnosis result analysis for the purposes of subsequent repair of SoC components. In this case a diagnosis is determined by the output response vector and the fault detection table ($M = [M_{tr}], t = \overline{1,p}; r = \overline{1,q+n}$ of dimension $p \times n$, p is a number of test-vectors, n is a number of stages of the boundary scan register). The result is a set of faulty lines and elements on a current input pattern. To provide computational processes, which result in exact diagnosis, metrics and representation method of the initial information are very important.

The interesting solution of a diagnosis problem can be obtained by application of the Boolean algebra and the fault detection table M that is Cartesian product of the test T on the specified fault set F together with the output response vector V, at that there is the most exact result of the covering problem solution in the form of DNF and every term is considered as a possible variant of fault existence in a device.

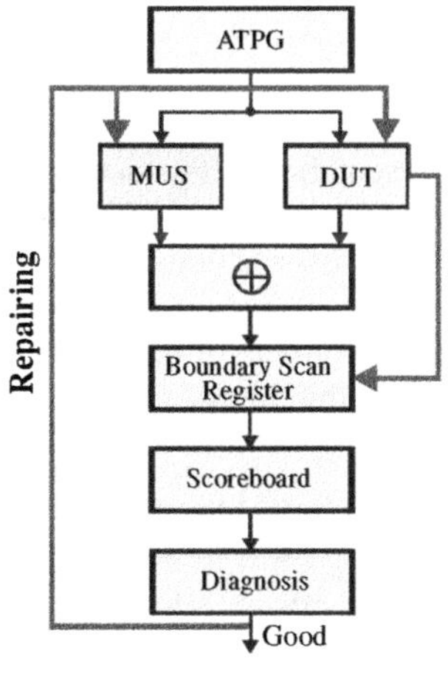

Fig. 12.12 Diagnosis process model for F-IP

Thus the diagnosis process model is represented by components:

$$
\begin{aligned}
&A = < T, F, M, V >,\\
&T = (T_1, T_2, ..., T_i, ..., T_n);\ F = (F_1, F_2, ..., F_j, ..., F_m);\\
&M = |M_{ij}|, i = \overline{1,n};\ j = \overline{1,m};\ V = (V_1, V_2, ..., V_i, ..., V_n);\\
&V_i = R(T_i) \oplus R^*(T_i);\ \{V_i, T_i, M_{ij}, F_j\} \in \{0,1\}.
\end{aligned}
\tag{12.19}
$$

The coordinate value of the vector V is a result of Xor operation at a generalized model output response and actual one.

Diagnosis problem solution comes to the fault detection table analysis that is ensued from fault simulation by means of subsequent forming of logical product of disjunctions (CNF), written by unit rows of the fault detection table

$$
F = \bigwedge_{\forall V_i = 1}^{i=\overline{1,n}} \Big(\bigvee_{\forall M_{ij}=1}^{j=\overline{1,m}} F_j \Big). \tag{12.20}
$$

The conjunctive normal form, derived from the fault detection table, is transformed to the disjunctive normal form (DNF) by means of equivalent transformations (conjunction, minimization and absorption) [4]. Therefore we have the Boolean function, where terms are the logical products, which represent full solution set in the fault combination form (they give the output response vector V at SoC outputs or its component):

$$
F = \bigwedge_{\forall V_i = 1}^{i=\overline{1,n}} \Big(\bigvee_{\forall M_{ij}=1}^{j=\overline{1,m}} F_j \Big) = \left| \begin{bmatrix} a \vee ab = b \\ a \vee a = a \end{bmatrix} \right. = \bigvee_{i=1}^{2^m} \Big(\bigwedge_{j=1}^{m} k_j F_j \Big), k_j = \{0,1\}. \tag{12.21}
$$

Function (12.21) in general case forms a diagnosis in the form of some fault combination subset, which need refinement further by means of application an additional probing of internal points by boundary scan register. A number of “1” in the output response vector V forms quantity of CNF disjunctive terms (12.21). Every term is line-by-line writing of faults (by logic operation OR), which influence on functional outputs. Table representation in the analytical form (conjunctive normal form) makes possible to reduce the volume of diagnostic information for fault location essentially. Subsequent transformation of CNF to DNF on the basis of the Boolean algebra identities enables to reduce the Boolean function. The algebra-logical method is considered by example of the following fault detection table M_1 and it is represented by five algorithm items.

$M_1 =$

T_i / F_j	F_1	F_2	F_3	F_4	F_5	F_6	V
T_1	1			1			1
T_2		1			1		1
T_3			1	1		1	1
T_4	1		1				1
T_5		1			1	1	0

1. Detection of all FDT rows, which correspond to zero values of the output response vector for nulling all 1-coordinates of found rows. In this case it is the row T_5.
2. Detection of all columns, which have zero values of rows coordinates with zero state of the vector V. Nulling of unit values of found columns. In this case they are F_2, F_5, F_6.
3. Removal the rows and the columns, which have only zero coordinate values (found in items 1 and 2), from the fault detection table.

$M_1 =$

T_i / F_j	F_1	F_2	F_3	F_4	F_5	F_6	V
T_1	1			1			1
T_2		0			0		1
T_3			1	1		0	1
T_4	1		1				1
T_5		0			0	0	0

$=$

T_i / F_j	F_1	F_3	F_4	V
T_1	1		1	1
T_3		1	1	1
T_4	1	1		1

4. Making CNF by unit ORV values:

$$
\begin{aligned}
&F = (F_1 \vee F_4) \wedge (F_3 \vee F_4) \wedge (F_1 \vee F_3) = \\
&= (F_1F_3 \vee F_3F_4 \vee F_1F_4 \vee F_4F_4) \wedge (F_1 \vee F_3) = \\
&= F_1F_1F_3 \vee F_1F_3F_4 \vee F_1F_1F_4 \vee F_1F_4F_4 \vee F_1F_3F_3 \vee \\
&\vee F_3F_3F_4 \vee F_1F_3F_4 \vee F_3F_4F_4 = \\
&= F_1F_3 \vee F_1F_3F_4 \vee F_1F_4 \vee F_3F_4 \vee F_1F_3F_4 \vee F_3F_4 = \\
&= F_1F_3 \vee F_1F_4 \vee F_3F_4.
\end{aligned}
$$

5. Transformation of CNF to DNF with subsequent minimization of the function. In this case it results in gaining sought-for result in the fault combination form: $F = F_1F_3 \vee F_1F_4 \vee F_3F_4$.

The proposed algorithm is oriented on analysis of the fault detection table to decrease its size and amount of subsequent computing related to DNF making those forms all solutions of SoC functionalities diagnosis. Further refinement of a diagnosis is possible by application of the multiprobe on the basis of the boundary scan register [15].

Example 12.1. Make algebra-logical analysis of the fault detection table $M = T \times F$ that contains 10 faults. Test of the length 11 input patterns checks all faults. The output response vector V = (10001001001), formed in the process of diagnosis experiment, fixes discrepancy between unit outputs and the gold standard on four (1, 5, 8 and 11) test patterns.

$M = T \times F =$

$T_i \backslash F_j$	F_1	F_2	F_3	F_4	F_5	F_6	F_7	F_8	F_9	F_{10}	V
T_1				1						1	1
T_2		1					1				0
T_3			1			1			1		0
T_4	1										0
T_5					1			1			1
T_6	1	1									0
T_7			1								0
T_8				1							1
T_9					1	1					0
T_{10}							1				0
T_{11}								1	1	1	1

In compliance with quantity of units in the output response vector V, a number of disjunctive terms CNF that is equal to 4 is formed. Every term is line-by-line writing of faults by logic operation OR which influences distortion of functional output signals. Then CNF is transformed to DNF on the basis of the Boolean algebra rules. It enables to get following result:

$$\begin{aligned} &F = (F_4 \vee F_{10})(F_5 \vee F_8)(F_4)(F_8 \vee F_9 \vee F_{10}) = \\ &= (F_5 \vee F_8)(F_4)(F_8 \vee F_9 \vee F_{10}) = (F_4F_5 \vee F_4F_8) \\ &(F_8 \vee F_9 \vee F_{10}) = (F_4F_5F_8 \vee F_4F_5F_9 \vee F_4F_5F_{10} \vee \\ &\vee F_4F_8F_8 \vee F_4F_8F_9 \vee F_4F_8F_{10}) = (F_4F_5F_8 \vee F_4F_5F_9 \vee \\ &\vee F_4F_5F_{10} \vee F_4F_8 \vee F_4F_8F_9 \vee F_4F_8F_{10}) = \\ &= (F_4F_5F_9 \vee F_4F_5F_{10} \vee F_4F_8). \end{aligned} \tag{12.22}$$

The function represented in the form

$$F = (F_4F_5F_9 \vee F_4F_5F_{10} \vee F_4F_8), \tag{12.23}$$

contains the fault F4 in all terms, it means that the fault is present in SoC functionality without fail. If to put forward hypothesis about existence of single fault or minimal quantity of multiple faults, the solution determinate by third term $F = F_4F_8$ is preferable (in a circuit there exist two faults, which form the output response vector that is equal to V= (10001001001)).

12.6 Simulation for Diagnosis Refinement

Disjunctive form represented in (12.21) is the main model of fault localization. It not always defines functional fault definitely, therefore it is necessary the diagnosis refinement procedures. All rows $M = T \times F$, which are marked by zero ORV values, have to be joined to non-existent fault disjunction (12.21). Obtainment of such form from given FDT enables to determine all faults, which can not exist in a circuit:

$$\begin{aligned}\overline{F} &= (F_2 \vee F_7) \vee (F_3 \vee F_6 \vee F_9) \vee (F_1) \vee (F_1 \vee F_2) \vee \\ &\vee (F_3) \vee (F_5 \vee F_6) \vee (F_7) = \\ &= (F_2 \vee F_7 \vee F_3 \vee F_6 \vee F_9 \vee F_1 \vee F_5) = \\ &= (F_1 \vee F_2 \vee F_3 \vee F_5 \vee F_6 \vee F_7 \vee F_9).\end{aligned} \tag{12.24}$$

Analysis of expression (12.23) and (12.24) enables to make interesting conclusions: 1. Faults, which can not exist in a circuit, are defined in DNF terms, obtained by zero rows relative to the output response vector. 2. Faults, which are in DNF (12.24), must be removed from function (12.23). 3. Removal of fault F_5 results in breaking of two terms $F_4F_5F_9 \vee F_4F_5F_{10}$, so far as every fault separately can not form given output response vector without fault F_5. 4. Thus the sole conclusion is made: double fault that determined by term $F = (F_4F_8)$ exists in a circuit. 5. The computational complexity of obtainment of exact and complete solution set is defined by expression $Q = 2^{m+1}(2m+1)$, where m is a number of faults.

Mark absence of the concrete fault as $F_i = 0$; input conditions for DNF (12.23) can be formed for purposes of subsequent simulation of a function under the following initial conditions:

$(F_1, F_2, F_3, F_5, F_6, F_7, F_9) = (0000000)$. Then simulation result of the function $F = (F_4F_5F_9 \vee F_4F_5F_{10} \vee F_4F_8)$ is equal to $F = (F_4 0 F_9 \vee F_4 0 F_{10} \vee F_4F_8) = F_4F_8$.

Actually, if faults $(F_1, F_2, F_3, F_5, F_6, F_7, F_9)$, which are checked on test patterns theoretically, give negative result (all of them don't distort output states) it means that they don't exist in a circuit. Justification of this fact is confirmed by the following proving.

Lemma 12.1. Complete set of all possible fault combinations, which are checked by the test T, is defined as DNF that obtained by transformation of conjunctive form

$$F = \bigwedge_{\forall V_i = 1}^{i=\overline{1,n}} \Big(\bigvee_{\forall M_{ij}=1}^{j=\overline{1,m}} F_j \Big) = \bigvee_{i=1}^{2^m} \Big(\bigwedge_{j=1}^{m} k_j F_j \Big),$$

every term of which is written by unit row values of FDT $M = T \times F$ that has ORV state $V_i = 1$.

Initial information, written in compliance with unit ORV values, is complete model of faulty behavior of a real object, which forms ORV with fixed quantity of "1" that is equal to k (FDT rows). Every row forms a fault disjunction, written by OR functions. A number of such disjunctions are equal to k; they are logical multiplied and form complete and consistent event set (a set of faults, which are in a circuit simultaneously). To obtain DNF that includes all possible combinations, written as elementary conjunctions, it is necessary to multiply elementary disjunctions and to simplify the expression using axioms $a \vee ab = b;\ a \vee a = a$. Fulfilled transformations are identical, so obtained function is equivalent to initial CNF logically, and it is technological notation of all solutions (all fault combinations, which are in a circuit) actually.

Lemma 12.2. In a real object there are not faults, checked in the fault detection table $M = T \times F$ rows and marked by zero ORV values $V_i = 0$.

Really the fault detection table $M = T \times F$ has two kinds of rows: unit 1 and zero ones relative to the actual ORV value: $[M_p(0110) \rightarrow V_p = 1] \& [M_q(0101) \rightarrow V_q = 0]$.

The row p detects two faults $F_2 \vee F_3$ in a circuit. The row q indicates of theoretical check the faults $F_2 \vee F_4$, if the vector is equal to "1": $V_q = 1$. But actually the signal $V_q = 0$ identifies the faults $F_2 \vee F_4$ inessentiality for circuit output distortion. Otherwise there are not the faults in a tested device. Substitute zero signals $F_2 \vee F_4 = 0$ in the function $F = F_2 \vee F_3$ and obtain result: $F = F_2 \vee F_3 \big|_{F_2 = F_4 = 0} = F_3$. Similarly all faults, defined in the rows, which have zero ORV value, are absent in a circuit. If it is true, they must be removed from DNF, written by unit ORV values. So, if there are DNF terms and a set of faults, which can not exist in a circuit for given ORV, the substitution procedure of zero signals in elementary conjunction variables of DNF function can be carried out. But in consideration of the fact $0 \wedge a \wedge b \wedge c... = 0$ the result of substitution and subsequent transformations for obtainment of the minimal function will contain the terms only, which don't have variables (faults with zero signal values). It means that all faults corresponding to zero FDT rows relatively ORV will be removed from DNF.

Theorem 12.1. Minimal set of all possible fault combinations, which are defined by the fault detection table $M = T \times F$, is computed by means of DNF simulation on an initial conditions set

$$F = \bigvee_{i=1}^{2^m} \left(\bigwedge_{j=1}^{m} k_j F_j\right)\Bigg|_{(\forall F_q = 0) \leftarrow (\exists M_{pq} = 1) \& (V_p = 0)},$$

which are determined by zero values of all checked faults, corresponding to zero ORV signals.

In compliance with lemma 12.1 complete set of all possible fault combinations under test is defined by DNF

$$F = \bigvee_{i=1}^{2^m} \left(\bigwedge_{j=1}^{m} k_j F_j\right),$$

that forms all solutions, satisfying to unit ORV values $V_q = 1$. It can be decreased by removal the faults, which are detected by a test theoretically, but actually they don't distort output states; it means that they are absent in a real circuit. So, they can be removed from DNF terms, which form complete set of all possible combinations. The mechanism of such removal according to lemma 12.2 consists of substitution of zero variable values in DNF terms and subsequent simulation of a function. If a term has "0"-component of some variable F_i according to the Boolean algebra, the whole term is turned into "0"; it means that it is removed from DNF. Thus after minimization subject to the conditions of lemma 12.2 the minimal DNF contains least quantity of possible fault combinations (a single and multiple ones) that can not be reduced without using of additional diagnostic information incoming from a troubleshooter on basis of the boundary scan register.

In practice the following statements are useful for diagnosis refinement:

Statement 1. If F_j are in all DNF terms, this fault exists in a circuit and it is not necessary to test it. Otherwise if one supposes that check result is equal to zero, all terms are turned into zero; and it contradicts to an existence condition of nonzero ORV values (V).

Statement 2. There are a single fault combination in a circuit that defined by one DNF term. If there are one confirmed solution in the form of DNF, other terms must be removed from consideration by means of their vanishing.

So, the check point minimization problem comes to fulfillment of two alternative strategies: 1) analysis of variables in minimal length terms to confirm all faults in a logical product by means of their probing; 2) check of variables, which turn maximal quantity of DNF terms into zero.

Thus proposed algebra-logical diagnosis method uses the Boolean calculus as base apparatus for solution of the covering problem by means of obtainment the disjunctive form that is minimized further by removal of terms, which have fault variables relating to rows with zero ORV values. An advantage of algebra-logical method is obtainment of DNF terms, which form all possible coverage's (multiple faults) of unit ORV coordinates. The computational complexity of the method has exponential relation from fault quantity: $Q = 2^n$. For small quantity of faults in a digital system the computational complexity enables to realize fault localization in real time.

12.7 Structure-Logical Fault Diagnosis Method

The fault detection table that turned out at fulfillment the fault simulation procedure doesn't have information about external outputs, which were distorted on test patterns. This fact can reduce the diagnosis depth essentially. On the other hand additional use of the information above in the aggregate with FDT has positive influence on diagnosis result and reduces the power of fault component set.

Lets there is FDT with outputs Y_1, Y_2, Y_3, which keep information about distortions on the test patterns T_1, T_4, T_8, T_{11} as in (12.25). The output response vector V was obtained by means of fulfillment OR operation at the output results of a diagnosis experiment $V = \bigvee_{j=1}^{n} Y_j$ that consists in input a test and subsequent comparison the reference response R and experimental one R* on every circuit output Y_j: $Y_j = R(T_i) \oplus R^*(T_i)$. The operation $V = Y_1 \vee Y_2 \vee Y_3$ was applied to the fault detection table (12.25); as a result ORV was obtained; at that the information content and the diagnosis depth was reduced.

T_i / F_j	F_1	F_2	F_3	F_4	F_5	F_6	F_7	F_8	F_9	F_{10}	Y_1	Y_2	Y_3	V
T_1				1						1	0	0	1	1
T_2		1					1				0	0	0	0
T_3			1			1			1		0	0	0	0
T_4	1										0	0	0	0
T_5					1			1			1	1	1	1
T_6	1	1									0	0	0	0
T_7			1								0	0	0	0
T_8				1							0	1	0	1
T_9					1	1					0	0	0	0
T_{10}							1				0	0	0	0
T_{11}								1	1	1	1	0	0	1
$\vec{Y}_1$	1			1	1			1		1	1			
$\vec{Y}_2$	1	1		1			1	1	1			1		
$\vec{Y}_3$	1		1		1	1			1	1			1	

(12.25)

It is necessary to have capability of diagnosis refinement by taking into account the circuit structure and exact analysis of device output states. For that additional information in the form of vector reachability matrix for all circuit component outputs, such as fallible lines and elements, is formed:

$$Y = \left|\vec{Y}_j\right|, i = \overline{1,n}; j = \overline{1,m},$$

where n is a number of outputs or reachability matrix vectors; m is the dimension of every vector that is equal to circuit fault quantity.

For given example the reachability matrix of three circuit outputs (scalars) Y_1, Y_2, Y_3 is represented by three last rows of expression (12.25) on the hand of potentially faulty components. The outputs are considered as independent ones, but in general case it occurs not always.

The diagnosis procedure with an allowance for circuit structure is modified on CNF forming stage. The mask operation by the appropriate row $\vec{Y}_j$ of the reachability matrix is applied to every FDT row for which $M_i(Y_j = 1)$:

$$M^1 = M^1 \vee [\vec{Y}_j \bigwedge_{\forall Y_j = 1} M_i(Y_j)]. \qquad (12.26)$$

If all circuit outputs have unit values of the experiment results $\forall Y_j = 1$, current FDT row is masked by all reachability matrix vectors with results union by OR function; it means that the mask procedure is not necessary.

In respect to the example above the mask procedure (12.26) is carried out at FDT (12.25) rows 1,5,8,11 by the vectors $\vec{Y}_1, \vec{Y}_2, \vec{Y}_3$, which are written in lower part of the fault detection table. It results in decrease a number of unit coordinate values of FDT:

T_i / F_j	F_1	F_2	F_3	F_4	F_5	F_6	F_7	F_8	F_9	F_{10}	Y_1	Y_2	Y_3	V
T_1										1	0	0	1	1
T_2		1					1				0	0	0	0
T_3			1			1			1		0	0	0	0
T_4	1										0	0	0	0
T_5					1			1			1	1	1	1
T_6	1	1									0	0	0	0
T_7			1								0	0	0	0
T_8				1							0	1	0	1
T_9					1	1					0	0	0	0
T_{10}							1				0	0	0	0
T_{11}								1		1	1	1	0	1

Here the unit coordinates $\{(T_1, F_4), (T_{11}, F_9)\}$ were removed from FDT. It results in more simple CNF and makes DNF obtainment procedure less laborious. Making of fault DNF is represented below:

$$F = (F_{10})(F_5 \vee F_8)(F_4)(F_8 \vee F_{10}) = (F_4)(F_{10})(F_5 \vee F_8)(F_8 \vee F_{10}) =$$

$$= F_4F_{10}F_5F_8 \vee F_4F_{10}F_5F_{10} \vee F_4F_{10}F_8F_8 \vee F_4F_{10}F_8F_{10} =$$

$$= F_4F_5F_8F_{10} \vee F_4F_5F_{10} \vee F_4F_8F_{10} \vee F_4F_8F_{10} =$$

$$= F_4F_5F_{10} \vee F_4F_8F_{10} = F_4F_{10}(F_5 \vee F_8).$$

Compare the result with the solution above

$$F = (F_4F_5F_9 \vee F_4F_5F_{10} \vee F_4F_8),$$

it is obvious that the solution F_4F_8 is absent, because it doesn't cover the unit ORV coordinates (1,5,8,11). The obtained subset $F = F_4F_{10}(F_5 \vee F_8)$ contains two fault combinations and indicates failure of two components 4 and 10; this subset is added by one fault 5 or 8. To refine the diagnosis it is necessary to carry out the probe procedure for one of lines 5 or 8; it result in obtainment the single term solution that defines a fault subset actually existing in a circuit.

12.8 Vector-Logical Diagnosis Method by the Fault Detection Table

To make diagnosis the fault detection table processing is carried out by an algorithm based on use the vector operations of conjunction, disjunction and negation over the fault detection table rows. Conjunction of the generalized vector that corresponds to unit coordinate values of the output response vector (ORV) and the inverted generalized vector by zero ORV coordinates:

$$F = M^1 \wedge \overline{M}^0 = \left(\bigvee_{V_i = 1} M_i \right) \wedge \left(\overline{\bigvee_{V_i = 0} M_i} \right). \quad (12.27)$$

A single fault diagnosis differs by fulfillment of the conjunction scenario (instead of disjunction one) of all vectors, which correspond to unit ORV coordinate values, on the first step:

$$F = M^1 \wedge \overline{M}^0 = \left(\bigwedge_{V_i = 1} M_i \right) \wedge \left(\overline{\bigvee_{V_i = 0} M_i} \right). \quad (12.28)$$

Example 12.2. Fulfill the multiple faults diagnosis in a circuit by the vector-logical method; the fault detection table and the output response vector are specified (12.29).

The fault detection table processing in compliance with formula (12.27) gives the result that is represented in four lower rows (12.29). Last FDT row fixes the faults presence in a circuit; faults are represented in vector or set-theory form

$$F = (0001000101\,) = \{F_4, F_8, F_{10}\}.$$

To transform the obtained result into DNF form the fault detection table structure and a set of faults, fixed in the last table row, are used. Synthesis of disjunctive form gives the following result and (12.29):

$$\begin{aligned} F &= (F_4 \vee F_{10})(F_8)(F_4)(F_4 \vee F_{10}) = \\ &= F_4F_8F_4F_4 \vee F_{10}F_8F_4F_4 \vee F_4F_8F_4F_{10} \vee F_{10}F_8F_4F_{10} = \\ &= F_4F_8 \vee F_{10}F_8F_4 \vee F_4F_8F_{10} \vee F_{10}F_8F_4 = F_4F_8. \end{aligned}$$

It is interested that due to writing of faults in the form of DNF terms, covering all unit ORV coordinate values, there is capability to remove the fault $F_{10} \in F$ from a fault list. The similar result has been obtained before when the algebra-logical fault diagnosis method is considered.

T_i / F_j	F_1	F_2	F_3	F_4	F_5	F_6	F_7	F_8	F_9	F_{10}	V
T_1				1						1	1
T_2		1					1				0
T_3			1			1			1		0
T_4	1										0
T_5					1			1			1
T_6	1	1									0
T_7			1								0
T_8				1							1
T_9					1	1					0
T_{10}							1				0
T_{11}								1	1	1	1
M^1				1	1			1	1	1	1
M^0	1	1	1		1	1	1		1		0
$\overline{M}^0$	0	0	0	1	0	0	0	1	0	1	1
F	0	0	0	1	0	0	0	1	0	1	1

(12.29)

Advantage of the vector-logical method is the technological analysis of the fault detection table; the analysis computational complexity–fault quantity–test power relation is multiplicative: $Q = n \times m$. It is recommended to use the method when there is the predominance of unit coordinate values in the fault detection table. Disadvantage of the method is the impossibility of making all fault combinations, which form the terms for covering of unit ORV coordinate values.

Example 12.3. Fulfill the vector-logical fault diagnosis of circuit lines subject to the circuit structure; a circuit is represented in Fig. 12.13.

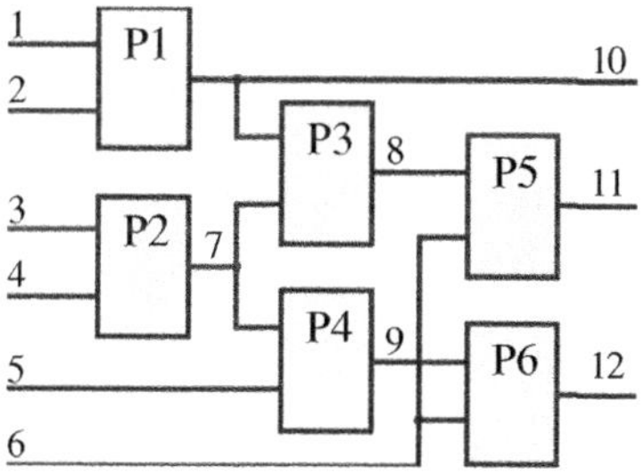

Fig. 12.13 A circuit example for diagnosis

The fault detection table (first 5 rows) and the output response vector V correspond to the circuit structure:

T_i / F_j	F_1	F_2	F_3	F_4	F_5	F_6	F_7	F_8	F_9	F_{10}	F_{11}	F_{12}	Y_{10}	Y_{11}	Y_{12}	V
T_1	1	1		1	1	1	1		1	1	1	1	1	0	0	1
T_2		1	1		1			1	1	1	1	1	0	1		1
T_3		1					1		1	1	1	1	0	0	0	0
T_4									1	1	1	1	0	0	0	0
T_5	1	1	1	1			1	1		1	1	1	0	0	1	1
$\vec{Y}_{10}$	1	1								1						
$\vec{Y}_{11}$	1	1	1	1		1	1	1			1					
$\vec{Y}_{12}$			1	1	1	1	1		1			1				
M^1	1	1	1	1	1	1	1	1	1	1	1	1				
M^0		1					1		1	1	1	1				
$\overline{M}^0$	1		1	1	1	1		1								
F	1		1	1	1	1		1								
$M^1(\vec{Y})$	1	1	1	1			1	1	1	1	1	1				
$F(\vec{Y})$	1		1	1				1								

(12.30)

Use the vector-logical method subject to expression (12.27) gives capability to get a result without taking into account the circuit structure: $F = (1011110100\ 00) = \{F_1, F_3, F_4, F_5, F_6, F_8\}$. Synthesis of disjunctive form by the fault detection table, masked by the obtained faults $F = \{F_1, F_3, F_4, F_5, F_6, F_8\}$, gives solution:

$$
\begin{aligned}
F &= (F_1 \vee F_4 \vee F_5 \vee F_6)(F_3 \vee F_5 \vee F_8)(F_1 \vee F_3 \vee F_4 \vee F_8) = \\
&= (F_1 \vee F_4 \vee F_5 \vee F_6)(F_1F_3 \vee F_1F_5 \vee F_1F_8 \vee F_3F_3 \vee F_3F_5 \vee F_3F_8 \vee \\
&\vee F_4F_3 \vee F_4F_5 \vee F_4F_8 \vee F_3F_8 \vee F_5F_8 \vee F_8F_8) = \\
&= F_1F_1F_3 \vee F_1F_1F_5 \vee F_1F_1F_8 \vee F_1F_3F_3 \vee F_1F_3F_5 \vee F_1F_3F_8 \vee \\
&\vee F_1F_4F_3 \vee F_1F_4F_5 \vee F_1F_4F_8 \vee F_1F_3F_8 \vee F_1F_5F_8 \vee F_1F_8F_8 \vee \\
&\vee F_1F_3F_4 \vee F_1F_4F_5 \vee F_1F_4F_8 \vee F_3F_3F_4 \vee F_3F_4F_5 \vee F_3F_4F_8 \vee \\
&\vee F_3F_4F_4 \vee F_4F_4F_5 \vee F_4F_4F_8 \vee F_3F_4F_8 \vee F_4F_5F_8 \vee F_4F_8F_8 \vee \\
&\vee F_1F_3F_5 \vee F_1F_5F_5 \vee F_1F_5F_8 \vee F_3F_3F_5 \vee F_3F_5F_5 \vee F_3F_5F_8 \vee \\
&\vee F_3F_4F_5 \vee F_4F_5F_5 \vee F_4F_5F_8 \vee F_3F_5F_8 \vee F_5F_5F_8 \vee F_5F_8F_8 \vee \\
&\vee F_1F_3F_6 \vee F_1F_5F_6 \vee F_1F_6F_8 \vee F_3F_3F_6 \vee F_3F_5F_6 \vee F_3F_6F_8 \vee \\
&\vee F_3F_4F_6 \vee F_4F_5F_6 \vee F_4F_6F_8 \vee F_3F_6F_8 \vee F_5F_6F_8 \vee F_6F_8F_8 = \\
&= F_1F_3 \vee F_1F_5 \vee F_1F_8 \vee F_3F_4 \vee F_3F_5 \vee F_3F_6 \vee F_4F_5 \vee F_4F_8 \vee F_5F_8 \vee F_6F_8.
\end{aligned}
$$

In case of additional use the circuit structure (Fig. 12.13) in the form of the table rows Y_{10}, Y_{11}, Y_{12} (12.30) that is applied to FDT a result can not be worst. Determination of the vector $M^1(\vec{Y})$ in (12.30) is realized by application of the reachability vectors $\{\vec{Y}_{10}, \vec{Y}_{11}, \vec{Y}_{12}\}$ masks, which correspond to faulty circuit outputs on test patterns:

$$M^1(\vec{Y}) = [\vec{Y}_{10} \wedge M_1(Y_{10})] \vee [(\vec{Y}_{11}) \wedge M_2(Y_{11})] \vee \\ \vee [(\vec{Y}_{12}) \wedge M_5(Y_{12})]. \tag{12.31}$$

Substitution of the concrete vectors from FDT to expression (12.27) gives the result:

$$M^1(\vec{Y}) = [1100000001\ 00] \vee [0110000100\ 10] \vee \\ \vee [0011001010\ 01] = [1111001111\ 11] = \\ = \{F_1, F_2, F_3, F_4, F_7, F_8, F_9, F_{10}, F_{12}\}.$$

Subsequent computation related to the vector operations $F(\vec{Y}) = M^1(\vec{Y}) \wedge \overline{M}^0$ on the fault detection table form the final solution:

$$F(\vec{Y}) = (1111001111\ \ 11) \wedge (1011110100\ 00) = \\ = (1011000100\ 00) = \{F_1, F_3, F_4, F_8\}.$$

Last vector (111100111111) creates a mask to form disjunctive normal form that has the following terms:

$$F(\vec{Y}) = (F_1 \vee F_4)(F_3 \vee F_8)(F_1 \vee F_3 \vee F_4 \vee F_8) = \\ = (F_1 \vee F_4)(F_3 \vee F_8) = F_1F_3 \vee F_1F_8 \vee F_3F_4 \vee F_4F_8.$$

In general case use the circuit structure in the form of the reachability matrix enables to get more exact diagnosis due to removal the faults, which can not influence on faulty outputs.

There are proposed diagnosis methods: 1) algebra-logical, 2) vector-logical and 3) structure-logical ones. They give the mathematical apparatus to a specialist in the field of SoC design and testing. It enables to diagnose of faulty components if there is preliminary constructed fault detection table. At that set-theoretical solutions, efficiently obtained by second method, can be represented by all possible fault combinations in the form of DNF terms that is typical of the first method. Second method is efficient if a number of “1” in FDT is greater then 10-20%. Third method needs additional information in the form of a reachability matrix for all external outputs that enables to reduce essentially the power of a diagnosed fault set or quantity of terms, defined all possible fault combinations, which form the output response vector.

It is necessary to note that all three methods can generate a result in three forms: vector, set and combine one in the form of DNF terms. Fault simulation stage is interested for the purposes of SoC faulty area reduction. The algebra-logical method can be added by the fault reduction procedure by means of subtraction all non-existent faults.

12.9 Algebra-Logical Memory Repair Method

The exact method of memory elements diagnosis and repair by spare rows and columns, which enables to cover a set of faulty cells by minimal quantity of spares is represented. The method is oriented on implementation of the Infrastructure Intellectual Property for SoC functionality. The structure solutions for realization of the method of diagnosis and repair of memory matrix fault cells are proposed [37,38,40,41,45,46,51,52].

During the SoC manufacturing and operation any kinds of memory the guarantee of its technical compliance are necessary. In this regard, three procedures are carried out as given below: 1) Memory testing that consists of test patterns input, which oriented on identification of specific kinds of faults [23,37,38]; 2) In the case of fault appearance, it is necessary an additional diagnosis procedure that enables to determine location, cause and kind of fault; 3) After detection of a fault set, which blocks carrying out of the memory function, it is necessary to activate the repair process – replacement of faulty elements by spares, which initially are on a chip [9-12]. Thereby, aforementioned actions are oriented on the growth of yield without significant additional time and material costs. To repair, it is necessary to apply a special mechanism for memory repair, by means of replacement of faulty components by fault free ones from the chip reserve.

As a rule the testing procedure is realized by BIST-block (Built-In Self Test), which is hardware high performance generator of test patterns, as well as an analyzer (signature) of memory outputs responses on test patterns. Repair analysis consists of definition of covering possibility of faulty memory elements by available reserve components. Memory module has two parts: 1) functional cells, which are used for data and program storage, when a module is used in SoC; 2) reserve or spare cells, which are designed for memory repair in case of functional cells failure. Functional and spare cells are joined together in the form of columns and rows. When a fault is detected, a row (a column), which includes a faulty element, is disconnected from the functional structure of memory cells and a row (a column) from chip spare is connected on its place. The number of reserve components is limited, so it is necessary to apply a special mechanism of effective allocation of repair resource, for support of faulty memory elements covering by the minimally possible quantity of redundant rows and columns.

The search procedure of faulty cells covering by the minimal quantity of reserve rows and columns described above can be realized as on-chip repair module or external one. In the second case data about errors is received from external modules; they are processed and pass to the controller that provides memory repair. It results in considerable time loss. So, the preferable solution is on-chip module realization, when data about errors is passed from BIST directly. Such mechanism is called as BIRA [37,38,45,46,51,52] – Built-In Repair Analysis.

Memory repair is realized by disconnection of faulty elements (rows and columns of a matrix) by means of electrical fusion of metal links and connection of spare ones. The fuse process can be electrical or laser [18]. Electrical fuse equipment has smaller dimensions than laser one and it is used more frequently. Fuse is carried out by means of an instruction set, which can be stored in permanent memory inside chip (hard repair) or in random-access memory (soft repair) [37-38]. Soft repair has several advantages: when a defect appears, a new corrected instruction can be recorded to memory easily; there provide economic use of chip area and sufficient reliability. Hard repair enables to use a simplified manufacturing test and provides detection of errors, which can not be fixed by soft repair under certain circumstances.

The structure of on-chip memory analysis and soft repair processes (BISR) [40,45,46] is represented in Fig. 12.14. There are: 1) Chip activation, filling of the BISR register by zero values. 2) Run the BIST controller. Memory testing and accumulation of information about faulty cells in the BIRA register. 3) Transfer of information about faulty cells to the BISR register for subsequent fusion. 4) Scanning the BIRA registers for obtainment of faults information 5) Run the fuse controller in record mode and transfer the repair instructions from the BISR. 6) Chip restart. Recording the fuse information to the BISR register, replacement of faulty rows and columns by reserve components is fulfilled. 7) Run the BIST controller for repeated memory testing and verification of the repair result correctness.

The idea of an algebra-logical memory repair method is in optimal replacement of memory matrix faulty cells by means of solution of the fault covering task by spare rows and columns. The objective function is defined as minimization of the memory matrix spares (S), necessary for its repair in the process of SoC operation by means of synthesis of a disjunctive normal form of faulty elements covering and subsequent choice of the minimal conjunctive term $X^t(R^t, C^t) \in Y$ that satisfies the limitations on a number of spare rows and columns S^r_{max}, S^c_{max}, which enter into the logical product:

$$Z = \min_{t=\overline{1,n}} (|X^t|) \Big|_{|S^r|+|S^c| \le S_{max};\, |S^r| \le S^r_{max};\, |S^c| \le S^c_{max}},$$

$$X^t \in Y = \{X^1, X^2, \ldots, X^t, \ldots, X^n\},\ X^t = (X^t_1 \,\&\, X^t_2 \,\&, \ldots, \&X^t_i \,\&, \ldots, \&X^t_{m_t}).$$

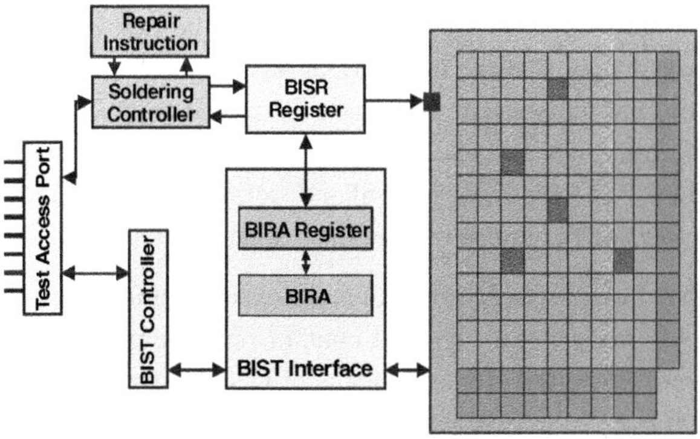

Fig. 12.14 Scheme of on-chip memory analysis and repair

There is every resultant conjunctive term of the function Y is formed from row and column identifiers $X^t = (R^t, C^t)$, which cover all faults in a memory matrix. The best solution is the minimal length term by Quine that contains rows and columns covering all faults. Particularly a solution can contain not a row (a column), when available columns (rows) from memory spare are enough for matrix repair. The model of determination process of minimal spares quantity, which covers all detected faults in a memory matrix, consists of the following items:

1. Transformation of two-dimensional model of memory matrix faults to a fault coverage table by spare rows and columns. To achieve the aim the topological memory model in the form of fault identification matrix is considered:

$$M = |M_{ij}|, M_{ij} = \begin{cases} 1 \leftarrow T \oplus f = 1; \\ 0 \leftarrow T \oplus f = 0. \end{cases} \quad (12.32)$$

Here matrix coordinate is marked by 1, if the fault-free behaviour function f of a cell and test reaction gives one value that corresponds to a fault. After detection all faults the fault coverage table is formed $Y = |Y_{ij}|, i = \overline{1, n}; j = \overline{1, m}$, where columns correspond to a set of detected faults m and rows are the numbers of columns and rows of a memory matrix, where faults are occurred:

$$Y = |Y_{ij}|, Y_{ij} = \begin{cases} 1 \leftarrow C_i(R_i) \cap F_j \neq \varnothing; \\ 0 \leftarrow C_i(R_i) \cap F_j = \varnothing. \end{cases} \quad (12.33)$$

Instead of the two-dimensional metrics components C and R it is used one-dimensional vector concatenated from two sequences C and R; its power is equal to $n = p + q$:

$$\begin{aligned} X = C * R = (C_1, C_2, ..., C_i, ..., C_p) * (R_1, R_2, ..., R_j, ..., R_q) = \\ = X^c * X^r = (X_1, X_2, ..., X_i, ..., X_p, X_{p+1}, X_{p+2}, ..., X_{p+j}, ..., X_{p+q}). \end{aligned} \quad (12.34)$$

At that there exists one-to-one correspondence between elements of the initial sets (C, R) and resultant vector X that is determined in the first column of the matrix Y. It is necessary to say that the transformation $X = C * R$ is carried out for ease of consideration and subsequent forming of disjunctive normal form within the bounds of uniformity of variables, which form the Boolean function. If the procedure is not carried out the function will be defined on two kinds of variables, which contain rows and columns of a memory matrix.

2. Forming CNF for analytical, complete and exact solution of the covering task. After forming of a covering matrix that contain zero and unit coordinates the synthesis of analytical covering form by means of CNF writing by columns is carried out. Here a number of conjunctive terms are equal to quantity of table columns and every disjunction is written by one values of the column:

$$Y = \mathop{\wedge}_{j=1}^{m} (Y_{pj} \vee Y_{qj})_{\{Y_{pj}, Y_{qj}\}=1} = \mathop{\wedge}_{j=1}^{m} (X_{pj} \vee X_{qj}). \qquad (12.35)$$

As follows from last expression every column has two coordinates only, which take on unit value, and quantity of logical products is equal to total quantity of faults m, detected in a memory matrix.

3. Transforming CNF to DNF enables to see all solutions of the covering task. For that it is necessary to apply the operation of logical multiplication and the minimization (absorption) rules to a conjunctive normal form to get disjunctive normal form:

$$Y = \mathop{\vee}_{j=1}^{w} (k_1^j X_1 \wedge k_2^j X_2 \wedge ... \wedge k_i^j X_i \wedge ... \wedge k_n^j X_n), k_i^j = \{0,1\}. \qquad (12.36)$$

The generalized notation of DNF is represented here, where a number of terms is equal to $w = 2^n$ in the limit, n is a number of rows in the set (C,R) or a number of the variables X in a matrix Y, on a set of which all solutions (fault coverings by spares) are formed; if k_i^j at X_i is equal to zero the variable X_i becomes inessential one.

4. Choice of minimal and exact solutions of the covering task. The procedure is related to the determination of minimal length conjunctive terms in obtained DNF. Subsequent transformation of a memory matrix to rows and columns on basis of use the correspondence defined above enables to write the minimal covering or set of ones by two-dimensional metrics of rows and columns that satisfies the objective function conditions (limitations) on spare quantity.

Example 12.4. Fulfill the process of a memory matrix repair in the part of determination of minimal quantity of spares, covering all faults. A memory matrix with faults and spares [11] is represented in Fig. 12.15.

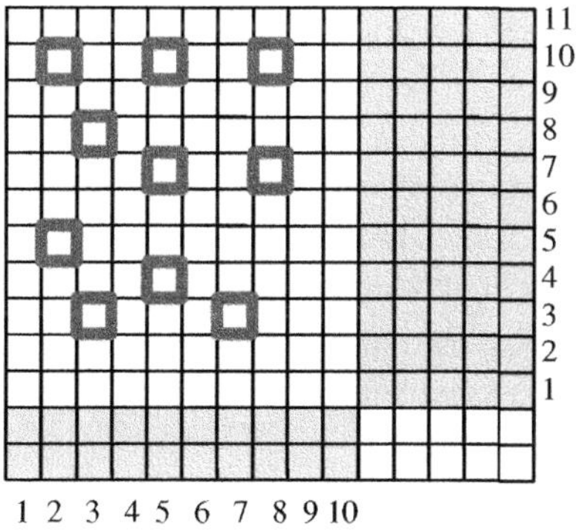

Fig. 12.15 Memory matrix with faults and spares

The matrix has limitations on diagnosis and repair possibilities for ten faulty cells, which are defined by two rows and five columns. In compliance with item 1 of the model of determination process of minimal quantity of spares, which cover all detected faults of a memory matrix, the coverage table of ten faults $F=(F_1,F_2,F_3,F_4,F_5,F_6,F_7,F_8,F_9,F_{10})$ by eleven rows is formed as in (12.37).

Y =

X_i / F_j	F_1	F_2	F_3	F_4	F_5	F_6	F_7	F_8	F_9	F_{10}
$C_2 \to X_1$				1						1
$C_3 \to X_2$		1					1			
$C_5 \to X_3$			1			1			1	
$C_7 \to X_4$	1									
$C_8 \to X_5$					1			1		
$R_3 \to X_6$	1	1								
$R_4 \to X_7$			1							
$R_5 \to X_8$				1						
$R_7 \to X_9$					1	1				
$R_8 \to X_{10}$							1			
$R_{10} \to X_{11}$								1	1	1

(12.37)

The rows are represented by concatenation of the columns C and rows R, which are in the one-to-one correspondence with the variable vector X:

$$C*R=(C_2,C_3,C_5,C_7,C_8)*(R_3,R_4,R_5,R_7,R_8,R_{10})\approx \\ \approx X=(X_1,X_2,X_3,X_4,X_5,X_6,X_7,X_8,X_9,X_{10},X_{11}). \quad (12.38)$$

In compliance with the coverage table-construction of DNF is performed, the terms are written by unit values of columns:

$$Y=(X_4 \vee X_6)(X_2 \vee X_6)(X_3 \vee X_7)(X_1 \vee X_8)(X_5 \vee X_9)\,\& \\ \&(X_3 \vee X_9)(X_2 \vee X_{10})(X_5 \vee X_{11})(X_3 \vee X_{11})(X_1 \vee X_{11}). \quad (12.39)$$

The following transformations related to obtainment of disjunctive normal form are based on application of the Boolean algebra identities, which enable to carry out: logical multiplication of all ten multiplicands and subsequent minimization of DNF terms by means of application the operator $(ab \vee a\bar{b}=a)$, absorption axiom and removal of the same terms. Skipped intermediate calculus, the final result is represented in the following form:

$$\begin{aligned}Y = {} & X_1 X_2 X_3 X_4 X_5 \vee X_2 X_3 X_4 X_5 X_8 X_{11} \vee \\ & \vee X_1 X_2 X_4 X_9 X_3 X_{11} \vee X_1 X_3 X_2 X_4 X_9 X_{10} X_{11} \vee \\ & \vee X_1 X_7 X_{10} X_{11} X_6 X_9 \vee X_6 X_9 X_7 X_8 X_{10} X_{11} \vee \\ & \vee X_2 X_4 X_9 X_3 X_8 X_{11} \vee X_1 X_2 X_4 X_9 X_7 X_{11} \vee \\ & \vee X_2 X_4 X_9 X_7 X_8 X_{11} \vee X_3 X_2 X_4 X_9 X_8 X_{10} X_{11} \vee \\ & \vee X_1 X_2 X_4 X_9 X_7 X_{10} X_{11} \vee X_1 X_2 X_3 X_5 X_6 \vee \\ & \vee X_1 X_3 X_5 X_6 X_{10} \vee X_2 X_3 X_5 X_6 X_8 X_{11} \vee \\ & \vee X_3 X_5 X_6 X_8 X_{10} X_{11} \vee X_1 X_2 X_3 X_{11} X_6 X_9 \vee \\ & \vee X_1 X_3 X_{10} X_{11} X_6 X_9 \vee X_2 X_3 X_8 X_{11} X_6 X_9 \vee \\ & \vee X_1 X_2 X_7 X_{11} X_6 X_9 \vee X_2 X_7 X_8 X_{11} X_6 X_9 \vee \\ & \vee X_3 X_8 X_{10} X_{11} X_6 X_9 .\end{aligned} \tag{12.40}$$

The choice of minimal length terms, which contain 5 variables in given case, forms a set of minimal solutions:

$$Y = X_1 X_2 X_3 X_4 X_5 \vee X_1 X_2 X_3 X_5 X_6 \vee X_1 X_3 X_5 X_6 X_{10} . \tag{12.41}$$

Transforming the obtained function to a coverage that contains variable designations in the form of rows and columns of a memory matrix, enables to represent solutions in the following form:

$$Y = C_2 C_3 C_5 C_7 C_8 \vee C_2 C_3 C_5 C_8 R_3 \vee C_2 C_5 C_8 R_3 R_8 . \tag{12.42}$$

All obtained minimal solutions satisfy the requirements (limitations) on spare quantity that is determined by the numbers:

$$(\left|C^r\right| \le 5) \,\&\, (\left|R^r\right| \le 2) .$$

Other solutions, determined in DNF, have no interest because they have not optimal covering of faulty cells that is determined by quantity of variables in the terms of (rows + columns), greater than five. Subsequent technology of embedded repair of faulty cells consists of electrical reprogramming of an address decoder of a column or a row of a memory matrix. In respect to memory, represented in Fig. 9, a procedure of writing or reading of information at access to any cell of column 2 will be readdressed to reserve column 11. In compliance with the last obtained solution (first term of DNF function Y) other faulty columns will be replaced on fault-free ones from memory reserve: 3 – on 12; 5 – on 13; 7 – on 14, 8 – on 15.

The computational complexity of algebra-logical memory repair method in the part of solving of the covering problem [35] is determined by the following expression:

$$Q = 2^{|F|} + |C + R| \times 2^{|F|} , \tag{12.43}$$

where $2^{|F|}$ is costs related to DNF synthesis by logical multiplication of two-component disjunctions (fault coordinate is defined by row and column numbers), where their quantity is equal to the quantity of faulty cells; $|C + R| \times 2^{|F|}$ is upper limit of computational costs, which are needed for minimization of the obtained DNF on maximum set of variables which is equal to the total quantity of rows and columns $|C+R|$.

In the worst case, when coordinates of all faulty cells are not correlated by rows and columns (they are unique), for instance, diagonal faults, the computational complexity of the matrix method is dependent only on the quantity of faulty cells and its analytic notation is transformed to the following view:

$$Q = 2^{|F|} + |C+R| \times 2^{|F|}\Big|_{|C+R| \le 2\times|F|} = 2^{|F|} + 2\times|F|\times 2^{|F|} = 2^{|F|}\times(1+2\times|F|). \quad (12.44)$$

If instead of fault set power to use quantity m of them, the previous expression can be represented more simplified form:

$$Q = 2^{m}\times(1+2\times m) = 2^{m}(2m+1). \quad (12.45)$$

According to the SoC Functional Intellectual Property Infrastructure, the matrix repair method on the basis of solving covering problem is implemented into a chip as one of I-IP components, designed for the functional support of SoC matrix memory and SoC in whole.

12.10 Conclusions

A deductive-parallel fault simulation method [14-17] focused on large dimension digital projects of gate and register levels to obtain descriptions of the fault detection table and assess the quality test coverage of prior given class defects. The unit under test model is represented in the form of structures, truth tables, Boolean equations, cubic coverage describing components of complex digital system implementing into in the SoC. The proposed method combines fault simulation dignity of deductive fault detection lists, effective from the point of view of mathematics, and the implementation of parallel procedures targeting high-speed digital devices with the gate, register and system descriptions levels.

Algebra logical method with the built-in diagnosing defects in the functional blocks SoC and its two modifications using analysis FDT, in order to reduce the volume and subsequent calculations related to the construction of DNF, which creates all decisions on diagnosis SoC functionalities in the real time is offered. The methods of approximation and exceptions include two procedures targeting the diagnosis of single or multiple defects. The proposed algebra logical method is a universal in terms of the defects number that is present in the scheme and allows for a single procedure to place an accurate diagnosis, indicating a single or multiple faults, all of which are required to cover all unit value coordinates of Output Response Vector (ORV).

A truncated infrastructure IP service for functionalities SoC is offered. It is a different from standard by minimum set of built-in-diagnosis procedure in real time and provides an opportunity to services: testing functions in operations on the basis of embedded set of test pattern generators, output analysis of test reactions;

diagnosis of a prior given depth of defects localization by using a boundary scan register of IEEE 1500 standard, fault simulation with a view to the providing the first two procedures executing by using the fault detection table.

A mapping process model of test synthesis is offered, which use different embedded test generators libraries for SoC functionalities, which allows substantially reduce the time construction of tests intended to verify the digital system in operation and fault detection.

In terms of the memory repair there is some scientific novelty. Memory SoC in the near future will be more than 90% of the SoC-based software. Urgently to develop not only the models and methods of rapid and accurate diagnosis, but also the creation of technologies for the repair of defective cells by using embedded service means in real time and at all stages of the product life cycle. This would essentially reduce the chip pins, increase yield, reduce time-to-market, and the cost of maintenance services as well as the exclude the external tools to diagnose and repair.

Algebra logical memory repair method based on the faulty cells coverage the spare elements by using apparatus of Boolean algebra. The method is quadratic computing complexity and can be implemented as software in the silicon outside, and inside in the form of a supplementary service module correction of defects that would automatically restore the memory elements in the operation.

The classical coverage challenge operates with two one-dimensional vectors (X, F), where the operator P allows you to find cover with minimal subset of the X components covering its functionality all the elements of F: $X_{min} = P(X,F) \leftarrow X \cap F = X_{min}$. The problem of one-dimensional vector F properties coverage by two-dimensional matrix $M = (C \times R)$ needs to bring both components to a single metric. This coordinate system has the common denominator for the both components. Naturally, such a metric for matrix M and vector F is a one-dimensional structure. Therefore, in this case, a priori need to complete the conversion of two-dimensional structures (the matrix defects memory) to a one-dimensional by the operation concatenation $X = (C * R)$ for the purpose of subsequent coverage task decision by applying formal actions defined operator $X_{min} = P(X,F)$.

A method of optimal repairing memory defects, which is different from the analogue technology of algebra logical defects coverage by two-dimensional matrix memory topology that provides the minimum and complete solutions for the repairing in real time, based on the use of redundant components in the form of rows and columns of memory.

The practical significance of the method lies in its implementation in the infrastructure IP service of SoC functional blocks. This allows significantly (by 5-10%) increase the percentage of yield to the market through electronic technology recovering defective memory elements in the production and operation phase, and increase the life cycle of memory through its repair in real time.

Built-in-repair focused on all components of the system is addressed: memory, multiprocessors, matrix processors. If you want to repair the other structures, they should be given the redesign with injection of addressable components.

Addressability and regularity of the components in the digital system transforms it to a reliable, robust, maintainable and fault tolerance one.

Algebra-logical presentation of the coverage task is very attractive to the optimal solution to the all tasks of synthesis and analysis of complex systems, where there is the problem of mapping (coverage): 1) specification – with a set of library components; 2) defects – with test sequences; 3) functionalities – with testbenches; 4) faulty elements – with the spare; 5) states of the UUT – with the line of observation.

In order to reduce the problem's dimension of mapping, the original model must be structured by hierarchy creating, which is typical and is widely used in design automation systems (ESL-, TLM-technology).

Priori looks very attractive FDT in the form of Boolean function in terms of compactness, which is transforms into a compact DNF terms, as all possible combinations of the faulty components to be repaired for the concrete output response vector.

Yervant Zorian (EWDT Symposium 2007, Yerevan): "Currently, one of the main problems for SoC design is the development of technologies and methods for embedded repair logic, which takes no more than 10% of the silicon area."

Further investigation in terms of the proposed material focused on the development maintainable structure of the System on a Chip and built-in hardware module BIRA for embedded repair of any component when defects are appear in the process of manufacturing and operating.

References

[1] Abramovici, M., Breuer, M.A., Friedman, A.D.: Digital System Testing and Testable Design, p. 652. Computer Science Press, Rockville (1998)

[2] Aitken, R.C.: Modeling the Unmodelable: Algorithmic Fault Diagnosis. IEEE Design and Test of Computers, 98–103 (1997)

[3] Armstrong, D.B.: A Deductive Method for Simulating Faults in Logic Circuits. IEEE Transactions on Computers, 464–471 (1972)

[4] Bayraktaroglu, I., Orailoglu, A.: The Construction of Optimal Deterministic Partitionings in Scan-Based BIST Fault Diagnosis: Mathematical Foundations and Cost-Effective Implementations. IEEE Transactions on Computers, 61–75 (2005)

[5] Benso, A., Carlo Stefano, D., Prinetto, P., Zorian, Y.: A Hierarchical Infrastructure for SoC Test Management. IEEE Design and Test of Computers, 32–39 (2003)

[6] Bergeron, J.: Writing testbenches: functional verification of HDL models. Springer, Heidelberg (2003)

[7] Bernardi, P., Veiras Bolzani, L.M., Rebaudengo, M., Sonza Reorda, M., Vargas, F.L., Violante, M.: A New Hybrid Fault Detection Technique for Systems-on-a-Chip. IEEE Transactions on Computers, 185–198 (2006)

[8] Cha, H., Rudnick, E.M., Patel, J.H., Iyer, R.K., Choi Gwan, S.: A Gate-Level Simulation Environment for Alpha-Particle-Induced Transient Faults. IEEE Transactions on Computers, 1248–1256 (1996)

[9] Chandramouli, R.: Infrastructure IP design for repair in nanometer technologies. IEEE Design and Test of Computers 17 (2005)

[10] Clark, C.J., Ricchetti, M.: Infrastructure IP for Configuration and Test of Boards and Systems. IEEE Design and Test of Computers, 78–87 (2003)

[11] Densmore, D., Passerone, R., Sangiovanni-Vincentelli, A.: A Platform-Based taxonomy for ESL design. Design&Test of computers, 359–373 (September-October 2006)
[12] DaSilva, F., Zorian, Y., Whetsel, L., Arabi, K., Kapur, R.: Overview of the IEEE P1500 Standard. In: ITC International Test Conference, pp. 988–997 (2003)
[13] Ghosh, S.: Behavioral-Level Fault Simulation. IEEE Design and Test of Computers, 31–42 (1988)
[14] Hahanov, V., Kteaman, H., Ghribi, W., Fomina, E.: HEDEFS – Hardware embedded deductive fault simulation. In: Proceedings of the 3rd IFAC Workshop, Rydzyna, Poland, pp. 25–29 (2006)
[15] Hahanov, V., Hahanova, I., Obrizan, V.: High-performance deductive fault simulation method. In: Proceedings of the 10 IEEE European test symposium, Tallinn, Estonia, pp. 91–96 (2006)
[16] Hahanov, V.I., Hahanova, I.V., Khan, S.U., Obrizan, V.I.: Topological fault simulation method. In: Proceedings of the 11th International Conference Mixdes Design of Integrated Circuits and Systems, Szczecin, pp. 211–214 (2004)
[17] Hahanov, V.I., Babich, A.V., Hyduke, S.M.: Test Generation and Fault Simulation Methods on the Basis of Cubic Algebra for Digital Devices. In: Proceedings of the Euromicro Symposium on Digital Systems Design DSD 2001, Warsaw, Poland, pp. 228–235 (2001)
[18] Hamdioui, S., Gaydadjiev, G.N., Van de Goor, A.J.: The State-of-the-art and Future Trends in Testing Embedded Memories. In: Records IEEE International Workshop on Memory Technology, Design, and Testing, San Jose, CA, pp. 54–59 (August 2004)
[19] Hansen, M.C., Yalcin, H., Hayes, J.P.: Unveiling the ISCAS-85 benchmarks: a case study in reverse engineering. IEEE Design & Test of Computers 16(3), 72–80 (1999)
[20] Harris, I.G.: Fault Models and Test Generation for Hardware-Software Covalidation. IEEE Design and Test of Computers, 40–47
[21] Huang, R.-F., Chen, C.-H., Wu, C.-W.: Economic Aspects of Memory Built-in Self-Repair. IEEE Design and Test of Computers, 164–172 (2007)
[22] Inoue, T., Miyazaki, S., Fujiwara, H.: Universal Fault Diagnosis for Lookup Table FPGAs. IEEE Design and Test of Computers, 39–44 (1998)
[23] Jain, S., Stroud, C.: Built-in Self Testing of Embedded Memories. IEEE Design and Test of Computers, 27–37 (1986)
[24] Khoche, A., Sherlekar, S.D., Venkatesh, G., Venkateswaran, R.: A Behavioral Fault Simulator for Ideal. IEEE Design and Test of Computers, 14–21 (1992)
[25] Levendel, Y.H., Menon, P.R.: Comparison of fault simulation methods – Treatment of unknown signal values. Journal of Digital Systems 4, 443–459 (1980)
[26] Menon, P.R., Chappell, S.G.: Deductive Fault Simulation with Functional Blocks. IEEE Transactions on Computers, 689–695 (1978)
[27] Niggemeyer, D., Rudnick, E.M.: Automatic Generation of Diagnostic Memory Tests Based on Fault Decomposition and Output Tracing. IEEE Transactions on Computers, 1134–1146 (2004)
[28] Nishida, T., Miyamoto, S., Kozawa, T., Satoh, K.: RFSIM: Reduced fault simulator. IEEE Transactions on Computer-Aided Design, CAD-6.3, 392–402 (1987)
[29] Novak, O., Gramatova, E., Ubar, R.: Handbook of testing electronic systems, p. 402. Czech Technical University Publishing House (2005)
[30] Ozguner, F.: Deductive Fault Simulation of Internal Faults of Inverter-Free Circuits and Programmable Logic Arrays. IEEE Transactions on Computers, 70–73 (1986)
[31] Pomeranz, I., Reddy, S.M.: On Fault Simulation for Synchronous Sequential Circuits. IEEE Transactions on Computers, 335–340 (1995)

[32] Pomeranz, I., Reddy, S.M.: Aliasing Computation Using Fault Simulation with Fault Dropping. IEEE Transactions on Computers, 139–144 (1995)
[33] Pomeranz, I., Reddy, S.M.: On Maximizing the Fault Coverage for a Given Test Length Limit in a Synchronous Sequential Circuit. IEEE Transactions on Computers, 1121–1133 (2004)
[34] Rashinkar, P., Paterson, P., Singh, L.: System-on-chip Verification: Methodology and Techniques. Kluwer Academic Publishers, Dordrecht (2002)
[35] Rossen, K.: Discrete Mathematics and its Applications, p. 824. McGraw-Hill, New York (2003)
[36] Silberman, G.M., Spillinger, I.: Functional Fault Simulation as a Guide for Biased-Random Test Pattern Generation. IEEE Transactions on Computers, 66–79 (1991)
[37] Shoukourian, S., Vardanian, V., Zorian, Y.: SoC Yield Optimization via an Embedded-Memory Test and Repair Infrastructure. IEEE Design and Test of Computers, 200–207 (2004)
[38] Shoukourian, S., Vardanian, V., Zorian, Y.: SoC Yield Optimization via an Embedded-Memory Test and Repair Infrastructure. IEEE Design and Test of Computers, 200–207 (2004)
[39] Tabatabaei, S., Ivanov, A.: Embedded Timing Analysis: A SoC Infrastructure. IEEE Design and Test of Computers, 24–36 (2002)
[40] Treuer, R., Agarwal, V.K.: Built-In Self-Diagnosis for Repairable Embedded RAMs. IEEE Design and Test of Computers, 24–33 (1993)
[41] Srikanth, V., Drummonds, S.B.: Poirot: Applications of a Logic Fault Diagnosis Tool. IEEE Design and Test of Computers, 19–30 (2001)
[42] Waicukauski, J., Lindbloom, E., Rosen, B., Iyengar, V.: Transition Fault Simulation. IEEE Design and Test of Computers, 32–38 (1987)
[43] Walczak, K.: Deductive Fault Simulation for Sequential Module Circuits. IEEE Transactions on Computers, 237–239 (1988)
[44] Wang, X., Hill, F.G., Mi, Z.: A sequential circuit fault simulation by surrogate fault propagation. In: Proc. 1989 IEEE International Test Conference, pp. 9–18. IEEE Computer Society, Los Alamitos (1989)
[45] Youngs, L., Paramanandam, S.: Mapping and Repairing Embedded-Memory Defects. IEEE Design and Test of Computers, 18–24 (1997)
[46] Youngs, L., Paramanandam, S.: Mapping and Repairing Embedded-Memory Defects. IEEE Design and Test of Computers, 18–24 (1997)
[47] Zhao, J., Meyer, F.J., Lombardi, F.: Analyzing and Diagnosing Interconnect Faults in Bus-Structured Systems. IEEE Design and Test of Computers, 54–64 (2002)
[48] Zhong, Y., Dropsho, S.G., Shen, X., Studer, A., Ding, C.: Miss Rate Prediction Across Program Inputs and Cache Configurations. IEEE Transactions on Computers, 328–343 (2007)
[49] Zorian, Y.: What is Infrastructure IP?. IEEE Design & Test of Computers, 5–7 (2002)
[50] Zorian, Y., Gizopoulos, D.: Guest editors' introduction: Design for Yield and reliability. IEEE Design & Test of Computers, 177–182 (2004)
[51] Zorian, Y., Shoukourian, S.: Embedded-Memory Test and Repair: Infrastructure IP for SoC Yield. IEEE Design and Test of Computers, 58–66 (2003)
[52] Zorian, Y., Shoukourian, S.: Embedded-Memory Test and Repair: Infrastructure IP for SoC Yield. IEEE Design and Test of Computers, 58–66 (2003)
[53] Zorian, Y.: Guest Editor's Introduction: Advances in Infrastructure IP. IEEE Design and Test of Computers 49 (2003)

13 Evolutionary Test Generation Methods for Digital Devices

Yuriy A. Skobtsov[1] and Vadim Y. Skobtsov[2]

[1] Donetsk National Technical University, Department of Automated Control Systems
[2] Institute of Applied Mathematics and Mechanics of NAS of Ukraine, Department of Control System Theory
e-mail: ya_skobtsov@list.ru

Abstract. In this chapter, we will discuss how evolutionary methods can be used for test generation of digital circuits. In present time it is strongly investigated the new direction in theory and practice of artificial intelligence and information systems – evolutionary computations. This term is used to generic description of the search, optimizing or learning algorithms, based on some formal principles of natural evolutional selection, which are sufficiently applied in solving various problems of machine learning, data mining, databases etc [1]. Among this approaches following main paradigms can be picked out: *genetic algorithms* (GA), *evolutionary strategy* (ES), *evolutional programming* (EP), *genetic programming* (GP). The differences of these approaches mainly consist in the way of target solution representation and in different set of evolutional operators used in evolutional simulation. Classical GA uses the binary encoding of problem solution and basic genetic operators are crossover and mutation. In ES solution is represented by real numbers vector and basic operator is mutation. EP uses FSM as solution representation and mutation operator. In GP problem solution is represented by program, crossover and mutation operators are applied. Now this classification is enough relative and interaction of basic evolutionary paradigms each other takes place.

13.1 Genetic Algorithms and Their Modifications

GA are random directed search algorithms, which emulate natural evolution process, to construct (sub)optimal solution of given problem. Any solution is represented with a chromosome or individual string of elements (genes). Classical "simple" GA [2] uses binary strings (for example, 0011101) to represent an individual. Therefore it looks very attractive to use GA techniques for a solution of ATPG problems for DS at structural or functional description levels. On the solution set the fitness (goal) function is determined. Fitness function allows to evaluate the closeness of each individual to the optimal solution – the ability of survival. Classical "simple" GA uses three basic operators: reproduction, crossover and mutation. Using these operators, the population (the set of individuals-solutions of considered problem) evolves from one generation to another. Classical steady state GA may be represented as the following sequence of operations shown in Fig.13.1.

M. Adamski et al. (Eds.): Design of Digital Systems and Devices, LNEE 79, pp. 331–361.
springerlink.com

Here parent individuals are selected with best fitness values. Then crossover is performed with a high probability P_c. The formed offspring are mutated with a low probability P_m and inserted in current population. To maintain the steady individuals number, the population reducing is performed.

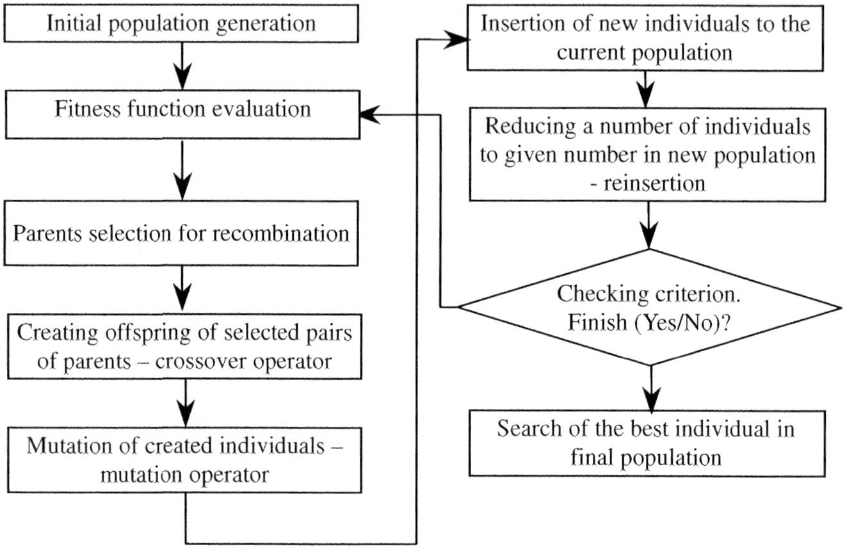

Fig. 13.1 Classical "simple" GA flow chart

At present there were suggested numerous modifications and generalizations of GA: 1) new variants of each GA step implementations (Fig.13.1); 2) essential modification of algorithm structure[3]. Here we can mark up different methods of parent selection, population reduction. Different genetic operators of crossover, of mutation. Further we briefly consider different variants of every GA step implementation and generalization of GA.

13.1.1 Parents Selection

At this step the individuals producing offspring are selected. The first step is a fitness assignment. Each individual in the selection pool receives a reproduction probability depending on the own objective value and the objective value of all other individuals in the selection pool. This fitness is used for the actual selection step afterwards.

In selection, the individuals producing offspring are chosen. As the result of selection intermediate population $\widetilde{P}^t$ from current population P^t (t is the generation number) is generated: $P^t \to \widetilde{P}^t$. Selection operator is based on fitness function values. Various selection methods are used fitness value information

differently and it significantly influences on GA effectiveness. Each individual a_i^t in the selection pool receives a reproduction probability $P_s(a_i^t)$ depending on the own fitness value and the fitness value of all other individuals in the population. And selection of individual a_i^t from current population P^t to intermediate population $\widetilde{P}^t$ is executed basing on the probability $P_s(a_i^t)$. The calculation methods of the probability $P_s(a_i^t)$ determines different selection methods:
roulette wheel selection [3]

$$P_s(a_i^t) = \frac{f(a_i^t)}{\sum_{j=1}^{N} f(a_j^t)} \tag{13.1}$$

where $f(a_i^t)$ - fitness function value, N – population size;
linear rank-based selection [3]

$$P_s(a_i^t) = \frac{1}{N}(\alpha - (\alpha - \beta)\frac{i-1}{N-1}), \tag{13.2}$$

where $1 \le \alpha \le 2$ is chosen randomly, $\beta = 2 - \alpha$.

On *tournament selection* [3], *m* individuals are chosen randomly, then the best of them is selected as parents. This procedure is continued until intermediate population $\widetilde{P}^t$ is not formed. Here selection parameter is $2 \le m \le N$.

13.1.2 Crossover Operators

Once the parents are selected, the crossover operator is used to generate offspring with a high probability P_c. The basic genetic operators and their properties can now be explained. In *single-point crossover* one crossover position $k \in \{1,2,...,L-1\}$, L is length of an individual, is selected uniformly at random and the substrings exchanged between the individuals about this point, then two new offspring are produced (Fig. 13.2).

A:	0	1	1	1	0	|	1	0	1	0
B:	1	0	0	1	1	|	0	1	0	1
						|		⇕		
A':	0	1	1	1	0	|	0	1	0	1
B':	1	0	0	1	1	|	1	0	1	0

Fig. 13.2 Single-point binary crossover

For *multi-point crossover*, *m* crossover positions $k_i \in \{1,2,...,L-1\}$, $i = \overline{1,m}$, are chosen at random with no duplicates and sorted in ascending order. Then, the substrings between successive crossover points are exchanged between the two parents to produce two new offspring (Fig. 13.3).

A:	0	1	1	1	0	0	1	1	0	1	0
B:	1	0	1	0	1	1	0	0	1	0	1
				⇕							⇕
A':	0	1	1	0	1	1	1	1	0	1	1
B':	1	0	1	1	0	0	0	0	1	0	0

Fig. 13.3 Multi-point binary crossover

The section between the first variable and the first crossover point is not exchanged between individuals [2,3].

Uniform crossover [2,3] makes every locus a potential crossover point. A crossover mask, the same length as the individual structure is created at random and the parity of the bits in the mask indicate which parent will supply the offspring with which bits, for example, 1 is the first parent, 0 is the second parent (Fig. 13.4).

Binary mask	1	0	0	1	0	1	1	1	0	0
First parent	1	0	1	0	0	0	1	1	1	0
	↓			↓		↓	↓	↓		
Offspring	1	1	0	0	0	0	1	1	1	1
		↑	↑		↑				↑	↑
Second parent	0	1	0	1	0	1	0	0	1	1

Fig. 13.4 Uniform crossover

13.1.3 Mutation

As new offspring are generated, each gene is mutated with low probability P_m. Usually the probability of mutating a gene is set to be inversely proportional to the number of genes in chromosome (dimensions). The more dimensions one individual has, the smaller is the mutation probability.

For binary individuals mutation means flipping of variable values. For every individual the variable value to change is chosen uniform at random with low probability $P_m \in [0{,}001;0{,}01]$ (Fig. 13.5).

0	1	1	[1]	0	0	1	1	0	1	0
0	1	1	[0]	0	0	1	1	0	1	0

Fig. 13.5 Binary mutation

In some cases *inversion mutation operator* is used. Two bits are chosen in individual at random and then chosen bits are exchanged (Fig. 13.6).

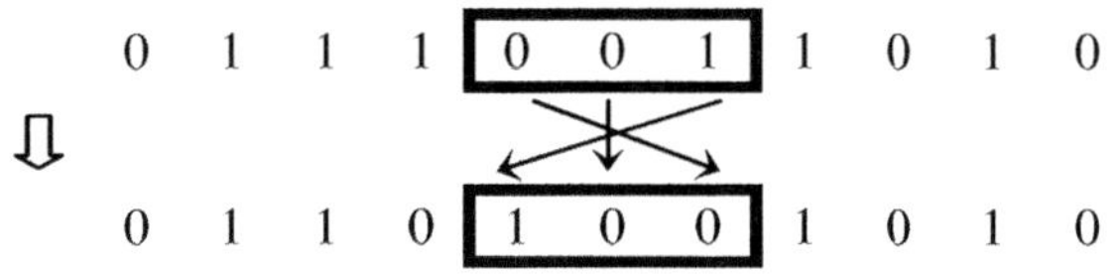

Fig. 13.6 Inversion mutation

Note that mutation serves the crucial role for providing the gene values that were not present in the current population. It enables new individual properties acquisition. Thus mutation makes the entire search space reachable, despite population finiteness. In spite of the fact that crossover has the most efficient search mechanism, it does not guarantee the reachability for each point of search space.

So, for solving any problem with genetic algorithm we must first of all define: *individual* and *population*, *genetic operators*, *fitness function*.

In ATPG problem solutions are represented as binary patterns or sequences of patterns also. Therefore it looks very attractive to use GA techniques for a decision of ATPG problems for DS at structural or functional description levels [2,3]. Further we will use different variants implementation and generalization of GA for test generation problem of digital circuits.

13.2 Genetic Test Generation Algorithm for Digital Circuits

The objective of digital circuits testing is to generate a compact sequence of binary test vectors that has high coverage of manufacturing defects. The test sequence applied should be able to uncover all possible defects that could occur in manufacturing process. That is, the output response of defective chip (or board) should be different from the outputs of a good chip. At the same time real physical defects are modeled with faults, such as stuck at fault, short, bridge fault, transistor stuck open, transistor stuck close and so on. Mostly stuck at fault modeling is used in digital testing. Here nodes are assumed to be stuck at constant either ‘0’ or ‘1’ for the purpose fault modeling. So each node may have two types of this fault, namely, *s-a-*0 and *s-a-*1. The approach usually used is to try to generate a test sequence that detects all single stuck-at faults in circuit under test. We would like to ensure, that generated test sequence contains a test for each single stuck-at fault in circuit under test. After a high fault coverage for single stuck-at faults is achieved, the additional test sequences may be generated that target other fault models, such as transistor stuck, delay fault model so on.

Generally the test generation process consists of two phases. At the first phase there are used the methods that do not direct to specific single fault but take the whole class of single stuck-at faults in circuit under test. Here test sequence is generated for faults that are checked enough easily with using not big computer power. Before the pseudorandom methods were used at this phase but now also genetic algorithms are exploited. Then the fault simulation determines the fault coverage and unchecked fault list for which it is necessary to generate test sequences.

So at second phase for each unchecked fault test sequence is generated using deterministic (or genetic) algorithm. Then again the fault simulation is used for reducing unchecked fault list. This process cycles until the high fault coverage is reached. We will discuss the genetic algorithm using basically at the second phase of test generation, where test sequence is generated for specific unchecked fault.

13.2.1 Test Generation Genetic Algorithms for Combinational Circuits

At first, the genetic algorithms were used for test generation of combinational circuits where output signals depend only on input signals and do not depend on state variables are usually represented with flip-flops. Here, as a rule, the individual corresponds to single binary test vector $X=(x_1,x_2,\ldots, x_n)$. The test generation problem may be formulated analytically for given single fault in one output combinational circuit[4]. Let $f(X)$ is a Boolean function implemented with correct combinational circuit and $\varphi(X)$ - Boolean function implemented with fault circuit. Then Boolean expression $D(X)=f(X)\oplus\varphi(X)$ is called a difference function. It is obvious that $D(X)=f(X)\oplus\varphi(X)=1$ defines the values of test vector X. So the test generation problem is reduced to search of Boolean equation $D(X)=1$. In the case multi outputs combinational circuit the difference function may be generalized in the following way:

$$D(X)=(f_1(X)\oplus\varphi_1(X))\vee(f_2(X)\oplus\varphi_2(X))\vee\ldots\vee(f_m(X)\oplus\varphi_m(X)).$$

Also it is obvious, that a solution of Boolean equation $D(X)=1$ gives the test vector for given fault.

This problem may be efficiently solved with using genetic algorithm. In this case the individual represents the binary test vector $X=(x_1,x_2,\ldots, x_n,)$, where $x_i=0,1$ and n equals circuit inputs number. So the population is composed of binary test vectors and standard genetic operators of crossover and mutation may be used in this case. Usually the number of individuals in population is proportional to inputs number n (for example 3n)[5]. We will consider the genetic algorithm using for test generation for example of circuit shown in Fig. 13.7.

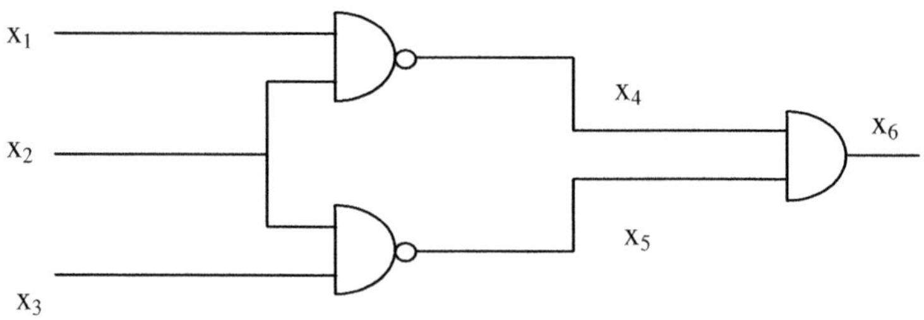

Fig. 13.7 Combinational circuit

For the convenience, all signal values for any inputs vectors are represented in Table13.1.

Table 13.1 Signal values for inputs vectors

X_1	X_2	X_3	X_4	X_5	X_6
0	0	0	1	1	1
0	0	1	1	1	1
0	1	0	1	1	1
0	1	1	1	0	0
1	0	0	1	1	1
1	0	1	1	0	0
1	1	0	0	1	0
1	1	1	0	1	0

From the beginning the initial test vectors population is generated in a random way. For example, the initial population for the circuit in Fig. 13.7 is shown in Table 13.2.

Here in second column the detected single stuck-at faults are shown for each population individual (binary test vector). Obviously, that test vector detecting more faults should have more chance to be inserted in test sequence. So, at initial stage we take the fitness function $h=F_n*r$, where F_n is the number of newly detected faults with corresponding individual (test vector) and r is "bonus" for each detected fault (for our example $r=10$). In real program system, the number of newly detected faults F_d is determined with using of fault simulation. Obviously the best individual (binary vectors 011 and 101) with maximum fitness function value $h=F_d*r$ must be inserted in test. Let for example the binary vector (101) be inserted in test. Further, for next population generation it is necessary to apply the genetic operators such as one(two)-point crossover and mutation. Once two individuals are selected, the crossover operator is used to generate two offsprings.

Table 13.2 Initial population for the circuit in Fig. 13.7

Input vector $(x_1, x_2, x_3,)$	Detected faults	Fitness function value $h= F_n *r$
000	$x_4= 0$, $x_5=0$, $x_6=0$	$3*10$
011	$x_2=0$, $x_3=0$, $x_5=1$, $x_6=1$	$4*10$
101	$x_2=1$, $x_4=0$, $x_5=0$, $x_6=0$	$4*10$
111	$x_6=1$	$1*10$

As new individuals are generated, each bit is mutated with given small probability P_m. In simplest case of binary-coded GA mutation may be done by flipping a random selected bit. While in a non-binary coded, mutation involves randomly generated a new value in a specified position. So mutation produces incremental random changes in the offspring generated through crossover and brings new properties for individuals. Each new individual (test vector) must be evaluated with fitness function. Obviously, the fitness function must take into account newly detected faults number for given test vector. Let after two generations current test set consists of two test vectors (101, 011), which detect (non detect) the faults shown in Table 13.3. The current population of vectors is presented in Table 13.4.

Table 13.3 The detected and undetected faults (based on two vectors)

Test vectors	Detected faults with current test sequence	Undetected faults with current test sequence
101	$X_2 \equiv 1$, $x_4 \equiv 0$, $x_5 \equiv 0$, $x_6 \equiv 0$	$X_1 \equiv 0$, $x_1 \equiv 1$, $x_2 \equiv 0$, $x_3 \equiv 0$, $x_3 \equiv 1$, $x_4 \equiv 0$, $x_4 \equiv 1$, $x_5 \equiv 1$, $x_6 \equiv 1$
011	$X_2 \equiv 0$, $x_3 \equiv 0$,$x_5 \equiv 1$, $x_6 \equiv 1$	$X_1 \equiv 0$, $x_1 \equiv 1$, $x_3 \equiv 1$, $x_4 \equiv 1$

Table 13.4 The detected faults and fitness function value

Input vector (x_1, x_2, x_3,)	Detected faults	Fitness function value $h = F_d *s + F_n *r$
000	$x_4 \equiv 0$, $x_5 \equiv 0$, $x_6 \equiv 0$	0*10+3*1=3
001	$x_2 \equiv 1$, $x_4 \equiv 0$, $x_5 \equiv 0$, $x_6 \equiv 0$	0*10+4*1=4
100	$x_2 \equiv 1$, $x_4 \equiv 0$, $x_5 \equiv 0$, $x_6 \equiv 0$	0*10+4*1=4
110	$x_1 \equiv 0$, $x_2 \equiv 0$, $x_4 \equiv 1$, $x_6 \equiv 1$	2*10+2*1=22

Note that here we have fitness function $h = F_d * s + F_n * r$ where F_n is the number of newly detected faults and F_d is the number of earlier detected faults with corresponding individual (test vector). Here r=10 is bonus of each newly detected fault and s=1 is bonus of each early detected fault. In concordance with Table 13.4 data the test vector (110) must be inserted in test sequence because it has maximum fitness function value. In the next Table 13.5 the situation is shown for current test that consists of three vectors and has only two undetected faults.

Table 13.5 The detected and undetected faults (based on three vectors)

Test vectors	Detected faults	Undetected faults with current test sequence
101	$X_2 \equiv 1$, $x_4 \equiv 0$, $x_5 \equiv 0$, $x_6 \equiv 0$	$X_1 \equiv 0$, $x_1 \equiv 1$, $x_2 \equiv 0$, $x_3 \equiv 0$, $x_3 \equiv 1$, $x_4 \equiv 0$, $x_4 \equiv 1$, $x_5 \equiv 1$, $x_6 \equiv 1$
011	$X_2 \equiv 0$, $x_3 \equiv 0$,$x_5 \equiv 1$, $x_6 \equiv 1$	$X_1 \equiv 0$, $x_1 \equiv 1$, $x_3 \equiv 1$, $x_4 \equiv 1$
110	$X_1 \equiv 0$, $x_4 \equiv 1$	$x_1 \equiv 1$, $x_3 \equiv 1$

Similarly may be shown that at the next step the test vector (010) must be inserted in test, because it detects the last faults $x_1 \equiv 1$, $x_3 \equiv 1$. Overall test generation genetic algorithm on base described approach of may be represented by way of following pseudocode:

```
Test generation(circuit)
  {
  Circuit initialization;
  Initial vectors population generation;
  While(stopping criteria met)
    {
    fault simulation;
    fitness function evaluation;
    insertion the best vector in test;
    genetic operators execution;
    reproduction;
    crossover;
    mutation;
    new population generation;
    }
  test sequence output;
  }
```

Here at initialization stage the fault list is generated and other auxiliary operations are executed. Usually the initial vectors population is generated in a random way, but a priori and available information about good vectors may be used also. Fitness function evaluation is based on fault simulation. Note that in described approach the genetic algorithm solves at each step the local problem of next test vector search (not whole test sequence) in contrast basic GA, described in section 13.1 which is used as a rule for global problem solution. In next section we shall consider the global GA application for test generation for sequential circuits, where the individual represent the whole test sequence but not single test vector.

13.2.2 Test Generation Genetic Algorithms for Sequential Circuits

The test generation problem for sequential circuits is much more complex and its target setting depends on observation time test strategy [4]. Let good sequential circuit realizes finite state machine (FSM) $A=(X,Y,Z,\delta,\lambda)$, where X is the input set, Y is the set of states, Z is the output set, $\delta: Y\times X\rightarrow Y$ is the next state function, $\lambda: Y\times X\rightarrow Z$ is the output function. Since we consider the structure model of sequential circuit then functions δ and λ are implemented with combinational circuits accordingly to Hafmen model:

$$Y=(y_1,...,y_k),\ \text{where}\ y_i=(0,1)\ \text{for}\ i=\overline{1,k}; \tag{13.3}$$

$$X=(x_1,\dots,x_n),\ \text{where}\ x_j=(0,1)\ \text{for}\ j=\overline{1,n}; \tag{13.4}$$

$$Z=(z_1,\dots,z_m),\ \text{where}\ z_j=(0,1)\ \text{for}\ j=\overline{1,m}. \tag{13.5}$$

Further we use the following notations[6]: $X(1), X(2),\dots, X(p)$ denotes an input sequence of length p; $Y(y_0,0), Y(y_0,1),\dots, Y(y_0,p)$ denotes the state sequence defined by initial state y_0; $Z(y_0,1),\dots, Z(y_0,p)$ denotes the output sequence defined by initial state y_0 and input sequence $X(1), X(2),\dots, X(p)$; $z_j(y_0,t)$ is the value at the j-th primary output after simulation step t. Using these notations the next state is defined by the following function:

$$Y(y_0,t)=\begin{cases} y_0 \, for=0 \\ \delta(X(t),Y(y_0,t-1)) \, for t \neq 0 \end{cases}. \quad (13.6)$$

Similarly, the output $Z(y_0,t)$ is defined by the function λ. A fault f transforms a state machine M into a machine $A^f = (Y,X,Z,\delta^f,\lambda^f)$, where functions δ^f,λ^f are defined analogically. Further we consider the different observation (respectively detection) time test strategy for sequential circuits.

Definition 13.1. A single stuck-at fault is detectable by input sequence X(1), X(2),…, X(p) with respect the single observation time test strategy (SOT) [6,7], if $\exists b \in \{0,1\}, \exists t \le p, \exists i \le k, \forall (r,q) : (z_i(r,t) = b \wedge z_i^f(q,t) = \overline{b})$, with r an initial state of fault-free circuit and q an initial state of faulty circuit.

According to above definition, a fault is SOT-detectable if there is a unique moment t such that independent of the initial states r and q of good and faulty machines the output values on a particular output are different. For sequential circuits sometimes the other strategy is used that allows more precisely to define the fault detectability.

Definition 13.2. A single stuck-at fault is detectable by input sequence X(1), X(2),…, X(p) with respect the multiple observation time test strategy (MOT) [6,7] if:

$$\forall (r,q) \exists t \le p, \exists i \le k, \exists b \in \{0,1\} : (z_i(r,t) = b \wedge z_i^f(q,t) = \overline{b}).$$

The fundamental difference between these two strategies is the following one. According to MOT, the is an individual time moment for each possible initial state pair (r,q), such that output signals on particular output are different. Obviously, that MOT strategy is more general then SOT. Some fault may be MOT-detectable but not SOT-detectable. So the test generation goal for sequential circuit is to find a input sequence *X*(1), *X*(2),…, *X*(p) for that it holds true Def.13.1 or Def.13.2 depending on using strategy. It is natural that the second strategy requires more computer resources. So, we use basically the SOT strategy. But the genetic based test generation algorithms allow generalizing the obtained results to MOT strategy in contrast to structural methods where it is problematic.

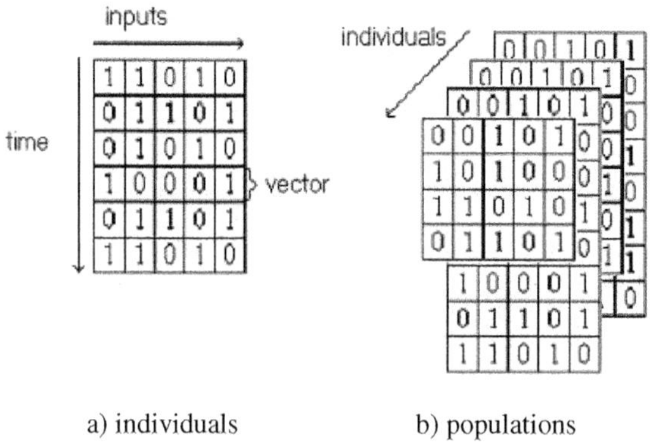

Fig. 13.8 Encoding of individuals and population in GA

Further for GA test generation of sequential circuits we will use as an individual a test sequence that is represented by binary table (Fig. 13.8a). Here the column number is determined with circuit inputs number and the test length determines the row number. In this case the population consists of the fixed number of test sequences, possibly, different length (Fig. 13.8b).For the chosen encoding of individuals and populations the following problem oriented genetic operators of crossover can be used [4,6,8]:

1. The classic one point crossover. In this case the table is interpreted as a single binary string.
2. The horizontal crossover where parents are crossed with subtables after some time point t<p as is shown at fig.13.9a.
3. The vertical crossover where parents are crossed with random selected colomns as is shown at fig 13.9b.
4. The free vertical crossover. It is executed in such a way, when crossover point is selected for each row and each pair is crossed by corresponding substrings. Note that this modification is the generalization of above vertical crossover.
5. The uniform crossover where each offspring gene is copied from one the parents accordingly to random binary mask as it is shown in Fig.13.4.
6. Structural crossover is the generalization of vertical crossover where also the parents are crossed by columns. Here it is used the exchange by columns groups corresponding to one treelike subcircuit. At that approach the exchange is directed to internal circuit check points that increases the test generation effectiveness for internal faults. Note that this crossover may be applied dynamically that is the partition of circuit to treelike subcircuits is doing for specific fault of the given circuit.

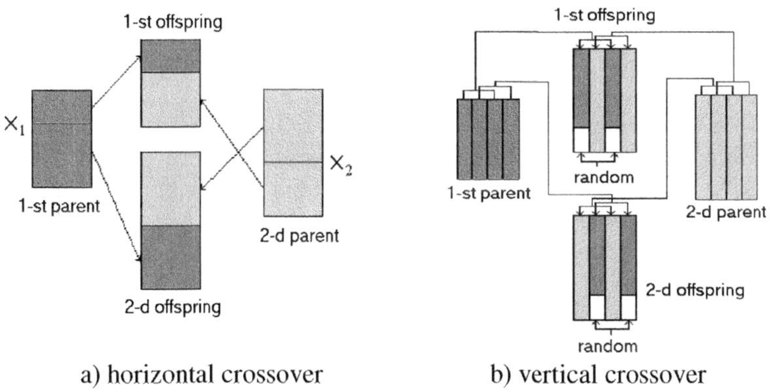

Fig. 13.9 Operations of the horizontal and vertical crossing GA

So, crossover is implemented with using above six independent operators that are selected randomly with probability P_1, P_2,..., P_6, which are derived experimentally under condition $P_1 + P_2 + P_3 + P_4 + P_5 + P_6 = 1$.

Then, as usually the generated offspring are mutated and three types of this operator accordingly are used with probabilities P_{m_1}, P_{m_2} and P_{m_3}:

1. Delete of one input vector from the random chosen position. Application of this operation allows to reduce the length of the generated test sequence in that case, when a remote vector does not worsen test properties of sequence;
2. Addition of one input vector in random position, that also allows to extend the search area of decisions;
3. Random replacement of bits in a test sequence.

Similarly, the random selection is used between these operators.

13.2.3 Problem-Oriented Fitness Functions for Test Generation

The fitness function type plays key role in the GA-based search process. Therefore, it is important to consider different types of fitness and evaluation functions, which are used in GA-based test generation methods.

The goal of testing process is to obtain different output values of correct and faulty devices. Therefore, the fitness function may be defined as measure of signal value changes in the simplest case [4]. In this case a fault free logical simulation may be used. Another and more accurate approach is to define fitness function as measure of detected faults. In this case more complex fault simulation is used, but such approach allows obtaining quite good results. Obviously, that the number of signal value changes and the number of detected faults are important parameters having influence on the effectiveness of test generation process. There are certain parameters, which are important to the evaluation function definition and to the effectiveness of test generation for sequential circuits in modern test generation systems:

1. N is the number of nodes in circuit
2. N_d is the number of nodes with different values in the fault free and in faulty circuits
3. T is the total number of flip–flops
4. T_d is the number of flip–flops that changed state
5. E is the number of events in the fault free and in faulty circuits
6. L is the length of test sequence
7. F is the total fault number;
8. F_d is the number of detected faults;
9. F_{dt} is the number of faults propagated to flip–flops;
10. D is fault detectability;
11. W is sequence power;
12. O is flip-flop observability;
13. E_f is the number of events in the faulty circuit;
14. T_s is the number of flip–flops which are hardly to set.

In addition to mentioned above parameters, the effectiveness of test generation algorithms depends on basic components of genetic algorithms such as population size, crossover and mutation probability, generation numbers, etc.

In CRIS [9] the hierarchical simulation technique is used that allows to reduce memory expenses and to deal with very large circuits. The classical GA is used, in which population evolves from generation to generation through reproduction, crossover and mutation. Each individual represents the test sequence. System CRIS is based on continued mutation of test sequence and its analysis with simulation procedure. Given system demonstrates good results (it is fast and produces compact test sequences with high fault coverage for combinational and sequential circuits) but has essential drawback, namely, the manual tuning GA parameters for each circuit.

System GATEST [10,11] is oriented on the sequential digital circuits and based on two–level GA. The first level GA generates single test vectors; the second level GA generates test sequences from these obtained vectors. Accordingly in the first level GA individuals correspond to single test vectors, while in the second level GA they are test sequences. GA uses different crossovers:1–pointed, 2–pointed, uniform.

The first–level GA is subdivided by three phases. Thus, GA has four phases. The evaluation functions are different according to the algorithm phase:

- in the first phase, the algorithm goal is the flip–flop initialization and so evaluation function is defined as follows

$$h_1 = T + \frac{T_d}{T}, \tag{13.7}$$

  here evaluation process uses only the fault-free logical simulation;
- in the second phase, all flip–flops are assumed to be initialized, and the goal is to find new test vectors able to detect additional faults; so evaluation function is represented as:

$$h_2 = F_d + \frac{T_d}{FT}; \tag{13.8}$$

- the third phase comes if the generated vector does not detect additional faults and uses the following evaluation function

$$h_3 = F_d + \frac{F_{dt}}{FT} + \frac{E}{NF}; \tag{13.9}$$

If the generated vector detects additional faults, then algorithm returns back to phase 2; otherwise if number of the unused vectors exceeds the definite limit than algorithm comes to phase 4;

- in the fourth phase test sequences are generated from designed vectors and GA uses the following evaluation function

$$h_4 = F_d + \frac{F_d}{F\,TL}. \tag{13.10}$$

In phases 2–4, evaluation uses the fault simulator that slows down the test generation process. This package shows good results: high fault coverage and low execution times for sequential benchmark circuits. But it also has the same drawback as the manual tuning GA parameters (alphabet size, iterations number, population size, mutation rate).

The interesting approach is used in package DIGATE [11,12]. It is organized in two phases:

- the first phase selects a target fault and GA activates it to a flip–flop;
- in the second phase GA searches the sequence that makes the target fault observable at the circuit primary outputs. It uses the distinguishing sequences that able to propagate a fault effect from flip–flop to primary outputs. The distinguishing sequences are pre–computed and stored for future use. The test sequences are constructed as concatenation of activating and distinguishing sequences. Evidently, here each individual is a sequence. The evaluation uses the fault simulation. Accordingly the phases 1 and 2 have the following evaluation functions

$$h_1 = 0.2D + 0.7(W + O) + 0.1(E_f + T_s + T_d), \tag{13.11}$$

$$h_2 = 0.8D + 0.1(W + O) + 0.1(E_f + T_s + T_d). \tag{13.12}$$

The weighted coefficients were found heuristically for each phase, but they are universal for any circuit. It is advantageous difference of considering method from previous packages.

In another system GATTO the individuals are input sequences too [13]. In this package the basic effort is directed to determination of the evaluation function as the measure of closeness to the optimum solution. The individuals are evaluated with fault simulation according to their activity (the more lines with different signal values in correct and fault circuits the more value of detectability probability). So, the evaluation function depends on three basic parameters: the weighted number of gates with different signals in correct and faulty circuit, the weighted flip–flop number with different signals in correct and faulty circuit, and the sequence length. The weights are empirical measures of gate and flip–flop observability accordingly. The last parameter is used for improvement of test sequence compactness. So, evaluation function for a single input vector combines the above parameters

$$h(v, f) = c_1 f_1(v, f) + c_2 f_2(v, f), \tag{13.13}$$

here f is the fault being considered and v is the input vector; c_1 and c_2 are the normalization constants, while f_1 and f_2 represent the weighted sums mentioned above.

The evaluation function H for the entire sequence s is computed according to the best vector it contains

$$H(s, f) = \max_{v_i \in s}(LH^i * h(v_i, f)). \tag{13.14}$$

Here constant $LH \in (0;1)$; i is a position of the vector v_i in the sequence s. Due using this evaluation function, shorter sequences are preferred and the final test length is reduced.

This approach does not contain the handle tuning of GA parameters for each circuit also.

The most effective is the test generation method that combines advantages of both structural deterministic and GA-based approaches. In this case the phase of entering fault and primary output activation is executed with deterministic method and multi-valued logic, and the phase of justification (the search of circuit inputs justified requirements obtained at first phase) is carried out with GA. Also the perspective approach is application of parallel GA where several populations of solutions are simultaneously evolved, basically independently of each other. But time to time the exchange of the best solutions is executed between populations in different methods. The nature of test generation problem is hierarchical, therefore it is reasonably to use hierarchical GA, where at every level different GA is applied.

13.2.4 GA Test Generation Implementation

The general approach to genetic test generation lies as follows. We use the individual encoding with binary tables as shown above at previous section in fig.13.8. For such kind of individual encoding and population representation, the above described special problem-oriented genetic operators are used. The test generation process contains three phases. The first phase goal is the fault sensitization. Here the signal difference correct and fault circuit is propagated to pseudooutputs. Then the second phase is executed for test properties improvement for sequence generated at first phase. This phase is algorithm kernel and use the genetic algorithm. After that at second phase, the test sequence is generated, the fault simulation is necessary to determine the undetectable faults.

The general pseudocode of test pattern generation is shown bellow.

```
Test_generation(sequential circuit)
{
 fault set generation();
 while(fault coverage < given threshold)
 {
  //Phase1
  goal = fault sensitization();
  if  (goal == empty set)
     exit;
  //Phase2
  sequence = GA test sequence generation(goal);
  // Phase3
  if (sequence != empty)
     fault simulation(sequence)
  else          // test sequence for goal fault not found
        mark fault as untested();
 }
}

GA test sequence generation(goal=f)
{
 For ( i=0 ; i<MAX_GENERATIONS; i++)
```

```
{
 For ( each individual s in population P )
  Fitness_evaluation(s,f );
 new P=∅;
 for (k=0 ; k< MAX_NEW_INDIVIDUALS ; k++ )
 {
  selection of 2 sequences s1 and s2 in P;
  crossover(s1,s2); //генерируются две особи
  mutation(si);
  newP=newP ∪s;
 }
 P=(best MAX_individuals from newP and P );
 for( each individual s in population P)
  if( s detect f )
   return s;
}
return( no_sequence )
}
```

Two fault simulation algorithms were integrated in order to accelerate test generation. The first one is a single pattern parallel fault propagation (SPPFP) method that is used in phase 1 for checking activation of any given fault by randomly generated sequence. Here the fitness functions defined with formulas (13.13, 13.14) are used. The evaluation function is computed with the help of the second fault simulation algorithm that belongs to the group of parallel pattern single fault propagation (PPSFP) methods. The second one was developed especially to using in GA based test generation algorithm.

13.3 Distributed Test Generation Methods

Today numerous modifications and generalizations of GA are suggested [4]. The parallel GA (PGA) are roughly upcoming and very promising from the theoretical investigation and practical application points of view [14-16].

13.3.1 GA Parallelization

Inherent GA "internal" parallelism and possibility of the distributed calculations promote to development of parallel GA. The first papers in this direction were appeared in 60-ties of XX century, but only in 80-ties, when accessible tools of parallel realization were developed, the PGA investigations adopted systematic mass character and practical orientation. Numerous models and realizations were developed in this direction, some of which are represented below.

First of all, it is necessary to note that the basis of PGA is population structuring such as decomposition to few subpopulations (subsets). This decomposition can be fulfilled with different methods, which define different types of PGA. Further we shall consider the modern methods of the PGA realization (Fig. 13.10).

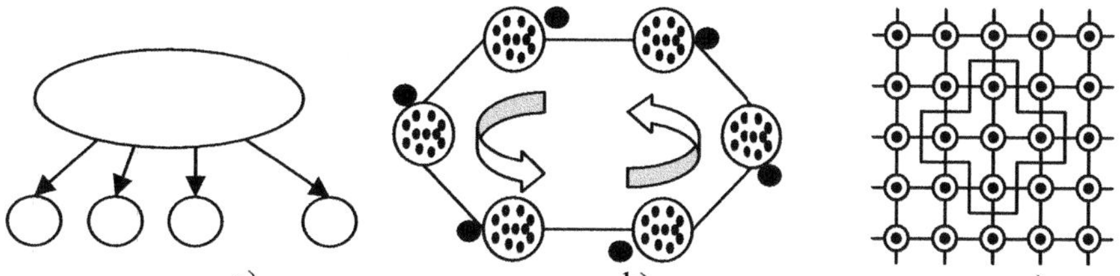

Fig. 13.10 Different types of parallel GA: a) global PGA, b) distributed PGA, c) cellular PGA

Most known is global parallelism which is represented on Fig. 13.10a. This model is based on simple (classic) GA in which the fitness function calculations are performed in parallel. This approach is faster, than classic sequential GA, and does not usually require balance on the load as on different processors. This model often named "**master-slave**". Many researchers use the pool of processors for the increase of speed execution of the algorithm. At the same time the independent programs running of algorithm at different processors are executed essentially faster than at one processor. It must be noted, that in this case there is no co-operation between different runs of algorithm. It is extraordinarily simple method of parallelization and it can be very useful. For example, it can be used for the solution of the same task with different initial conditions. GA allow effectively use this method by virtue of their probabilistic nature. At the same time we have minimum program changes, but advantages are considerable.

In *distributed PGA* (Fig.13.10b) a population is divided by a set of subpopulations, which evolve independently (accordingly to simple GA) and can communicate with neighbor subpopulations in certain manner after some "isolation time". This parallel paradigm is often implemented in an extraordinarily popular "*model of islands*" (coarse grain), where great number of subalgorithms simultaneously work in parallel, exchanging in the search process by some individuals. This model assumes direct realization on the computing systems with MIMD- architecture. Thus every "island" corresponds to its own processor.

In *cellular PGA* (fine grain) (Fig. 13.10c) there is a set of subpopulations consisting of only one individual. Given individual-subpopulation can communicate only with neighbor individuals-subpopulations at once. A neighbor relationship is defined as certain regular structure named as grid (Fig.13.10c). For cellular PGA parallelism is usually implemented on the computer systems with SIMD-architecture, where every processor represents subpopulation-individual. Although another works are known where authors use single possessor computers and systems with MIMD-architecture.

13.3.2 Parallel Test Generation Method Based on the "Master-Slave" Model

In this section for parallelization of GA we use a model «master-slave», because it requires the small changes in the existing software implementation of test generation GA and gives quite good results.

In this approach, every processor has its own copy of population. The calculation expenses of fitness-functions values (witch use a logical simulation) are evenly distributed to all processors. For all processors, the same list of faults is used. Therefore, for n individuals and P processors, we take the P / n individuals to every processor. The values of fitness-functions are calculated by the slave processors and are sent to one selected processor-master, which collects all information and passes it to all processors. Every processor has information about the values of fitness-function for all individuals and can create next population generation on this basis.

So, the processor-master executes central part (kernel) of test generation algorithm, while the logical simulation (fault-free and fault) of digital circuits are implemented on processors-slaves. The fault simulation is most critical from point of view of calculation expenses. Different methods of the distributed fault simulation are known, which are mainly based on decomposition: 1) circuits on subcircuits; 2) test sequence on subsequences. We will take combined approach of these two methods (Fig. 13.11).

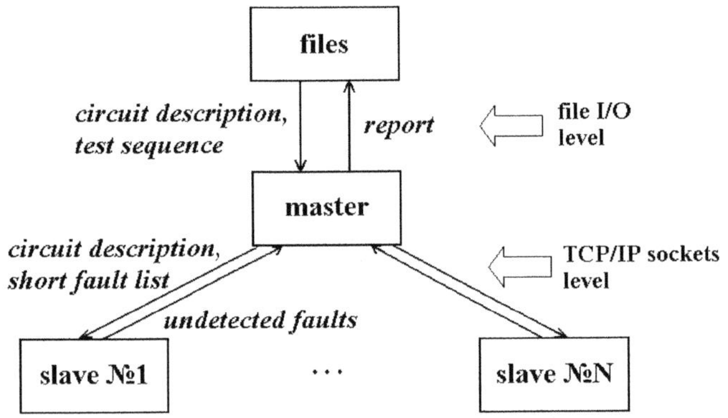

Fig. 13.11 Data flow diagram for distributed test generation and fault simulation algorithms

On the first and second stages the simulated input sequences are distributed between working processors. On the first stage every working processor is loaded by the generation (simulation) of one subsequence. For balance the list of undetected faults is broken up on approximately identical subgroups.

At the end of each of three stages, the points of synchronization are placed. When a processor-master arrives at these points, it goes to the wait mode, while all working processors will not make off the tasks that guarantee global correctness of algorithm. Thus, work is distributed between a processor-master and workers as follows. Processor-master:

- Performs all input-output operations with an user and file system: it reads circuit description and fault list, and writes the generated input test sequence;
- Initially runs «slave» processes on available resources;
- Distributes the copies (internal form) of circuits and fault lists to every working processor;

- Organizes the process control of test generation: as soon as input sequence has to be fault simulated, it sends the proper message for activating of working processors; when working processors finish their work, processor- master receives results and accordingly changes global data structures (general fault list, values of fitness-functions for individuals etc).

A processor-slave keeps the local copy of circuit (in internal format) and fault list. Every «slave» takes an input sequence from the «master» and determines the faults are detected by this sequence by the logical simulation and calculates the values of fitness-function for individuals. It sends the obtained results to the master and wait next task. As the population size is much larger than a number of processors, good balance in the load of processors is achieved. For every working processor the change of local fault list with the detected and undetected faults from other working processors requires a lot of resources and it is critical.

Final results (test input sequences and fault coverage) are near to the results obtained on the single possessor computer system with the use of a similar algorithm. Quality of decision (test fault coverage) is not here lost and is in most cases got better, and time of test generation grows short substantially.

13.3.3 Distributed Fault Simulation

Described above distributed genetic algorithm of test generation is based mainly on the distributed fault simulation algorithm. Now we will shortly describe also this method.

One of the central problems of digital device diagnostic is fault simulation of digital circuits. Persistent increasing of modern device complexity makes the task of reducing fault simulation time still very important. One of possible ways to speed up fault simulation procedure is adaptation of existing algorithms for multiprocessors computing systems (clusters) implementation.

Distributed fault simulation is organized in similar way and is based also on the «master-slave» approach like distributed test generation. One processor here is selected as master and remained processors – as slaves. There exist several approaches to implementation of distributed fault simulation: partitioning of circuit and partitioning of fault list. Our algorithm is based on the fault list partitioning.

Data flow chart for this scheme of computational process is showed on fig.13.11.

Every slave processor performs fault simulation on the data received from the master: circuit description and fault sublist. The pseudocode of this process is given below.

```
slave_process_fault_simulation()
{
 search_of_master_process();
 if( master_was_found )
 {
  receive_circuit_description();
  receive_fault_sublist();
  parallel_fault_simulation()
  send_list_of_undetected_faults();
 }
}
```

The kernel of this process is the procedure of *«parallel_fault_simulation»*, which is a regular fault simulator that used in single processor implementation. In our case we used home built PROOFS-based fault simulator, described in [17]. Mark the main advantages of this algorithm that makes it very successful: 1) dynamic fault-list processing: detected fault is eliminated from fault list in the same time it was detected, no simulation performs for this fault further; 2) fault sorting which allows to include in one group the faults that cause the same simulation events; 3) the technique of functional fault injection.

Common data flow chart diagram that describes interaction among master and slaves processes is shown on Fig. 13.11. It is necessary to notice that master process performs two types of exchange operation. File input/output operations are necessary to obtain both circuit description and test sequence to be simulated. In contrast all data interchange among master and slave process is performed via TCP/IP sockets. This fact enables to construct computing cluster on the common used computers. Authors used 100 Mbit local intranet as such cluster.

Data flowchart diagram shows that master processor does not perform any simulation but organizes the computing:

- Reads the circuit description to be simulated and input test sequence;
- Sends this description and test sequence for all client processors;
- Receives from slaves fault simulation results and makes common report.

Algorithm for master process for distributed simulation is given below.

```
distributed_simulation(circuit,test)
{
 number_of_slaves = search_of_slaves();
 if( number_of_slaves != 0 )
 {
  input_circuit_description();
  input_test();
  make_full_fault_list();
  partiting_the_fault_list(number_of_slaves);
  for( i=0 ; i< number_of_slaves ; i++ )
  {
   send_to_client_i_circuit_description ();
   send_to_client_i_part_of_fault_list();
   send_to_client_i_test_sequence();
  }
  for( i=0 ; i< number_of_slaves ; i++ )
  {
   receive_list_of_undetected_faults();
  }
  make_report();
 }
}
```

Master process starts with a search procedure of calculation clients. Further it divides full fault list into sublists prorate number of found clients. Then for all clients the following operation sequence is fulfilled in cycle: sends circuit description in internal format; sends test sequence and corresponding short fault list.

After that master transfers to state of waiting data from clients. At the next step, master receives the results of fault simulation from each client and makes general reports: fault coverage, common simulation time, time of simulation on every clients. The constructed in described way distributed fault simulation algorithm allows a high parallelization of simulation process.

13.3.4 Distributed Test Generation Based on the "Model of Islands"

In this section the "model of islands" is used for GA parallelization. Here separate subpopulation, which is initialized randomly and evolved independently, is realized on each processor. In given iteration, number subpopulations are exchanged by some individuals in certain way. Each processor selects the best individuals of own subpopulation and send them to neighbor processors subpopulations (neighborhood concept is a parameter of method). These individuals are accepted in neighbor processors subpopulations and then independent evolution on each processor-"island" is continued.

In this approach there are more chances to obtain high-quality solution, since different areas of search space are investigated on different processors [19]. Moreover, in this case it is possible the reducing of search time due to the best individual migration.

In contrast to previous method ("master-slave" model), where GA works only on the central processor-master and processors-slaves are used only for fitness function computing, in this approach full GA is implemented on every processor. In other words, each processor executes full cycle of GA evolution operations: fault-free and fault logical simulation, test sequences generation. Each processor works with full circuit and fault list. At the same time there are at least two reasons of speeding-up test generation process. Every processor operates with subpopulation of less dimension, then less time is required. Due to the best test sequences migration each processor can detect faults quicker then in case of independent operation in subpopulations. One of the most important parameters of this model is population power (individual number) of subpopulation. The influence and selection of this parameter will be considered below.

The main factors that affect on migration in "model of islands" (hence it affects the effectiveness) are as following:

Migration rate, that is the number of exchanged individuals;

Selection method of individuals for migration;

Isolation time which defines generation number between migration phases;

Strategy of individual replacement with migrated individuals from neighbor subpopulations; here also different approaches exist the worst individuals are replaced, random individuals are replaced in subpopulation etc.;

Replication strategy of migrating individuals. Under first approach migrating individual also stays in starting subpopulation. The second approach demands removal of migrating individual from starting subpopulation. The first strategy can result in domination the same strong individuals in different subpopulations. Under the second strategy, an individual can return back to start subpopulation in some time that results in extra computing expenses;

Topology which defines neighborhood relationship between subpopulations, here exchange is fulfilled only between neighbor subpopulations.

Several standard methods of selected individuals exchange between subpopulations exist. Time expenses to individual migration between subpopulation depend of used exchange method.

- Exchange by the ring:

  In this method individuals can migrate to one neighbor subpopulation. In this case the number of $m = n - 1$, where n is the number of computers;

- Two-way exchange by the ring:

  Here, likewise to previous method, exchange of individuals is executed between the closest neighbors, and neighborhood relation is defined by two-dimensional structure.

13.3.5 Implementation and Experimental Investigations of Distributed Genetic Algorithms of Test Generation and Simulation

Developed algorithms were implemented with using blocking sockets technology in C++ Builder programming environment. For computer experiments the computing cluster on the base of 100 Mbit local intranet was used. The cluster nodes have the following parameters: Intel Celeron 2 GHz processor, RAM 256 MB, OS Windows XP.

For research of effectiveness of suggested algorithms following time parameters were calculated during computer experiments: whole time of simulation process, events number in fault-free and fault simulation, whole number of events. For comparison the experimental results of algorithms from [2] were taken.

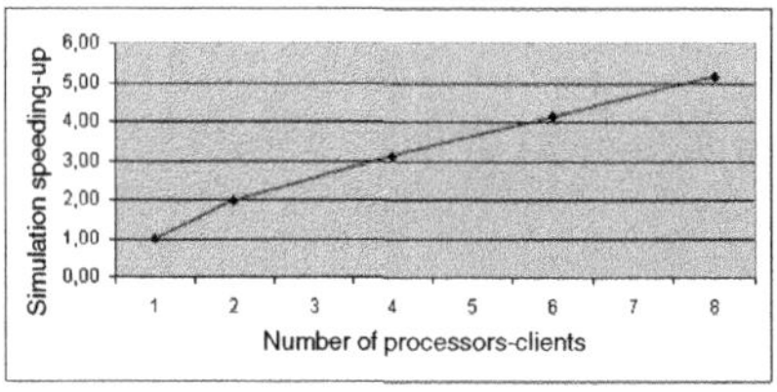

Fig. 13.12 Speeding-up of fault simulation for circuit 35938 according to worker number

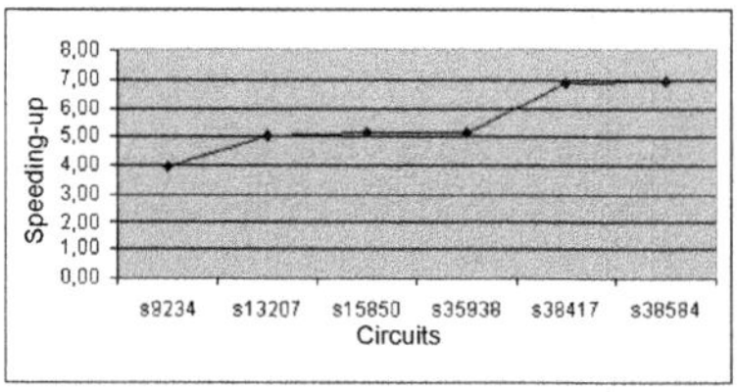

Fig. 13.13 Speeding-up according to circuit complexity

At first let consider experimental results obtained for distributed GA implementation based on the "worker-farmer" model. The diagram of simulation speeding up for circuit s35938 (ISCAS89), under condition of change of processors-client number from 1 to 8, is represented in Fig. 13.12. Given experimental results confirm the effectiveness of suggested parallelization method of simulation algorithm.

Finally, in the Fig. 13.13 the simulation results for large circuits of ISCAS benchmark are represented. These data show the relative speeding-up with increase of circuit size. It is explained by that fact that for large circuits the expenses of parallelization are reduced compare with fault simulation expenses.

Further let consider the results of implementation of test generation distributed GA which is based on the "islands model". In table13.6 there are represented experimental results, which show the speeding-up and test quality for several circuits from ISCAS89. In this case 8 processors and ring migration method were applied.

Table 13.6 The speeding-up and test quality

Circuit from ISCAS89 benchmark	Speeding-up relatively to one processor	Fault coverage increase
S1196	1.59	+0.8%
S1238	1.8	+0/6%
S1423	1.1	+12.8%
S1488	6.1	+7.1%
S5378	5.16	+1.3%
S35932	5.35	+1.6%

Obtained results confirm the effectiveness of test generation and fault simulation algorithms parallelization. The comparison of experimental results show that "farmer-worker" model gives more speeding-up in comparison with "island model" relatively to one-processor system and essentially easier in software implementation. But "island model" raises fault coverage of generated tests especially for large circuits. Therefore parallelization based on the "island model" has a reason only in case when generated tests have unsatisfactory fault coverage for "farmer-worker" model.

13.6 Hierarchical GA of Test Generation for Highly Sequential Circuits

Usually the sequential circuits have the (re) set sequences that allow installing the memory elements to some determined states. In this case the test generation is essentially simpler. The general and hard case of testing sequential machines is testing sequential machines without possibility to set it in initial state. This kind of machines with memory is often called highly sequential machines or circuits. In this case other approaches are applied [11,18].

Hierarchical approach can be effectively applied and implemented for sequential circuits at structural (gate) level. In this case at low level the some characteristic sequences are generated and then are used under test generation at high level.

Given approach is applied to test generation in highly sequential circuits for hard-to-test faults. Here two-phase strategy is used for test generation: 1) fault activation; 2) fault propagation. The iterative combinational circuit is used as a model of sequential circuit (Fig. 13.14). At the first phase, an attempt is made to derive a sequence that activates the chosen fault and propagates its effect to primary outputs (POs) or flip-flops (FFs). At the second phase, the fault effect is propagated from FFs to POs of iterative combinational circuit with assistance of distinguishing sequences basically. So, the basic problem in fault activation is transition of circuit under test (correct and faulty) to specified set conditioned by obtained FFs input values (pseudo inputs).

A transition sequence is generated with assistance of genetic algorithm. In order to generate such kind of sequences with assistance of the dynamic state traversal algorithm, a table of visited states is mapped to the list of input vectors in the test set [18]. However, if ending state was not visited, then transition sequence is generated with help of GA. In this case, initial population consists of random sequences of given length and the sequences, which solve problem partially. For example, it can be input sequences that set only some FFs to necessary values. In this case different input sequences can set different FFs to necessary values and, obviously, can be useful in transition sequences generation.

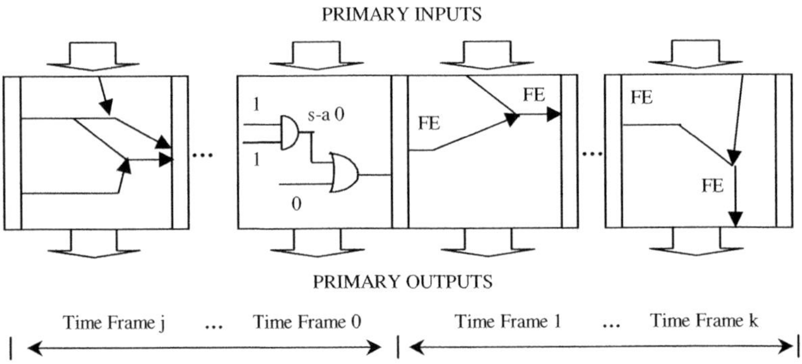

Fig. 13.14 Two-phase strategy test generation in iterative combinational circuit

In this case, some characteristic sequences are generated at low level that permits to set the flip-flops at some deterministic values that simplify a problem test generation for sequential circuits.

For test generation at high level there are useful the following input characteristic sequences. First of all we define set and reset sequences as follows:

1. **S_i-sequences.** A flip-flop **set sequence** is a sequence that sets the i-th flop to a 1-state;
2. **R_i-sequence.** Similarly a flip-flop **reset sequence** is a sequence that resets the *i*-th flip-flop to a 0-state.

These sequences associated with flip-flops are intended to (re)set the flip-flops starting from an unknown state. Such S_i (R_i) – sequences are called **type A** and generated at preprocessing step of test generation. The sequence type A length is restricted with 4D, where D is a sequential depth of circuit. If S_i (R_i) – sequences require some flip-flops must be (re) set to specific (not arbitrary) states that these sequences are called **type B**. These sequences are generated dynamically in case of need during the test generation process.

3. A **pseudoregister justification sequence** is a sequence that is able to justify (set or reset) the required flip-flops states for particular pseudoregister. Here the pseudoregister is the group of flip-flops.

At second test generation phase a distinguishing sequences propagating fault effect from FFs to Pos are required. In this case three types of the distinguishing sequences are used [18] (Fig. 13.15).

4. The **distinguishing sequence of type A** for FF *i* is defined as a sequence that produces two distinct output responses when applied to the fault-free DD for two initial states, and initial states differ in the *i*-th position and are independent of all other FF values.
5. The **B-type distinguishing sequence** for FF *i* is a sequence that, when applied to the fault-free DD with *i*-th FF = 0 (or 1) and applied to the faulty DD with the same FF = 1 (or 0) for two initial states, produces two distinct output responses independent of all other FF values.
6. The **C-type distinguishing sequence** is similar to type B except that the subset of FFs are assigned to specific logic values.

For every distinguishing sequence, the "distinguishing power" is assigned. It evaluates the possibility of given distinguishing sequence to propagate fault effect from according FF to PO. A distinguishing sequence has major "distinguishing power" if it is necessary to set specified values to small number of FFs. Also distinguishing sequences, which are able to propagate effects of several faults, have greater "distinguishing power".

Type A		Type B		Type C	
Fault-free DD	Fault-free DD	Fault-free DD	Faulty DD	Fault-free DD	Faulty DD
FFs	FFs	FFs	FFs	FFs	FFs
Uuu1uuu	uuu0uuu	uuu1uuu	uuu0uuu	uuu1Suu	uuu1Suu

Fig. 13.15 Types of distinguishing sequences

The described characteristic sequences are generated with using the low level genetic algorithm. In this GA individuals are represented with binary tables and the problem oriented genetic operators (crossover and mutation) adjust to these tables as shown in subsection 13.2.2. In this case initial population consists of random sequences of given length and the sequences, which solve problem partially. For example, it can be input sequences that set only some FFs to necessary values. But different input sequences can set different FFs to necessary values and, obviously, can be useful in transition sequences generation.

During test generation, the high level genetic algorithm uses as fabricated parts the characteristic sequences which are generated at low level. It makes the evolutionary search more directional and effective. At the high level the modified genetic algorithm is used. In the first place the initial population includes not only random binary tables, but also the generated characteristic sequences. In the second place more extensive set of genetic operators is used during test generation.

Note that the different fitness functions are used at the various level and phases. Since fault activation and fault propagation phases target different goals, their corresponding fitness functions are differed. The used parameters are as follows:

- P_1 is a fault detection;
- P_2 is a sum of dynamic controllabilities;
- P_3 is a matches of FFs values;
- P_4 is a sum of distinguishing powers;
- P_5 is a induced faulty circuit activity;
- P_6 is a number of new visited states.

Parameter P_1 is self-explanatory, in particular during the fault propagation phase. It is included in fault activation phase to cover faults that are propagated directly to the POs in the time frame in which are excited. P_2 indicates the quality of states to be justified. P_3 guides the GA to match the required FFs values in the state to be justified during state justification, from the least controllable to the most controllable FF value. P_4 measures the quality of the set of FFs reached by the fault effects. P_5 evaluates the number of events generated in the faulty circuit, with events on more observable gates weighted more heavily. P_6 is used to expand the search space. Thus, on basis of considered parameters following types of fitness functions are used.

Fault activation phase:

- Multiple time frames

$$F_1=0,2P_1 + 0,8P_4; \tag{13.15}$$

- Single time frame

$$F_2=0,1P_1 + 0,5P_2 +0,2(P_4 + P_5 +P_6); \tag{13.16}$$

- state justification

$$F_3=0,1P_1 + 0,2(k - P_2) + 0,5P_3 + 0,2(P_5 + P_6), \tag{13.17}$$

where k is a constant;

Fault propagation phase:

$$F_4=0,8P_1 + 0,2(P_4 + P_5 + P_6). \tag{13.18}$$

Note that large value of weight coefficient of the P_4 is used in fitness function at the activation phase. If activation sequence for target fault cannot be generated directly, then this problem is solved in few stages: at first the fault activation is fulfilled within a one iteration of combinational iterative circuits, and then the sequence for setting flip-flops to target state is generated. Obviously, that the main parameter is the number of detected faults at the propagation phase. Therefore, coefficient of P_1 has enough large value. Note that GA cannot find out undetectable (redundant) faults. Therefore it is desirable to use a deterministic test generation method for residuary undetectable faults.

13.7 Genetic Programming in Test Generation of Microprocessor Systems

Testing of microprocessor-based systems is a very serious problem. The most complicated task is that of generation of test sequence. Traditional structural methods of test generation, which normally require the description of the logic circuit structure on the gate level, are not applicable for such systems owing to very high task dimension. The generation of test-programs of microprocessor systems (MS) usually was carried out at function level practically "manually". At that the test represents an assembler-program unlike binary sequences for logic circuit.

One of the most perspective approaches to the MS test pattern generation is approach based on the genetic programming (GP). Checking sequence for MS is a test program consists of assembler language operators. Classical GP uses for individual representation tree-like structures that does not allow operate with arbitrary programs. Therefore, in given case graph-based program representation, especially directed acyclic graph (DAG), is applied (Fig. 13.16) [19].

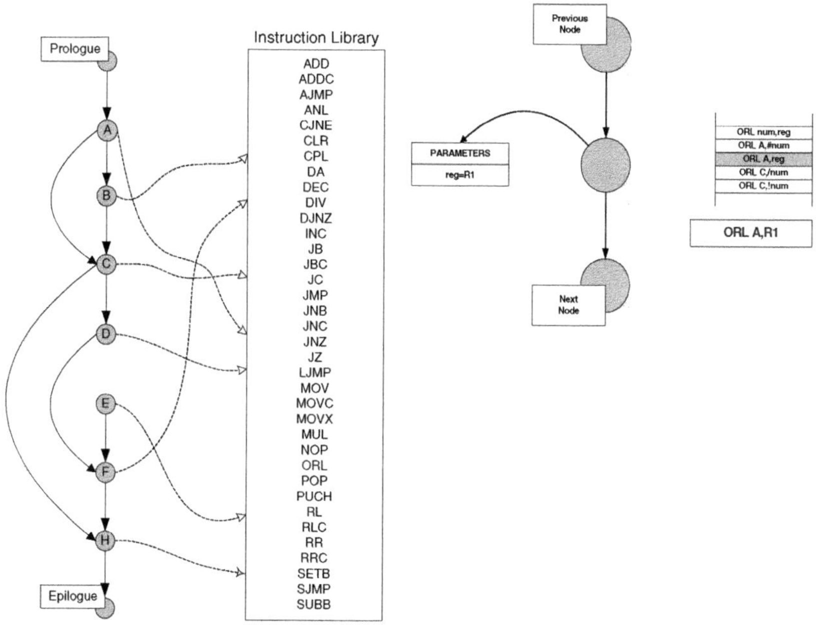

Fig. 13.16 DAG and Instruction Library (on the left), a sequential instruction and its parameters (on the right)

Each node of the DAG (Fig.13.16) contains a pointer to the instruction library and, if necessary to instruction parameters (i.e., immediate values or register specifications). The instruction library describes the assembly syntax, listing each possible instruction with the syntactically correct operands. Although instruction library may also contain macros instead of instructions, with the exception of prologue and epilogue, all entries correspond to individual assembly instruction.

For instance, Fig. 13.16 shows a sequential node that will be translated into an "*ORL A, R1*", i.e., a bit-wise OR between accumulator and register R1. *DAGs* are built with four kinds of nodes:

- *Prologue* and *epilogue* nodes represent required operations, such as initializations. They depend both on the processor and on the operating environment, and they may be empty.
- *Sequential-instruction* nodes represent common operations, such as arithmetic or logic ones (e.g., node **B, F** (Fig. 13.16)). *Unconditional* branches are considered as sequential, since execution flow does not split (e.g., node **D** (Fig. 13.16)).
- *Conditional-branch* nodes are translated to assembly-level conditional-branch instructions (e.g., node **A** (Fig. 13.16)). All common assembly languages implement some jump-if-condition mechanisms. All conditional branches implemented in the target assembly languages must be included in the library.

Test programs are induced by modifying *DAG* topology and by mutation parameters inside *DAG* nodes. Following genetic operators (mutation and crossover) are applied:

- ***Mutation 1 - Add node*:** a new node is inserted into the *DAG* in a random position. The new node can be either a sequential instruction or a conditional branch. In both cases, the instruction referred by the node is randomly chosen. If the inserted node is a branch, ether unconditional or conditional, one of the subsequent nodes is randomly chosen as the destination. Remarkable, when an unconditional branch is inserted, some nodes in the *DAG* may become unreachable.
- ***Mutation 2 - Remove node*:** an existing internal node (except prologue or epilogue) is removed from *DAG*. If the removed node was the target of one or more branch, parents' edges are updated.
- ***Mutation 3 - Modify node*:** all parameters of an existing internal node are randomly changed.
- **Crossover:** two different programs are mated to generate a new one. First, parents are analyzed to detect potential cutting points, i.e., vertices in the *DAG* that if removed create disjoint sub-graphs. Then a standard 1-point crossover is applied to generate the offspring.

Fitness-function of the second level is build on the basis of coverage measure of VHDL operators. Thus fitness-function exploits the data obtained by means of Active VHDL (code coverage).

The following fitness-function is used during construction of tests for microprocessor system:

$$F = c_o N_{ao} / N_o + c_b N_{ab} / N_b + c_b N_{ac} / N_c \,, \tag{13.19}$$

where N_{ao} is the number of linear statements VHDL have been activated by test-program; N_{ab} –the number if statements have been activated by test program; N_{ac}

is the number of case statements have been activated by test program, N_o, N_b, N_c the common number of *linear*, *if*, *case* statements accordingly; c_o, c_b, c_c are normalizing constants ($c_o + c_b + c_c = 1$).

The program implementation (Fig.13.17) is carried out in the Active VHDL environment in accordance with the following scheme. The present population of test-program (in assembler) is being generated by the method based on the genetic programming which is implemented beyond Active VHDL.

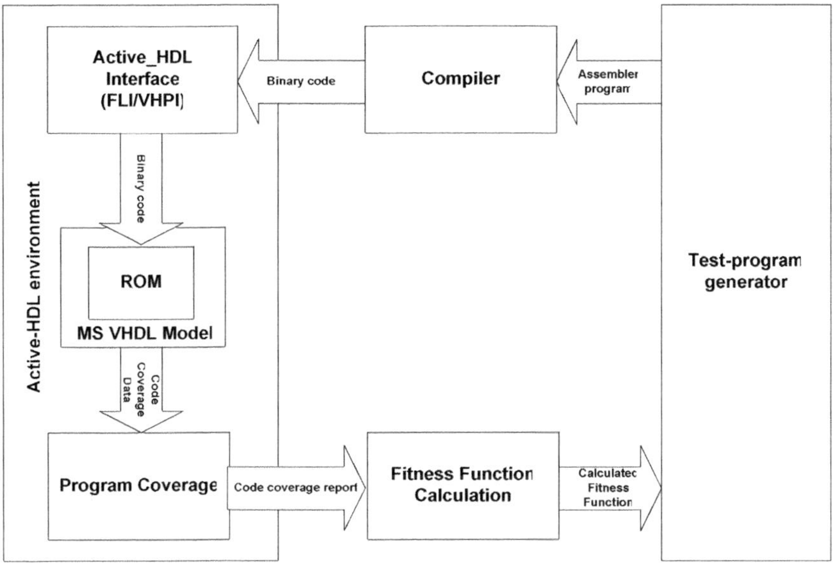

Fig. 13.17 Program implementation

The algorithm of test program generation is presented below as pseudo code:

```
generation of test-program initial population;
While (not attained maximum number of generation)
// loop  according to generations
Generation of  various paths for each test program;
   While (not attained stop condition )
   //loop according to paths
        {
          Test-program generation according to correspondent path
           Compilation of test-program to binary code
           Entry to Active VHDL environment
          Loading of binary code to ROM of microprocessor
          system VHDL  model
  Estimation of test program coverage using Active VHDL
   Exit from Active VHDL environment;
   Calculation of fitness-function according to correspondent  path;
  } // end of loop according to paths
```

Calculation of fitness function for test-program (graph);
//creation of the next generation;
Selection of parents according to fitness-function value;
Crossover;
Mutation;
Reduction of population;
} //end of loop according to generations

Generation of initial population is implemented on base of graph of test program representation by link list of neighbor nodes. For each node of graph the correspondent link list of adjacent nodes have been processed. The graph complexity may be vary by tuning the following parameters: number of nodes in graph, number of macros and number of successor-nodes for each node.

The approbation of the presented approach is done for microcontroller 8051, the model of which is given on the function level in the VHDL language. The analysis and comparison of simulation data of circuits at the function and logic level show that generated test-programs have high fault coverage. At the same time the generation of checking tests on function level is being done essentially faster. In Fig .13.18 the results of genetic algorithm of test program generation's implementation for microcontroller 8051 are shown graphically. As we can see the fitness function has reached its maximum value equal to 97.36 %.

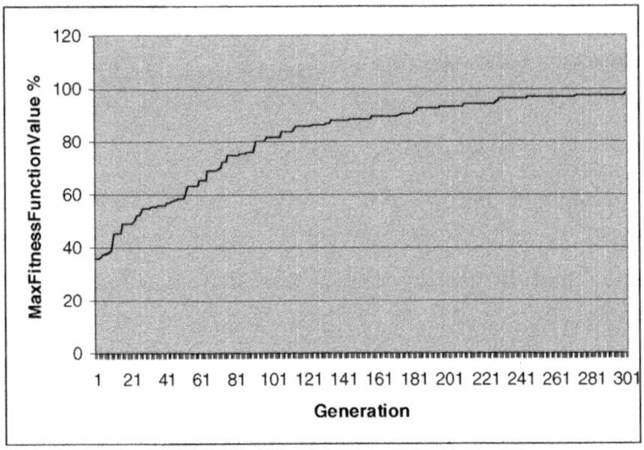

Fig. 13.18 Fault coverage

13.8 Conclusions

This paper presents new perspective approach to DS test generation that is based on using evolutionary algorithms and hierarchical solution. It was shown that given approach could be effectively applied to test generation at basic DS representation levels: FSM and structural, for highly sequential circuits, 2-levels hierarchical genetic algorithm is applicable;– GP-based approach is applied for MS level.

References

[1] Holland, J.P.: Adaptation in Natural and Artificial Systems. In: An Introductionary Analysis With Application to Biology? Control and Artificial Intelligence 210 (1975)

[2] Goldberg, D.E.: Genetic Algorithms in Search, Optimization and Machine Learning. Addison-Wesley, Reading (1989)

[3] Skobtsov, Y.A., Skobtsov, V.Y.: Modern modifications and generalizations for genetic algorithms. Tavrichesky Bulletin of Information Science and Mathematics 1, C60–C71 (2004) (in Russian)

[4] Skobtsov, Y.A., Skobtsov, V.Y.: The logical simulation and testing of digital devices.-Donetsk:IAMM NASU 436 (2005) (in Russian)

[5] O'Dare, M.J., Arslan, T.: Hierarchical test pattern generation using a genetic algorithm with a dynamic global reference table. In: First IEE/IEEE international Conference on – Genetic Algorithms in Engineering Systems: Innovations and Applications, vol. 414(12-14), pp. 517–523 (1995)

[6] Becker, B., Keim, M.: Hybrid fault Simulation for Synchronous Sequential Circuits. Journal of Electronic Testing: Theory and Applications 15, 219–238 (1999)

[7] Pomeranz, I., Reddy, S.M.: The multiple observation time strategy. IEEE Transactions on Computers 41(5), 627–637 (1992)

[8] Prinetto, P., Rebaudengo, M., Sonza, R.M.: An Automatic Test Pattern Generator for Large Sequential Circuits based on Genetic Algorithms. In: Proc. Int. Test Conf., pp. 240–249 (1994)

[9] Saab, D.G., Saab, Y.G., Abraham, J.: CRIS: A Test Cultivation Program for Sequential VLSI Circuits. In: Proc. Int. Conf. on Computer Aided Design, pp. 216–219 (1992)

[10] Rudnick, E.M., Holm, J.G., Saab, D.G., Patel, J.H.: Application of Simple Genetic Algorithm to Sequential Circuit Test Generation. In: Proc. European Design & Test Conf., pp. 40–45 (1994)

[11] Muzuder, P., Rudnic, E.M.: Genetic algorithms for VLSI design, layout & test automation, p. 336. Prentice Hall PTR, Englewood Cliffs (1999)

[12] Hsiao, M.S., Rudnick, E.M., Patel, J.H.: Automatic test generation using genetically-engineered distinguishing sequences. In: Proc. IEEE Test Symp., pp. 216–223 (1996)

[13] Prinetto, P., Rebaudengo, M., Sonza, R.M.: An Automatic Test Pattern Generator for Large Sequential Circuits based on Genetic Algorithms. In: Proc. Int. Test Conf., pp. 240–249 (1994)

[14] Mahfoud, S.W.: A comparison of parallel and sequential niching methods. In: Proceedings of VI International Conference on Genetic Algorithms, pp. 136–143 (1995)

[15] Corno, F., Prinetto, P., Rebauden, M., Sonza Reorda, M., Veiluva, E.: A PVM tool for automatic test generation on parallel and distributed systems. In: Proc. Int. Conf. on High-Performance Computing and Networking, pp. 39–44 (1995)

[16] Krishnaswamy, D., Hsiao, M.S., Saxena, V., Rudnik, E.M., Patel, J.H.: Parallel genetic algorithms for simulation-based sequential circuit test-generation. In: IEEE VLSI Design Conf., pp. 475–481 (1997)

[17] Abramovici, M.: Digital System Testing and Testable Design, vol. 652. Computer Science Press, New York (1990)

[18] Hsiao, M.S., Rudnic, E.M., Patel, J.H.: Automatic test generation using genetically-engineered distinguishing sequences. In: Proc. VLSI Test. Symp., pp. 216–223 (1996)

[19] Corno, F., Cumani, G., Sonza Reorda, M., Squillero, G.: Fully Automatic Test Program Generation for Microprocessor Cores. In: DATE 2003: Design, Automation and Test in Europe, Munich, Germany, pp. 1006–1011 (2003)

Index

GPSR Compliance

The European Union's (EU) General Product Safety Regulation (GPSR) is a set of rules that requires consumer products to be safe and our obligations to ensure this.

If you have any concerns about our products, you can contact us on ProductSafety@springernature.com

In case Publisher is established outside the EU, the EU authorized representative is:

Springer Nature Customer Service Center GmbH
Europaplatz 3
69115 Heidelberg, Germany

Batch number: 10755284

Printed by Printforce, the Netherlands